Computational Modelling and Simulation of Aircraft and the Environment

Computational Modelling and Simulation of Aircraft and the Environment

Volume 1: Platform Kinematics and Synthetic Environment

Dominic J. Diston
University of Manchester, UK

A John Wiley and Sons, Ltd., Publication

This edition first published 2009
© 2009 John Wiley & Sons, Ltd.

Registered office
John Wiley & Sons Ltd, The Atrium, Southern Gate, Chichester, West Sussex,
PO19 8SQ, United Kingdom

For details of our global editorial offices, for customer services and for information about how to apply for
permission to reuse the copyright material in this book please see our website at www.wiley.com.

MATLAB® MATLAB and any associated trademarks used in this book are the registered trademarks of
The MathWorks, Inc.

Library of Congress Cataloguing-in-Publication Data

Diston, Dominic.
   Computational modelling of aircraft and the environment / Dominic Diston.
     v. cm.
   Includes bibliographical references and index.
   Contents: Volume 1. Platform kinematics and synthetic environment
   ISBN 978-0-470-01840-8 (cloth) (v. 1)
   1. Airplanes—Computer simulation.   2. Airplanes—Performance—Mathematical models.
   3. Aeronautics—Systems engineering—Data processing.   4. Navigation (Aeronautics)—
   Computer simulation.   5. Atmosphere—Computer simulation.   6. Gravitational fields—
   Computer simulation.   7. Spherical astronomy—Data processing.   I. Title.
   TL671.4.D57 2009
   629.1301'13—dc22

                                             2009008333

A catalogue record for this book is available from the British Library.

ISBN: 978-0-470-01840-8

Set in 9/11pt Photina by Integra Software Services Pvt. Ltd., Pondicherry, India
Printed in the UK by Antony Rowe Ltd, Chippenham, Wiltshire.

IN MEMORY OF
John Joseph Diston
Master Mariner
1925–2000

Contents

Preface

The growth in computational processing over the past thirty years has been nothing short of revolutionary. What is available now was unthinkable in 1979 when I began my career. Back then, computing was a long-winded activity involving cards, punch-tape and teleprinters. Processing was usually performed overnight on a mainframe computer and chargeable by CPU time and memory usage. Now, computers are ubiquitous, tiny and very powerful. Internet access, portable computers, mobile phones and personal music players are taken for granted.

At the same time Modelling and Simulation has found its way into popular culture, most recognisably in the form of computer games (in particular through animation and visual scene generation). This too is taken for granted, to the extent that there is far greater interest in the consumption of gaming scenarios than in the production of the underlying models and simulations. In the professional engineering world, computational modelling has enabled a radical approach to product development, via digital prototypes and dynamic simulation. Complete systems can be represented with a small computer, providing the basis for extensive analysis and experimentation (within the limitations of the underlying models). Gradually, industry has changed its approach to this multi-discipline, through integrated development processes underpinned by databases and design tools.

The educational opportunities are widespread as this multi-disciplinary subject is assimilated into corporate culture and expectations become enshrined in corporate doctrine. At a time when major aerospace programmes are investing heavily in the skills and technologies for modelling and simulation, there is a need for a broad-based text covering the applicable mathematics and science in the key domains that are needed in order to build the new generation of simulators, i.e. incorporating environments and vehicles. This is what motivates this particular text; it is in two volumes that cover environments and vehicles, respectively.

In a very real sense, this is the text that I would have liked to have had seventeen years ago when I began to get heavily involved in diverse simulation projects. I have extensive experience across many aircraft projects, as a control engineer and then as a modelling and simulation specialist. I have written a lot of simulation code for aircraft, gas turbines, onboard systems (especially fuel systems and hydraulic systems), control systems, signal processing and various aspects of the natural environment. All of this was undertaken for real applications, to allow synthetic aircraft to fly with synthetic systems inside a synthetic environment.

The underlying physics can be as sophisticated as anyone wishes it to be. However, the art of simulation is to draw upon expert knowledge in order to form a combined perspective on a system of interest. To this end, I have concentrated on the framework of computational models that are required in order to establish a reasonably compete operational space, containing operational platforms. Some of the mathematics is quite involved, as befits a text that can support teaching on university programmes. I hope that you find this material interesting and informative and that it will make a difference to how you see the world around you.

Happy Reading!

Acknowledgements

Terry Hillyer, who taught me maths at the Marist College in Hull from 1974 to 1976. This was the start of my love of all things mathematical.

Allan Seabridge (formerly head of Flight Systems at BAE SYSTEMS), who is the editor of this book series and the instigator of this particular book.

The Mathworks Inc., for their generous support via the MATLAB® Book Program. Software and graphics that appear in this book were developed using MATLAB and parts of the MATLAB product family that were supplied for this purpose.

Many former colleagues and friends at BAE SYSTEMS (formerly British Aerospace), for help, advice and inspiration over two and half decades.

My family . . . an island of turmoil in a sea of calm.

List of Abbreviations

2-D	Two-Dimensional
3-D	Three-Dimensional
6-DOF	Six Degrees-Of-Freedom
ADC	Astronomical Data Center
ADDS	Aviation Digital Data Service
ALT	Altitude
AOA	Angle of Attack
AOS	Angle of Sideslip
ASCC	Air Standardization Coordinating Committee
ASTM	American Society for Testing and Materials
ATM	Air Traffic Management
AZ	Azimuth
BIPM	Bureau International des Poids et Mesures
BGS	British Geological Survey
BSC	Bright Star Catalogue
BST	British Summer Time
CAD	Computer-Aided Design
CAE	Computer-Aided Engineering
CCIR	International Radio Consultative Committee
CGPM	General Conference on Weights and Measures
CIA	Central Intelligence Agency
CIPM	International Conference on Weights and Measures
COESA	Committee on Extension to the Standard Atmosphere
DCW	Digital Chart of the World
DEC	Declination
DEM	Digital Elevation Model
DMA	Defense Mapping Agency
DMSO	Defense Modeling and Simulation Office
DTED	Digital Terrain Elevation Data
EASA	European Aviation Safety Agency
EATCHIP	European Air Traffic Control Harmonisation and Integration Programme
ECAC	European Civil Aviation Conference
ECEF	Earth-Centred Earth-Fixed (Axis System)
ECI	Earth-Centred Inertial (Axis System)
EDC	Earth Resources Observation Systems (EROS) Data Center
EGM96	Earth Gravitational Model (1996)
EME2000	Earth's Mean Equator and Equinox for Julian year 2000
ENU	East-North-Up (Axis System)
EROS	Earth Resources Observation Systems

ESDI	European Spatial Data Infrastructure
ESDU	Engineering Sciences Data Unit
ET	Ephemeris Time
EU	European Union
EUREF	Reference Frame Sub-Commission for Europe,
FGDC	Federal Geographic Data Committee
FK5	Fifth Catalogue of Fundamental Stars
GHA	Greenwich Hour Angle
GIS	Geographical information Systems
GMC	Geometric Mean Chord
GMT	Greenwich Mean Time
GPS	Global Positioning System
GRS67	Geodetic Reference System (1967)
GRS80	Geodetic Reference System (1980)
GSDI	Global Spatial Data Infrastructure
GSHHS	Global Self-consistent Hierarchical High-resolution Shoreline
GST	Greenwich Sidereal Time
GTOPO30	Global Topography (30-arc-minutes resolution)
HA	Hour Angle
HMNAO	HM Nautical Almanac Office
IAG	International Association of Geodesy
IAU	International Astronomical Union
IATA	International Air Transport Association
ICAO	International Civil Aviation Organisation
ICD	Interface Control Document
ICRF	International Celestial Reference Frame
ICSU	International Council for Science (*formerly* International Council of Scientific Unions)
IEC	International Electrotechnical Commission
IERS	International Earth Rotation and Reference Systems Service (*formerly* International Earth Rotation Service)
Inf	Infinity
IP	Intermediate Pressure
ISA	International Standard Atmosphere
ISO	International Standardisation Organisation
ITRF	International Terrestrial Reference Frame
ITU	International Telecommunication Union
ITU-D	International Telecommunication Union – Development
ITU-R	International Telecommunication Union – Radiocommunication
ITU-T	International Telecommunication Union – Telecommunications
IUGG	International Union of Geodesy and Geophysics
JAA	Joint Aviation Authorities
JD	Julian Date
JPL	Jet Propulsion Laboratory
JSD	Joint Services Designation
KBO	Kuyper Belt Object
LAT	Latitude
LHA	Local Hour Angle
LON	Longitude
LST	Local Sidereal Time

MJD	Modified Julian Date
MSL	Mean Sea Level
NaN	Not a Number
NACA	National Advisory Committee on Aeronautics
NASA	National Aeronautics and Space Administration
NGA	National Geospatial-Intelligence Agency
NGDC	National Geophysical Data Center
NGS	National Geodetic Survey
NIMA	National Imagery and Mapping Agency
NMCAs	National Mapping and Cadastral Agencies
NOAA	National Oceanic and Atmospheric Administration
NSDI	National Spatial Data Infrastructure
OGC	Open Geospatial Consortium (OGC)
OS	Ordnance Survey
PA	Parallactic Angle
PLF	Pressure Loss Factor
PR	Pressure Ratio
RA	Right Ascension
SEDRIS	Synthetic Environment Data Representation and Interchange Standardisation
SI	Système International d'Unité
TAI	International Atomic Time
TCB	Barycentric Coordinate Time
TCG	Geocentric Coordinate Time
TLE	Two-Line Element
TMY	Typical Meteorological Year
TNO	Trans-Neptunian Object
TR	Temperature Ratio
TT	Terrestrial Time
UK	United Kingdom
UN	United Nations
US	United States (of America)
USA	United States of America
USGS	US Geological Survey
USNO	US Naval Observatory
UT	Universal Time
UTC	Coordinated Universal Time [*commonly thought of as GMT*]
UTM	Universal Transverse Mercator
VMAP	Vector Map
VTP	Virtual Terrain Project
WDB-II	World Data Bank II
WGS84	World Geodetic System (1984)
WMM2005	World Magnetic Model (2005)
WMO	World Meteorological Organisation
WN	Week Number (GPS)
WVS	World Vector Shoreline

How To Use This Book

This is a book about applied mathematics and the physics that underpin the creation of aerospace simulations. It is a textbook. It is the first volume of a two-volume set, which aim to provide a working manual for students and practitioners that is focused on relevant theory and its translation into computational algorithms. Volume 1 deals with the geometry and geophysics that defines the operational environment for flight around the Earth. This defines the synthetic environment and the viewpoint from a mobile platform. Volume 2 will deal with flying vehicles and their embedded systems, with a strong emphasis on equations of motion, force generation and energy transfer. It will conclude with an integrated approach to dynamic simulation.

Primarily what is provided in the two volumes is a package of explanations and derivations so that, if anyone were to required to programme a simulation from scratch, they would be able to do so. More generally, the explanations and derivations have an inherent educational value. They provide a technical overview of the multidisciplinary science that is involved in simulation and provide reference points for university courses in aerospace engineering, as well as for continuing professional development. The content herein is based on teaching materials used at The University of Manchester from 2005.

Specifically, this book contains six chapters (1-6) and three appendices (A-C):

1. **Computational Modelling**
2. **Platform Kinematics**
3. **Geospatial Reference Model**
4. **Positional Astronomy**
5. **Geopotential Fields**
6. **Atmosphere**
A. **Introduction to MATLAB**
B. **Data and Functions**
C. **Organisations**

Chapter 1 introduces the context and motivation for computational modelling, and establishes the underlying definitions and philosophies. It presents an overview of development processes and the role played by through-lifecycle modelling and simulation. It summarises the scope of a synthetic environment that could be assembled in order to provide a virtual Earth and identifies the main sources of relevant information and (importantly) mathematical models. There then follows a discussion of aerospace vehicles and the appropriate way of interpreting them as platforms with associated systems. Final comments relate to integration and inter-operability. Collectively this sets the scene for Volume 1 and provides the link with the companion volume.

Chapter 2 develops the subject of vehicle kinematics, which defines position and orientation and their rates of change without any concern for forces and moments. Most textbooks would bundle this together with equations of motion and with flight dynamics. However, for current purposes, it is important to have the mathematical definition of a mobile platform from which to observe the synthetic environment. Thus, if the platform trajectory and velocity profile are known, then a range of external parameters (e.g. atmospheric pressure, magnetic variation) can be calculated as the platform traverses the environment. Also, an outside-world visual scene can be generated at any instant for the observer's eye position and the associated field of view. Accordingly this chapter also considers the issues of establishing lines of sight and projecting them on to displays (as would be required in a flight simulator). [*Note: The key elements of this chapter will be reviewed in Volume 2, as part of the Equations of Motion.*]

Chapter 3 gives the geospatial reference model. This is a lengthy and wide-ranging discussion of technical principles that enable positions and trajectories to be determined around the Earth. This starts with a spherical earth model and its reference frame and then considers coordinate the methods for navigation using great circles and rhumb lines. It continues with an ellipsoidal earth model, specialised for various reference ellipsoids and datum definitions, and re-develops the navigational methods based on differential geometry. Two-dimensional map projections are introduced, with the aid of examples, and lead into the development of the global mapping systems based on Universal Transverse Mercator (UTM) and Universal Polar Stereographic (UPS) grids. In addition, the National Grid of Great Britain (so-called OSGB36) is defined as an examplar for regional grid reference systems. The chapter concludes with a brief overview of three-dimensional mapping and finally, for practical purposes, some airport locations around Europe.

Chapter 4 provides the main principles of positional astronomy. Again, this is quite lengthy and is needed for navigation. Thus, it follows on from the previous chapter and sets the Earth in the context of the solar system and the celestial background. It defines reference frames for observations and it examines various concepts of "time" . Specifically, it considers mean and actual solar time, the diurnal cycle and the overarching chronological scale. There is a concise mathematical summary of orbital mechanics and the related geometrical transformations. Keplerian orbital elements are defined in several ways that cover solar system dynamics and Earth satellite orbits. Particular emphasis is placed on the Global Positioning System (GPS) satellite constellation because of its importance in airborne navigation. Finally there is a very brief discussion of star catalogues and the construction of maps of the night sky.

Chapter 5 deals with potential field models that describe gravitation and geomagnetism. Both are based on Gauss' Theorem and are expressed using Laplace's Equation. This gives rise to mathematical models that are constructed from a series of spherical harmonics, which can be extended in order to improve accuracy. While aerospace applications may vary in their requirements for accuracy, it is important to explain the basis for calculation, especially as the sophistication of this subject area is not necessarily appreciated amongst the wider engineering community. The discussion is grounded in the common concepts of gravitational and magnetic potential, with relevant simple explanations. The chapter closes with a heavily mathematical treatment of computational methods based on Cunningham's Method.

Chapter 6 is an amalgamation of many interesting and useful aspects of atmospheric modelling, without delving into too much detail in any one aspect. The focus is on bulk properties, as these will provide the basis for calculations related to atmospheric flight (in Volume 2). The aim is to develop the relevant mathematics for standard atmosphere models and then to characterise the typical regional, seasonal and diurnal variations. There is reinforced by a reasonably detailed treatment of humidity. The closing section presents a brief discussion of weather systems.

The technical development within this book proceeds sequentially up to Chapter 3 and splits into parallel streams thereafter, i.e.

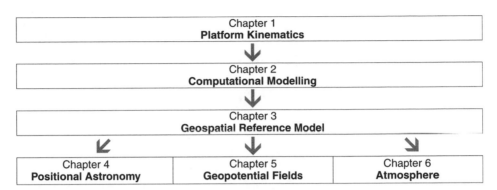

Appendix A gives an introduction to MATLAB and the general capabilities of its programming language, with an explanation of how it accesses program files and performs arithmetic. A number of basic examples are presented in order to show an end-to-end process from data definition, through calculation and finishing with data plotting.

Appendix B opens up some of the detail of the MATLAB programming language, examining data and functions. The main data types are defined and described, as applicable to general computational tasks. Likewise, program/function structure is defined and described, with various small-scale examples, in order to illustrate the range of possibilities that available. The final sections discuss general prinicples of program architecture and practical aspects of programming. This does not do justice to the product; inevitably there is much more that could be written on this subject but a condensed overview is sufficient to underpin the technical contents of this book.

Appendix C is a collection of useful information sources, arranged by organisational category, namely Specailist Agancies of the United Nations, International Organisations, US Government Organisations, UK Government Organisations, European Organisations and finally Open Projects and Consortia. This is by no means an exhaustive list but it does provide a set of starting points for detailed research, making the best of what the internet has to offer. Each organisation is named and described by a short paragraph. The wording is essentially a précis of what is presented on the respective web sites.

Series Preface

The field of aerospace is wide ranging and covers a variety of products, disciplines and domains, not merely in engineering but in many related supporting activities. These combine to enable the aerospace industry to produce exciting and technologically challenging products. A wealth of knowledge is contained by practitioners and professionals in the aerospace fields that is of benefit to other practitioners in the industry, and to those entering the industry from University.

The Aerospace Series aims to be a practical and topical series of books aimed at engineering professionals, operators, users and allied professions such as commercial and legal executives in the aerospace industry. The range of topics is intended to be wide ranging covering design and development, manufacture, operation and support of aircraft as well as topics such as infrastructure operations, and developments in research and technology. The intention is to provide a source of relevant information that will be of interest and benefit to all those people working in aerospace.

Modelling of systems is a valuable aid to understanding system behaviours and to assisting with trade-off and selection of candidate solutions. Using models and simulations to replace physical mock-ups and test rigs also makes a significant contribution to reducing project costs and time-scales. It is now second nature for engineers to develop quite sophisticated models and simulations of their systems and to use them as the basis of debate and agreement of system behaviours, and to develop an understanding of system interactions.

This book, *Computational Modelling and Simulation of Aircraft and the Environment: Volume 1 – Platform Kinematics and Synthetic Environment* provides an insight into the use of models and simulations, as well as the system environment they inhabit, to provide analysis and information which is used to support the design and certification of aircraft systems. The first stage in this process is to understand how to develop a consistent understanding of the outside world in which models and simulations exist. This is followed by the development of specific system models to integrate with this synthetic world to be published as a companion volume – *Aerospace Vehicles and Flight Dynamics*. The result of this combination is a set of tools and techniques which produce robust and trustworthy design information.

Ian Moir, Allan Seabridge and Roy Langton

Chapter 1

Introduction

1.1   Computational Modelling

Computational modelling is an increasingly important activity in the development of new aerospace systems. Its importance reflects the growth in system complexity as well as the emphasis on integration and interoperability. This approach can encompass virtually every aspect of system behaviour within an operational environment. The philosophy is to apply mathematical methods and relevant sciences in order to build synthetic systems, which are expressed as equations and algorithms and supported by relevant data. These can be exercised inside a computer for the purposes of analysis, optimisation, simulation and visualisation.

The term 'system' is ubiquitous. It is applied to an array or network of components that perform integrated functions and exhibit dynamic behaviour. By definition, a component is viewed as the lowest unit of decomposition for a particular system and, as such, it is treated as indivisible. Components are assembled in order to form a system and systems can be assembled into bigger system, i.e. a system of systems. Boundaries and partitions can be drawn anywhere that makes sense in the context of what a system is intended to do and how it is organised.

System representations are commonly called 'models' and can be given any form or content that is relevant to a specific purpose. Computational models are based on constitutive equations and relationships that describe the interconnection of system components and the interdependency of system parameters. What a model contains is dictated by what the model is intended to do and how it will be used. The level of detail establishes the capability of a complete model and, ultimately, how close it comes to being a replication of a real system.

The scale and complexity of systems can vary considerably, as can the capacity for information processing. Most systems involve some level of human interaction, such that human beings become part of the system when in use. Other systems are designed to be automatic or autonomous, which raises a whole set of issues regarding their safety and dependability in the absence of human involvement.

All systems have an external environment that defines the extent of the operational space and the variability of operational parameters. It is often appropriate to think of many environments coexisting in the same operational space, each supporting a different mode of transaction (e.g. information transfer, heat transfer). Thus, within all prescribed environments, a system must operate under ambient conditions and it must interoperate with other systems,

Computational Modelling and Simulation of Aircraft and the Environment   D.J. Diston
© 2009 John Wiley & sons, Ltd

artifacts or agents. This implies a real-world environment (where physical contact and geophysical processes take place) in conjunction with a network environment (whereby information exchange takes place).

Thus, a *system* can be just about anything, provided that it is an assembly of parts, it performs a function and it exhibits time-varying behaviour. Moreover, a *model* of a system can be just anything, provided that it represents the system in some credible manner (regardless of its format or level of detail). As a general expectation, the term 'system' conveys the dual assumption of complexity and (in the age of software-controlled systems) adaptability. The real challenge is to understand what is important in the construction of a system and how best to represent it.

In simple terms, computational modelling is usually undertaken with one or more of the following objectives:

- to confirm the concept of system operation;
- to understand the inherent functional mechanisms;
- to predict system behaviour;
- to investigate parametric uncertainty and physical constraints;
- to exercise control and monitoring functions;
- to propose and refine design requirements;
- to assess aspects of safety and interoperability.

The challenge is to combine objectives in order to build an appropriate facility for investigating system functionality, operability and interoperability. In the context of *Computational Modelling and Simulation of Aircraft and Environment*, the emphasis is placed on the principles that support the simulation of airborne systems, namely the technical descriptions of air vehicles and flight physics (which treat the system platform as a system in its own right) and the theoretical foundations for spatial reference models and geophysical processes (which treat the external environment as a system in its own right).

Thus, an aircraft can be considered as a *system of systems* that comprises a structural platform with external geometry that determines aerodynamic parameters and internal systems for vehicle management and operational tasks. Taking a wider perspective, the aircraft can be considered as one system in a larger *system of systems* such that it belongs to a network of vehicles and supporting infrastructure that collectively operates as a collaborative enterprise. Inherently, this idea underpins the concept of network-centric operations or network-enabled capability, in which the ability to acquire and distribute information is the critical factor in ensuring the optimum mix of resources and the maximum effectiveness of all participants.

Looking further outwards, there is the big environment within which all the operational platforms and services reside. This defines the universe or, at least, the small part of it that contains the Earth and a projection of major objects that are visible from Earth. This too could be considered as a *system of systems*, constructed as a set of interlocking physical processes, but it makes better sense to model this as a *synthetic environment*. This provides a single operational space for all participants, with a common definition of physical processes and interactions. In principle this can be extended in order to encompass a Virtual Earth within which activities and missions can be undertaken . . . and all by simulation.

1.2   Modelling and Simulation (M&S)

Modelling and Simulation encompasses a wide range of techniques for building representations of systems and using them in order to predict/replicate system behaviour over time. Often the distinction between these terms has been blurred and this is probably because of the interdependent nature of the two activities. This is reflected in the combined designation 'Modelling and Simulation' (or M&S) that has entered the common vocabulary of system developers.

'Modelling' is a process that collects and collates information about a system and then constructs representations of that system, based on mathematical methods and applicable sciences. Computational modelling has become the standard approach to building system prototypes because of the flexibility that is afforded by software in the building diverse representations (e.g. functional, physical, operational models) of any system that is amenable to mathematics or informatics, i.e. any system at all.

A computational model is an artifact that is constructed from equations, functions and data (regardless of its presentation). It is expressed in a form that allows computational experiments to be performed in order to answer questions about a system of interest. As with the term 'system', the term 'model' is ubiquitous and its meaning has become ambiguous, lying somewhere between the ideas of structure, content, purpose and mechanisation.

Generically, 'simulation' is a process that generates dynamic performance and operational features that are appropriate to one or more systems of interest, together with modes of interaction between systems, operators and environment. It uses models that are mechanised in software (i.e. machine-readable or machine-executable) and that are exercised via numerical algorithms that are capable of evaluating cause-and-effect relationships and aggregating the resultant changes over time, with updates associated with discrete time increments or discrete events. Models must be fit for their intended purpose and, inevitably, all will have limitations.

Specifically, a 'simulation' is a computational task or experiment that is initialised in a known condition and then is steered by stimuli (i.e. commands or environmental conditions) that are applied at system boundaries (which necessarily match the boundaries that delineate the associated models). The nature of simulation will dictate whether stimuli are routed via physical interfaces (which implies *real-time simulation*) or are predefined via time-based records (which implies *non-real-time simulation*).

A 'simulator' is an assemblage of computers, interfaces and related equipment that support the execution of specific types of simulation. Perhaps the example that is brought to mind most readily is a flight simulator. In its full configuration, this comprises a six degree-of-freedom motion platform, carrying a fully functional replica of a cockpit or flight deck surrounded by a high-fidelity display system for projecting images of a synthetic outside world. All airborne systems are simulated, together with navigational and atmospheric environments. For training and operational purposes, external operators provide air traffic management and induce faults and failures. In its minimal configuration, this comprises software inside a computer with a limited array of control and display functions. At this level, 'software only' simulators are applicable to all types of system.

The context in which simulators are used leads to a number of designations for the type of simulation being undertaken. There is some degree of variability and preference in the choice of designation but typical distinctions might include:

- **Constructive M&S** involving software-based simulations only, representing functional principles of the various systems of interest;
- **Hybrid M&S** involving software-based simulations plus equipment components and, representing a mix of functional principles and physical embodiment;
- **Live M&S** involving software-based simulations plus equipment components, with so-called Human In The Loop (HITL) operation.

By definition, Live and Hybrid M&S necessitate real-time simulation, while Constructive M&S can be real-time or non-real-time. Note that non-real-time simulation can be set at an arbitrary update rate (i.e. slower or faster than real time).

1.3   Development Processes

On major development programmes, M&S is an integral feature of lifecycle planning, as typified by Figure 1.1. This is based on the conventional 'V-Diagram' for system development, starting with requirements at top-left. Time runs from left to right. The left side of the 'V' follows the design path down to detailed specification and acquisition of system components; the right side of the 'V' follows the integration of the end-product and the related test and verification activities. This example considers the creation of a hypothetical air system (i.e. a complete airborne capability), through the progressive design of an air vehicle, its integrated systems and the constituent system components. The complete system is assembled to completion and reaches operational readiness. In this case, the requirements are implemented in an operational simulation, which implies that the operational capability of the end-product is completely predicted in advance and that performance can be confirmed and refined during development.

However, Figure 1.1 reinterprets the conventional lifecycle by subdividing it into mini-lifecycles that are applied in a hierarchy. Each 'V' starts with requirements at top-left, proceeds to the solution components at bottom-centre and concludes with an assembled product at top-right. It must be demonstrated that the product satisfies the requirements and any noncompliance would have to be addressed by either changing requirements or repeating development. So it is highly preferable that problems with requirements and/or solutions and/or products are rectified as soon as possible.

In this setting, models and simulations have the unrivalled potential of being able to represent architectures and predict behaviour during any level of system development. Stacking up the hierarchy, the complete set of mini-lifecycles can be tracked in a comprehensive and coordinated manner. Crucially, each mini-lifecycle contains integration and verification, performed via M&S as a prediction supported by analysis. Thus, components can be demonstrated in advance of their construction and assembly; they can be placed in the context of the full system before the full system exists. Subject to the quality and validity of computational models, M&S will predict the outcomes of physical integration and verification activities towards the end of the lifecycle. Early prediction will reduce the amount of physical testing that is required and will allow that testing to be targeted effectively. The underlying philosophy is that design will be validated by M&S early in the lifecycle and M&S will be validated by physical test later in the lifecycle.

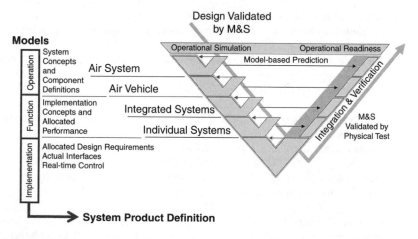

Figure 1.1 Model and simulation within the system development lifecycle.

The 'depth' of development is revealed in different class of models, as suggested in Figure 1.1. There are no rules for classification but it is reasonable to suppose the high-level development will focus on capabilities and outcomes, the mid-level development will focus on functionality and performance and low-level development will focus on implementation and build standards. Thus, three broad groupings are designated as Operational Models, Functional Models and Implementation Models. All contribution to the overall System Product Definition.

Simplistically, the V-Diagram can be seen as a linear process in two parts, namely design followed by integration. Realistically this is better characterised as a V-shaped bucket, containing numerous iterations that confirm design properties as they emerge in order to ensure that development is contributing progressively to a viable end-product. This is shown in Figure 1.2, as a combination of Design and Build activities, conducted at levels that are labelled as Product, Systems and Units. In addition, the development path continues to the right as the product enters service and is subsequently maintained.

As time runs from left to right, the problem with this traditional view of lifecycles is the large gap between top-left (where the requirements start) and the top right (where the product arrives). This is bridged by validation activities and, if problems are found here, they will be expensive to fix. The application of Verification and Validation (V&V) at each level of development is shown schematically in Figure 1.3. This involves testing on the upward path but, still, it is not desirable to find too many problems during the Build phase.

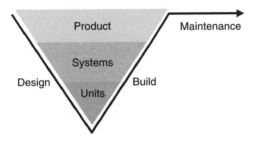

Figure 1.2 Simplified lifecycle.

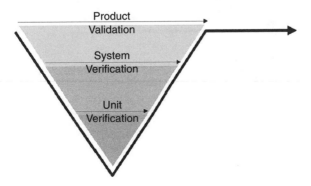

Figure 1.3 Verification and validation.

The principal benefit and attraction of modelling is the ability to predict how a product will work and what it will look like well in advance of actually building it. This requires a set of models at different levels of detail, each having its own lifecycle, as shown in Figure 1.4. These are superimposed on the product lifecycle and suggest that significant time can be saved if good models can be produced early enough. Clearly there is a trade-off between the content of a model and the time required to create it.

Top-level concept models can help confirm requirements before committing to full development. They provide a rehearsal for detailed design and testing, as well as allowing assessment of many options prior to down-select. System models look at implementation issues and detailed design. Importantly they can identify problem hotspots that require further modelling and analysis. The aim of a modelling framework to support the product lifecycle is to build up a family of models, as shown in Figure 1.5, each representing different aspects of the product. In principle, the bulk of product validation can be performed against models of what the product will be, leaving a set of acceptance test procedures to be applied to the final product. Note that this is not the same as model validation, which

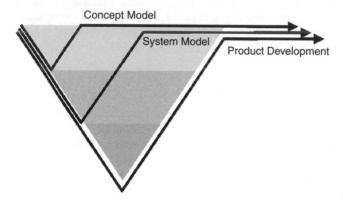

Figure 1.4 Model lifecycles.

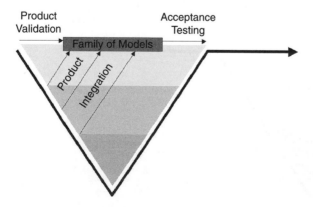

Figure 1.5 Model-based Integration.

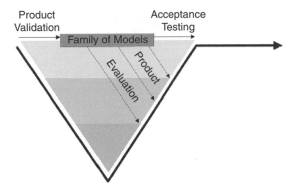

Figure 1.6 Model-based evaluation.

would have to be performed in order to demonstrate the model is an adequate and accurate representation of the product.

Having acquired models, they can be deployed for product evaluation, as shown in Figure 1.6. The models themselves can be used to generate results to be confirmed on a physical test stand. This allows extensive analysis and simulation, as well as definition of tasks and test scripts. Armed with this preparation, product build will progressively replace modelled units and systems with real units and systems. Note that the aim is to provide a product-wide context for testing such that each unit and system will experience realistic (dynamic) boundary conditions. Increasingly it is recognised that stand-alone testing is not adequate in itself for integrated systems.

The whole point of organising a product lifecycle is to be able to achieve a delivery target with minimal risk. The earlier problems are found and fixed, the smaller the risk of delay and the smaller the actual delay if the risk were realised. This is what Figure 1.7 attempts to show. Modelling can certainly facilitate early product validation. What it can also do is allow suppliers to be fully engaged at the concept stage, as in Figure 1.8. This results in a more complete product model and a more effective flow-down of product requirements through the

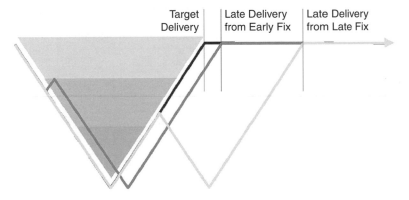

Figure 1.7 Possible late delivery.

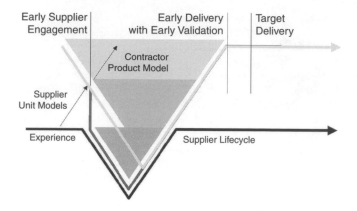

Figure 1.8 Possible early delivery.

value chain. There is an opportunity to shift unit development to the left, based on early supplier engagement and early access to supplier experience. In effect, the contractor sees a 'thin' and 'short' unit development compared with the standard lifecycle.

Cost and time can be reduced and quality can be improved in this way. However, this sort of enrichment is based on rapid design iterations and shared design information. Many complex products are developed by several risk-sharing partners and these are supported, in turn, by a large number of supplier organisations. Products evolve through many versions and variants before the final product family emerges.

1.4   Models

Quite simply, a model is defined simply as a *representation* of a system. It does not have to be an exact representation but it must contain the correct configuration and deliver the correct operational capability. Thus, it is an approximation but it must exhibit credible behaviour and performance. It can also be thought of as a framework that allows a system to be analysed or simulated. As such, the level of detail must be appropriate to the analysis or simulation but note that the level of detail does not have to be uniform across the model. Some technical features may be of greater interest than others and these may warrant greater detail.

A system can have many different representations and therefore a family of models may be required in order to cover the full range of technical features and to reflect the evolution of those features through the development lifecycle. Perhaps the two most important observations are as follows:

- All models serve a purpose and that purpose must be made explicit.
- All models have limitations and those limitations must be made explicit.

What is at issue here is that a model is applicable to the context within which it was created and are not necessarily applicable anywhere else; if it is applicable then either it is based on generic principles or it is a fluke.

Thus, it is essential that the authors specify (a) intended usage, (b) known limitations and (c) test coverage. In critical applications, independent review should establish the adequacy of this information, as well as the extent to which a model is a true representation of a system. What is needed is a qualification statement that, quite literally, states that a model is qualified to be used for a given purpose.

A *representation* invokes a set of classifications, associations and explanations. These are expressed using a *language*, which can be thought of as a descriptive framework of concepts, definitions and attributes. The process of converting between representations will be called a 'transformation'; the process of converting between languages will be called a 'translation' (Gawthrop & Ballance, 1998). These interrelationships are shown generically in Figure 1.9.

It is widely accepted that a model is either a physical replica or a mathematical abstraction. For an abstract model, there are many options for deciding on its content and for adopting suitable methods of construction (cf. Bennett, 1995). These might include:

- stochastic vs deterministic;
- event-driven vs time-driven;
- equation-based vs algorithm-based;
- theoretical vs empirical;
- physics-based vs effects-based.

Having obtained a model of a system, a number of activities can be performed. For instance, it can be used to predict steady-state performance, to establish analytical models (e.g. via linearisation) and to develop simulation codes.

It should be noted that the terms 'modelling' and 'simulation' are notoriously interchangeable. Under its correct interpretation, *modelling* is the process of mapping the structure of a physical system into a mathematical form and checking that it is correct and fit for purpose. By contrast, *simulation* is the process of experimentation whereby behaviour is predicted/reproduced by a computational algorithm that integrates rates of change in continuous states and induces transitions in discrete states, all along an advancing timeline.

It is intuitive to think of model development based on objects (cf. Rumbaugh *et al.*, 1991). Physical systems are built using physical objects, linked via physical connectors of various types. Systems can contain subsystems (i.e. systems in their own right), giving the property of *hierarchy*. Connections allow objects to interact and thus provide the interface medium between objects. The content of an object can be defined independently of its interfaces, giving the property of *encapsulation*. Many different object types could match a single interface definition, giving the property of *polymorphism*. A hierarchy of subsystem objects would access

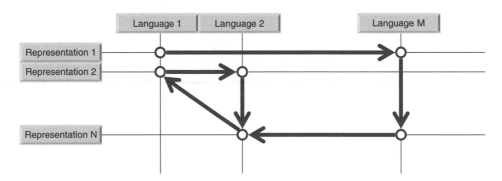

Figure 1.9 Matrix of model representations vs languages.

the top-level system interface, giving the property of *inheritance*. The adoption of this development paradigm confirms and formalises what modellers and simulationists have taken for granted over many decades.

The basic concept of object-oriented modelling is shown in Figure 1.10. This shows an arrangement of component objects within a defined architecture, with interfaces to an external environment. Associations between components are created by means of links which plug into component icons via sockets that are usually referred to as *ports*. Each port is defined as part of the component definition. The decomposition reveals that one subsystem happens to be a supertype of various component entities. One is a constitutive relationship (in this case, a mathematical expression involving states x, port variables p, internal variables θ and time t). Another is a composite component definition (which is a model in its own right). The question mark indicates that potentially any other entity could satisfy the same interface or, perhaps, the interface is left open (such that a model is only partially defined).

The issues and motivations behind polymorphic modelling are discussed by de Vries (1994). In his definition, this is 'the combined application of modularisation and subtyping during model building'. The first concept introduces an abstraction principle that focuses on the separation between essential and incidental properties of a subsystem. *Essential* properties are those that are necessary in order to classify the subsystem: *incidental* properties serve to add descriptive detail to a particular subsystem and, as such, they may differ depending on context. The second concept makes it possible to refine or specialise a generic type in various forms.

Critical aspects of modelling are verification and validation. These have a strong distinction, as recommended by the Society for Computer Simulation (Technical Committee for Model Credibility) (SCS 1979). With a number of specific points of clarification (e.g. Murray-Smith, 1995), *Verification* confirms that the *internal* structure of a model is correct and that its constituent parts are mutually consistent and *validation* confirms that the *external* behaviour is credible and that it satisfies the user requirements. Typically, the latter implies the use of test scenarios in order to demonstrate that a model can reproduce known benchmarks and it is important to distinguish between *theoretical* validation (which considers general principles),

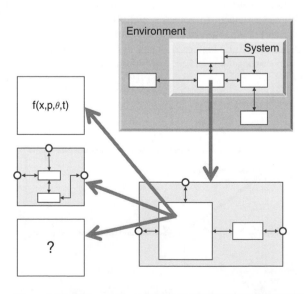

Figure 1.10 Polymorphic modelling with encapsulated components.

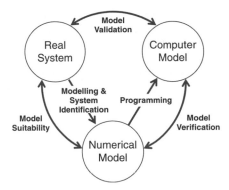

Figure 1.11 Modelling of a dynamic system.

functional validation (which deals with specific mechanisms contained in actual systems) and *empirical* validation (which compares model outputs with real measurements).

In overall terms, this view is depicted in Figure 1.11 (Buccholz *et al.*, 1995) highlighting the modelling and programming activities which lead from a real system through to a computer-based simulation. Validation and verification are shown as comparative exercises, along with a vague reference to model 'suitability'.

Adopting a different perspective, as discussed in the previous section, it is useful to separate the structure of a model from its parametric instantiation. In this way, the fundamental task is to map the constituents of a *real system* into an *object model*. A *structural* validation can be performed in order to confirm that both the component resolution and interface definition correctly reflect the functional organisation of the system and that the model is capable of delivering the information required of it. Instance data transforms an object model into a *parametric model* that can support analysis and simulation. A *parametric* validation is then applied in order to compare model prediction with actual measurement. This revised scheme is shown in Figure 1.12. In contrast, the numerical and computer models of Figure 1.11 are merely two parametric representations. The prime distinction is the explicit declaration of model structure which offers an interpretation of 'suitability' which can support a strategy for model testing.

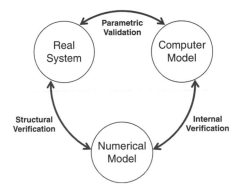

Figure 1.12 Validation and verification.

Test strategies for models fit into six broad categories although the terminology can vary considerably. These involve simulation and other analyses in order to confirm system performance within known tolerances. The main categories are, as follows:

- *replication* where consistent results are produced by independent means;
- *substitution* where part of a model is replaced by an equivalent model to confirm its behaviour;
- *approximation* where a model is reduced to its bare essentials to confirm dominant properties;
- *inversion* where a model is reconstructed in order to reproduce input stimuli from a known response;
- *identification* where internal model parameters are reconstructed from measured inputs/outputs;
- *sensitivity* where parametric variation is quantified against predefined design margins.

The rationale is to build confidence in the correct operation of a model. Choice of strategy and the level of testing will depend on the criticality of application and the perceived complexity of the system of interest. This will be reflected in the number of components and component interfaces; it will be influenced by carry-over experience from similar systems and previous application of relevant technologies.

With increasing scale and connectivity of systems, less reliance can be placed on testing and more has to be placed on the process of model development and the traceability of development activities. The basic principle is that, when a model is too big to test effectively, verification rests on adherence to a set of standard practices. Auditors will always look for evidence of this before looking at the technical content of any work that has been done.

There is a general recognition that risks associated with a complex system can never be zero but should be as low as reasonably practicable. In pragmatic terms, this says that risk reduction should be pursued until the cost grossly outweighs benefit. In the context of modelling, there is an additional recognition that it is not feasible to fully specify a model in advance of its design and implementation. Invariably, in all but the simplest of systems, there will be significant uncertainty about the detailed characteristics of the system of interest and, depending on the modelling requirements, the full extent of development problems may not be immediately apparent.

1.5   Meta-models

In order to illustrate some of the formalities that underpin model development, it is appropriate to consider how to specify the information content of a generic model in such a manner as to enable its storage within a database. An example is shown in Figure 1.13, using graphical notation that is based on ISO-10303.[1] This is a *model of a model*, which is otherwise known as a meta-model or a schema. The design of meta-models is open to preference and reflects the concepts and interpretations that are important to users and designers. Crucially this establishes a structure within to express information and necessarily it removes the use of ambiguous terminology.

With reference to Figure 1.13, the main objectives are, firstly, to associate model components, model connectivity, constitutive relationships and data items with distinct entities and, secondly, to rigorously separate definitions from instances. In this case, an *instance* relates to an

---

[1] ISO-10303 Product Data Representation and Exchange, otherwise known as the STandard for Exchange of Product model data (STEP).

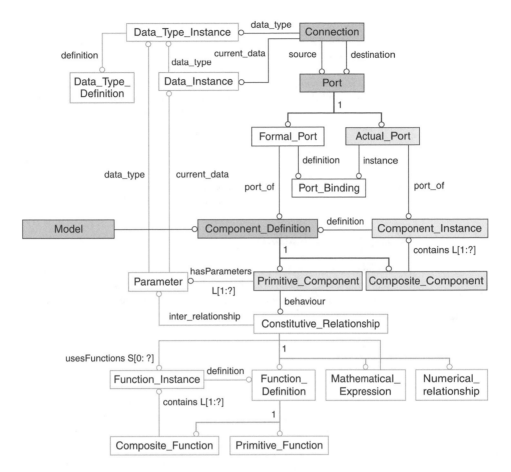

Figure 1.13 Extract of information model for system 'modelling'.

actual object, while a *definition* relates to the object type. Note that all entities are given singular names (i.e. never plural), relational links are traced along thin lines and inherited links are traced along thick lines (with the number '1' signifying mutual exclusion between subtypes). All links are traced towards the line-end that is marked by the symbol 'o'.

A *model* is a *Component_Definition* in this particular context (although the term could be given several interpretations). In turn this can be specialised as a *Composite_Component* or a *Primitive_Component*, depending on whether the model is reducible or not, respectively. A composite component is decomposed into a list of one or more *Component_Instance* entities (annotated as 'L(1:?)'). A primitive component has its behaviour defined by a *Constitutive_Relationship*. Note that, in this formulation, component definitions are nonrecursive.

Models are drawn using components, ports and connections. A *connection* has a *port* at each end, typically designated a source and a destination. The ports of a component definition are defined by the concept of a *Formal_Port* while those of a component instance are defined by the concept of an *Actual_Port*. This is akin to the concepts of actual and formal arguments in structured programming languages, which relate respectively to the use and definition of functions. When models are built, each connection to an actual port needs to be referenced to the equivalent formal port in the relevant component definition by means of a *Port_Binding*.

A constitutive relationship can be specialised in various forms, as shown. For illustration, one scenario has been developed in which functions can be defined using a hierarchy. Also, mathematical expressions (in other words, equations) can optionally contain functions. This topic is a major area of research in its own right[2] and will not be discussed further.

Finally, what about data? Primitive components and constitutive relationships contain *parameter* sets and the *connection* mechanism transfer information between components, via the port bindings. Both of these entities need to reference a *Data_Instance* in order to hold current values of data and a *Data_Type_Instance* in order to declare a parameter as being of a particular data type. Each data type instance then references a *Data_Type_Definition*.

With further development, a meta-model like this might form part of a much larger information model. To this end, useful insights can be drawn from ESPRIT Project 20496 entitled Systems Engineering Design Representation and Exchange Standardisation (SEDRES)[3] and a review of relevant work in the field of Computer-Aided Control Engineering (e.g. Varsamidis, 1998).

1.6   Aerospace Applications

In the two volumes of this text, a framework will be constructed for computational modelling that will support aerospace operations. This will concentrate on the technical and scientific principles that underpin the development of specific models and will introduce the software structures that enable their implementation. There are two steps in this construction, namely:

- *synthetic environment*;
- *aerospace vehicles*.

Thus, geophysics is defined first in order to define the components of an operational space. Also, the kinematics of generic platforms are considered here because the position and orientation of an observer is crucial to what is being observed. This is the scope of this volume (Volume 1). Generic aircraft and embedded systems are then developed in order to provide the operational platforms. There will then be a concluding section that brings everything together and presents an overview of practical simulation. This is the scope of the companion volume (Volume 2).

It is stressed that this is not a 'flight simulator' game and, in particular, it does not cover the generation of photo-realistic visual scenes; but, equally, it is not limited to 'appearances' because all the applicable physics and geometry is stated explicitly. In fact, an implicit theme throughout this book-set will be to open up the 'black box' approach to simulated environments and provide the means to implant new objects and information structures within it.

1.6.1   *Synthetic Environment*

The need for more realistic models of Planet Earth has been driven by a growth in computer-aided engineering and the introduction of simulation-based product assessments that extend far beyond the traditional scope of flight simulators. The focus is no longer just on an air vehicle and the pilot-vehicle interface but on an array of outside-world objects and environments. Necessarily this is a multidisciplinary endeavour, involving many participants pursuing diverse interests.

---

[2] Refer to http://www.OpenMath.org/intro.html, http://www.inria.fr/OpenMath/ or http://pdg.cecm. sfu.ca/OpenMath/Lib.
[3] http://www.ida.liu.se/projects/sedres.

Representations of the Earth, its surroundings and its natural processes are collected together within a *synthetic environment*. This can be configured in many ways depending on the task being undertaken, incorporating a range of relevant computational models. The big challenge is to acquire models that support multidisciplinary objectives. Inevitably this will involve the sharing of information between disciplines and the merging of information for different purposes. What is being proposed here is a collection of models that describe any, many or all of the geosciences (at varying levels of detail). As revealed by the slow tempo of standardisation efforts, this is highly ambitious.

For aerospace applications the problem of sharing geographical and geophysical information is particularly challenging. Aerospace operations extend to all points on the Earth's surface, across all terrains and oceans, in virtually all weather conditions. There are various aspects of computational modelling that could be addressed. For the purpose of this book, the greatest importance is attached to the *technical basis for a synthetic environment that will support the simulation of aerospace vehicles relative to a reference model of the Earth and its surrounding natural environment*. This involves (1) position and orientation relative to the Earth, (2) the geometry of the Earth and its relationship with astronomical bodies (especially the Sun and Moon), (3) forces exerted by the Earth and (4) atmospheric properties that are relevant to vehicle aerodynamics and propulsion.

Positions are defined by cartesian coordinates relative to a convenient origin, such as the centre of the Earth. A coordinate frame can be fixed that rotates with the Earth; this will establish positions relative to the surface but it is a non-inertial[4] frame. It is necessary to provide a fixed (or inertial) frame as the basis for dynamic calculations and, also, observations of celestial objects (many of which are used as navigational reference points). Thus, there is a need for a consistent set of reference frames and a discussion of how these frames interrelate.

The shape of the Earth can be represented as a slightly flattened sphere. The practicalities of position-fixing lend themselves to a grid system (or *graticule*) of longitude and latitude, based on angles measured north and south from the equator and east or west of a prime meridian, respectively. Apart from the obvious importance of knowing the position of object, this also implies the need to declare a system of time measurement, so that observations from any position are consistent with what should be observed at any given time. Crucially, cartesian coordinates must be consistent with longitude, latitude and altitude (measured with respect to any global or regional datum) and, in turn, these parameters must be consistent with grid coordinates defined by national mapping and cadastral[5] agencies.

International standardisation has adopted the *WGS84*[6] ellipsoid as the nominal shape of the Earth. The mass distribution generates a particular gravitational field, which can be implemented as a geopotential model (such as *EGM96*[7] or simplifications thereof). From this, a hypothetical surface can be calculated that approximates mean sea level and extends inland connecting points with the same gravitational potential. This surface is called the *geoid* and, given that local gravity is always perpendicular to it, the geoid defines a local horizontal datum. Because of undulations in geoid height, it is usual to apply reference ellipsoids (other

---

[4] In this context, 'non-inertial' means that the reference frame is accelerating and, thus, true acceleration cannot be calculated from force and inertia; it has to take account of frame acceleration (including rotation).

[5] 'Cadastral' originally related to a cadastre or public register of lands drawn up for fiscal purposes but is now applied more widely to large-scale surveying.

[6] World Geodetic System (1984).

[7] Earth Gravitational Model (1996) (which accompanies WGS84).

than WGS84) that approximate the geoid over a geographical area of interest. This provides the basis for national and regional surveys and the production of maps.

Mapping information will be required, based on a standard projection such as UTM[8]. This needs to be related to methods for determining flight paths along fixed headings and great circles. In addition, it is essential to incorporate earth rotation, standard time (i.e. UTC[9]) and time zones. As part of the general environment, some representation of the overflown land-mass will be needed in order to provide terrain reference information both for navigation and for visual observation. Digital terrain is available in many formats; it is available in the public domain at low resolution (typically 1 arc-minute spacing or 30 arc-second spacing). The better known databases are *DTED* Level 0 (published by the NGA[10]) and *DEM* and *GTOPO30* (published by the USGS[11]). A good introduction to the whole subject of terrain modelling and related topics is provided by the Virtual Terrain Project.[12]

Although not covered in this book, vector maps are available that define linear features such as road and rail networks, rivers, elevation contours, political boundaries and so on. Data of this type, augmented by terrain geometry, land-use data and more detailed cultural features are becoming increasingly common in simulations. The standard vector map in the public domain is NGA's *VMAP* Level 0 (formerly known as the Digital Chart of the World, or DCW). For accurate simulations in coastal regions, shoreline data is also available in vector formats, most notably NGA's World Vector Shoreline (or *WVS*), the CIA[13] World Data Bank II (or *WDB-II*) and, more recently, the combination of these two databases by the NOAA[14] to create *GSHHS*.[15] However, for present purposes, none of this is really important.

The relationship between 'Earth' and 'Sky' is critical in aerospace and maritime applications. It provides the oldest method of navigation, as well as determining the diurnal cycle and the seasonal variations. It is important to have celestial objects in their correct positions so that day and night can be separated and that conspicuous objects can be easily found (especially the Sun, the Moon, occasionally Venus and the bright stars in the night sky). This is all based on the basic principles of orbital mechanics, which applies equally to artificial satellites. Of particular interest in the context are the constellations of navigational satellites, such as GPS.[16]

As mentioned already, detailed gravity models are based on geopotential theory. Another application of the same theory is the representation of geomagnetism. This provides the practical concept of North and South for direction-finding (via a compass) in the absence of any more sophisticated navigational aid. Because magnetic devices are widely used, it is appropriate to model the Earth's magnetic field. The widely adopted standard is *WMM2005*,[17] which models the dominant effects of the Earth's magnetic core and the associated long-term variations over periods of years. The main importance in general aviation is the prediction of (1) the separation of magnetic north and true north (otherwise known as *magnetic variation*) and (2) the inclination of magnetic field lines (known as *magnetic dip*).

---

[8] Universal Transverse Mercator.
[9] Coordinated Universal Time (also known as Greenwich Mean Time).
[10] US National Geospatial-Intelligence Agency.
[11] US Geological Survey.
[12] www.vterrain.org.
[13] Central Intelligence Agency.
[14] National Oceanic and Atmospheric Administration.
[15] Global Self-consistent Hierarchical High-resolution Shoreline.
[16] Global Positioning System.
[17] World Magnetic Model (2005).

Air vehicles spend their operational lives travelling through air and therefore a model is needed for atmospheric properties. There are many such models, developed for more or less specialist applications. The most widely used model for aviation is the ICAO[18] International Standard Atmosphere (or *ISA*) although a number of nonstandard atmospheres are defined in order to summarise climatic variations. While it may be useful to develop a comprehensive treatment of weather modelling, this is not practicable in a book like this; the subject matter is complex and highly specialised. Also, it is very difficult to generalise requirements for weather modelling because, by definition, the usage is specific to particular places and times. The standard approach is to adopt a so-called Typical Meteorological Year (TMY). However, for present purposes, it will suffice to offer an introduction to the atmosphere of the Earth, supplemented by calculations that are relevant to atmospheric flight.

1.6.2   *Aerospace Vehicles*

In essence, an aerospace vehicles is a flying platform with installed systems. This book places its emphasis on conventional fixed-wing and rotary-wing aircraft but the same set of generic concepts govern the motion of all flying objects, namely distributed inertia, applied forces and equations of motion. Additionally, powered vehicles need an energy source (i.e. fuel), together with methods for propulsion and energy conversion, distribution and management. Lift generation is necessary in order to overcome the force of gravity. This requirement might be satisfied purely by a propulsion system (e.g. rocket-propelled launch vehicle) or take advantage of the atmosphere via the use of aerodynamic lifting surfaces (e.g. conventional fixed-wing aircraft). It should be noted that propulsion and lift generation can be combined in powered-lift systems (e.g. rotary-wing aircraft).

As with any dynamic system, the behaviour of a vehicle in flight is entirely determined by its energy. Of greatest interest are the velocities associated with the six degrees of freedom of the vehicle; three translational and three rotational. Note that vehicle motion is most usually described by so-called 6-DOF models, signifying the use of 'six degree-of-freedom' equations of motion.[19] What is important is that the conversion rate from chemical energy (stored in the fuel) to other forms has to be sufficient to acquire/maintain a given flight condition, to manoeuvre the vehicle and to power the on-board systems and services. The vehicle has to be able to vary its potential energy and its kinetic energy, as well as having excess power capability within its normal operational envelope so that it can accelerate and climb.

A fixed-wing aircraft uses a thrust source (either a jet or a propeller) as its means of forward propulsion. It uses lifting surfaces (in many possible configurations) to generate lift at a given orientation; it uses control surfaces to vary the amount of lift that is generated. The rate at which the airflow can react to changes in the orientation and/or geometry of the aircraft dictates how rapidly the force/moment variation can occur. In conjunction with mass and inertia, it dictates how the aircraft will respond to flight controls and, crucially, how energy is transferred to and from the aircraft, as evidenced in the changes to kinetic energy and potential energy.

---

[18] International Civil Aviation Organisation.

[19] Depending on the vehicle under consideration (or perhaps a particular aspect of its flight dynamics), its motion can be described by fewer than six degrees of freedom. The general designation of 'n-DOF' models, implies that $3 \leq n \leq 6$, where a reduced model can be justified on the basis of symmetric loading or some simplification in the method of operation. This is often done when supplying aircraft and guided weapon simulations in order to conceal details of operational performance from potential competitors or potential enemies.

A rotary-wing aircraft uses its lifting surfaces (i.e. its rotor blades) as the combined source of thrust and lift. Thus, this is a powered-lift aircraft with an ability to hover and to manoeuvre independently in each degree of freedom. Note that this is achieved far more efficiently than applying thrust-vectoring to a fixed-wing aircraft. Total rotor force is varied by changing the angle of each blade and, since this is done collectively, there is no nett moment generated. By changing the blade angle cyclically (i.e. following a sinusoidal profile as the rotor turns through a complete cycle) the centre of lift moves away from the rotor hub, thereby generating a moment. Aerodynamic flowfields are much more complicated for rotary-wing aircraft than for fixed-wing aircraft, as is the coordination of flight controls.

A vehicle can usefully be thought of as an integration of the following system elements:

- structural system
- environmental system
- power system
- control system
- information system.

Without being overly assertive about this particular categorisation, it is one of a number of ways of highlighting scientific and technical aspects that come together in design. In short, a vehicle is a mobile platform with a load-bearing structure of frames, panels and effectors (that determine shape, strength and stiffness) together with an array of enclosures, apertures and interfaces. It is surrounded by an environment that determines its configuration, its means of propulsion and so on. The remaining elements relate to the way in which the vehicle is mechanised for its intended role and involves the selection of appropriate technologies. In effect, these are *embedded* systems.

In this context 'power' involves an energy source, a powerplant,[20] a distribution network,[21] actuation equipment and various effectors. 'Control' involves sensors and measurements that are combined in order to determine how drive the various effectors. 'Information' involves anything that collects, communicates, records and displays data. Whereas it is easy to think of a generic airplane or a generic helicopter (in terms of layout and appearance), it is difficult to think of embedded systems in this way because of the level of design detail, the range of applicable technologies and the need to tailor a system solution for each aircraft type.

It is important to appreciate that dynamic behaviour is driven by the transfer of energy within and between systems, i.e. the vehicle structure, the embedded systems and the external environment. Actual dynamic response is determined by system pathways and the rate of energy propagation, which give rise to characteristic modes of energy redistribution. Operability and safety of any system will depend on its response to control inputs and environmental factors, as well as failures and malfunctions in system components. It is well known (and inescapably obvious) that the instantaneous state of a system has a profound effect on the rate at which energy is absorbed or dissipated. Ultimately this defines the parametric boundary outside of which normal operation[22] is physically impossible, i.e. the *operational envelope*. It also defines how rapidly system parameters can vary within that envelope.

---

[20] A 'powerplant' is comprise one or more engines for converting stored energy into usable power.

[21] A 'distribution network' is a set of pathways, each dedicated to one form of power, with appropriate methods of generation and modulation.

[22] Normal operation denotes the intended capability of the system product, under full control within a predefined and credible environment.

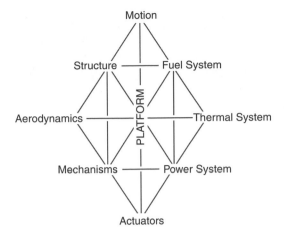

Figure 1.14 Generic platform schematic.

This overall perspective can be adjusted and refined, as shown schematically in Figure 1.14. This shows a platform with its embedded constituents and pathways, laid out in a symmetrical pattern. Conceptually, the *platform* holds everything else; it has position, orientation and velocity and it is the collection point for overall force/moment and mass/inertia calculations. Depending on how the model (and its meta-model) is arranged, the platform might also have direct associations with data belonging to components (e.g. overall geometry and appearance) so that it provides the 'external view'. The *structure, aerodynamics* and *mechanisms* denote a vehicle; the *fuel system, power system* and *thermal system* denote the installed utilities. *Actuators* are the main devices that generate mechanical forces onboard the platform (e.g. moving control surfaces or extending/retracting landing gear). *Motion* provides the equations of motion, covering accelerations due to applied forces and the related kinematics.

As with any generic schematic of this type, the labelling of placeholders and the allocation of system functionality are a matter of philosophy. In some systems, structure and mechanisms might be considered to be indivisible. Also, what about the engine? It is a power system but perhaps it deserves its own placeholder. Perhaps it should be considered as a combustor and a machine, thereby partitioning its principle processes between two places. So the arguments might continue and a compromise achieved. While it is desirable to maintain a common framework for description and categorisation, this is not an end in itself. If one framework is too awkward to apply or does not satisfy user needs then change it!

The missing pieces from Figure 1.14 are the external environment and the internal control functions. The former was introduced in Section 1.6.1 and will be developed in this volume; the latter will be discussed in Volume 2. The overall depicttion is shown in Figure 1.15 and this sets the scene for this text, namely the control and operation of platforms flying around inside a synthetic environment.

In its minimal configuration the environment is defined (with candidate models):

- *Geospatial Model* (WGS84 and GTOPO30)
- *Geomagnetic Model* (WMM2005)
- *Gravity Model* (EGM96)
- *Atmosphere Model* (ISA)

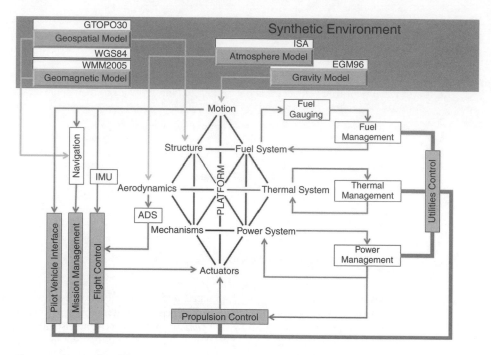

Figure 1.15 Generic platform plus control plus environment.

The first two components interface with a navigation function and the remaining two components interface with motion and aerodynamics. As it stands this configuration is adequate for atmospheric flight over wide areas but the resolution of the geospatial model is modest and there are cultural features (like airports). In this schematic, ground contact is implies by an association between the geospatial model and the platform structure (on the basis that it is the structure that react loads when platform is sitting on the ground).

The control functions are distributed between (a) embedded systems and (b) overall control of the platform. Fuel, power and thermal systems have management functions associated with them, which signifies a high degree of automation through measurement, control, scheduling and adaptation. These three functions are coordinated via a Utilities Control System (or UCS), which ensures optimal management of platform resources with respect to the requirements of other onboard systems.

The Flight Control System (or FCS) stabilises and steers the platform by moving control surfaces, thereby changing the external aerodynamics around the platform. In addition, it is convenient to include autopilot functions in this system because these are just control functions by another name. This uses motion data from an Inertial Measurement Unit (IMU) and atmospheric/airflow data from an Air Data System (ADS). For powered flight, thrust is generated by burning fuel and using the energy that is released to drive a propulsive machine (e.g. a propeller). The Propulsion Control System (or PCS) regulates the combustion process and modulates the thrust generation. For powered lift, flight and propulsion are inextricably linked for both fixed-wing and rotary-wing aircraft and, for many years, research has been focused on the development the Integrated Flight and Propulsion Control System (IFPCS).

The operation of the flying platform is shared between a Mission Management System (or MMS) and a Pilot-Vehicle Interface (or PVI). Generically, these interface with FCS, PCS and UCS, as indicated in Figure 1.15 by the thick-line interconnections. Practically, the interfacing would be implemented by a data network of some type, which is of no immediate interest here. The PVI encompasses all cockpit or flight deck functions and devices, as well as the dynamic environment that provides audible, visual and motion cues to the pilot. The array of controls and displays has become highly sophisticated as flying has moved away from mechanical control, towards workstation commands. So too has the array of sensors that are installed on platforms, as observation has evolved from discrete equipment with a limited field of regard to integrated multi-aperture, multi-spectral architectures giving all-round awareness by splicing and fusing available images. By extending the synthetic environment, platforms can then be equipped with sensors but this whole area of capability warrants a separate text. As mentioned already, this text (in its two volumes) will concentrate on the control and operation of platforms flying around inside a synthetic environment. Moreover, its emphasis will be on the underlying physics rather than enabling systems and sensors.

1.7   Integration and Interoperability

Much of the motivation for M&S, as applied to multiple platforms, combined operations and synthetic environments, is based on the philosophy of network-centric operations or network-enabled capability (as introduced in Section 1.1). In this context it is appropriate to consider the interlinked issues of integration and interoperability. The purpose of this text is provide methods for building platform/environment models as components in an operational simulation, the principal aim of which would be to demonstrate interoperability of one platform with many others, within an integrated *system of systems*.

Integration is a fundamental aspect of complex products and reflects the interdependent nature of the parts that comprise a system. While the term 'systems integration' often denotes the process whereby parts are connected together, what is really implied is an entire design philosophy that evolves an end-product that contains systems that are intended to operate in an integrated manner. In other words, systems are integrated by design, not by assembly.

This means that each system can be optimised in the context of its interfaces with other systems, rather than being acquired as a standalone entity and then installed alongside other systems. Integration allows modularisation in the sense that functionality can be allocated across an architecture that comprises specific and generic resources, plus redundant resources (if necessary) in order to ensure continuity of function in the presence of faults and failures.

Interoperability is a fundamental aspect of complex organisations and reflects the interdependent nature of systems that comprise a *system of systems*. From this perspective, each system is viewed as a discrete entity that has to play its own rôle in mission tasks and interact with other systems at some level of coordination, cooperation or collaboration. This concept is especially prominent in military thinking and has two distinct aspects. First, there is an exchange of services between 'systems, forces and units' that enable them to operate together effectively. Secondly, there is an exchange of information and the ability to use it intelligently. In concept, information is to be acquired by subscription to one or more information systems and is to be accessible via network infrastructure (covering one or more compatible networks). There is much debate on levels of interoperability, recognised that platforms and systems do not have the same capabilities for exchanging services or information. This problem is in stark relief when combining a brand new system with legacy systems that were introduced ten or twenty years previously. Such problems motivate the continuous programmes of systems upgrade and technology refresh.

In one possible interpretation, Level 1 interoperability implies common *data*, Level 2 implies common *information*, Level 3 implies common *processes* and Level 4 implies common *understanding*. Clearly, Level 4 is the ideal conclusion, supported by sophisticated processes for analysis, assessment, verification and decision-making. In turn, this depends on the richness of the information (including such qualities as completeness, correctness, accuracy, consistency, timeliness and relevance). However, Level 1 is always a challenge and international efforts in the standardisation of data exchange formats have very long gestation periods. The fundamental issues are that meta-models can be readily designed within a prescribed scenario but (1) meta-models are not unique, (2) meta-models are not easy to combine and (3) scenarios evolve and expand. Meta-models can be augmented incrementally but, as soon as a new structure is introduced that is not compatible, a new meta-model is needed.

Integration and interoperability dictate the need for open architectures, flexible acquisition and planned obsolescence. The creation of a completely new system is sometimes referred to as the Genesis Scenario; there is a sense of completion in relation to the requirements that apply at the time of creation. It is only when an existing system is in need of major upgrade that its design constraints become fully recognised. Change is then implemented in the light of new requirements that perhaps were never envisaged previously. There is no escape! Many systems are now characterised as *perpetual* (i.e. air traffic management systems) in the sense that they must provide a continuous service over an indefinite period of time. Note that activities as maintenance and training must not interrupt or compromise the level of service.

Collectively, M&S is a major activity in the development of integrated and interoperable systems. In new programmes, it will probably be the predominant method for assessing and demonstrating system functionality and its operational effectiveness. Moreover, it will probably underpin the bulk of verification activities because of its ability to represent systems and adapt to design change (without having to build and rebuild physical prototypes). It has an unrivalled potential to investigate system dynamics, to take different technical views and to vary the resolution of internal processes. A vast amount of data can be derived from a computational model; far more than could be obtained from a physical system apart from when it has been built, deployed and monitored over a long time period. Thus, M&S is the ideal (and only) method for encompassing complex systems and complex operations that span large geographical areas. Depending on the available computational resources, this could cope with global operations.

The inevitable conclusion is that all systems have to be designed from the viewpoint of integration and interoperability. In effect, all computational models are systems in their own right and therefore they must be designed from the same viewpoint. The structure of systems is reflected in the structure of models that represent them and the structure of models is defined by meta-models. Given the high level of sophistication that is possible in system design, the practice of modelling and simulation has also reached a high level of sophistication.

1.8   The End of the Beginning

Computational modelling has become a wide and varied multidiscipline, with the potential for consolidation of multi-platform, multi-system, multi-physics models and simulations for functional and operational assessments. These can be performed on single computers or networks of computers. Participation can extend to large numbers of people, as developers, users or observers. Thus, it is appropriate that the technical principles should be opened up in the form of a multidisciplinary exposition so that participants can appreciate the diverse contributions from science and engineering. Aerospace simulation provides a strong motivation for this as well as a large community of interest. This is the intended purpose of this text on *Computational Modelling and Simulation of Aircraft and Environment.*

Chapter 2

Platform Kinematics

2.1 Axis Systems

2.1.1 Platform Axis System

The motion of an observer (on a moving platform) in six degrees of freedom requires a frame of reference for measuring position and orientation and for tracking their rates of change. The mathematical description of is universally based on the axis system shown in Figure 2.1. This defines the *platform axes* for all calculations involving generalised forces and velocities (i.e. involving both translation and rotation). Its orientation is fixed with respect to the vertical, horizontal and transverse datum planes that are used to construct the platform geometry.

There are six degrees of freedom; three translational and three rotational. Applied forces are resolved to give components of Axial Force (X), Side Force (Y) and Vertical Force (Z). Applied moments are resolved into Rolling Moment (L), Pitching Moment (M) and Yawing Moment (N). The corresponding components of linear velocity are labelled u, v and w. Angular velocities are labelled p, q and r (otherwise known as roll rate, pitch rate and yaw rate). These parameters are summarised in Figure 2.2. Velocity information can be consolidated as a linear velocity vector v and an angular velocity vector $\boldsymbol{\omega}$, defined as follows:

$$\mathbf{v} = u\mathbf{x} + v\mathbf{y} + w\mathbf{z} \tag{2.1}$$

$$\boldsymbol{\omega} = p\mathbf{x} + q\mathbf{y} + r\mathbf{z} \tag{2.2}$$

The symbols $\mathbf{x}$, $\mathbf{y}$ and $\mathbf{z}$ define basis vectors for the x-axis, y-axis and z-axis, respectively.

2.1.2 Local Axis Systems

2.1.2.1 Ground Axis System

For excursions of short range and short duration it is sufficient to measure displacements from some convenient point on the Earth's surface (i.e. a local origin) with a horizontal datum plane that is appropriate to the geospatial reference model being used. Conceptually this is a tangent plane and is usually anchored to a point on the WGS84 ellipsoid.

Computational Modelling and Simulation of Aircraft and the Environment   D.J. Diston
© 2009 John Wiley & Sons, Ltd

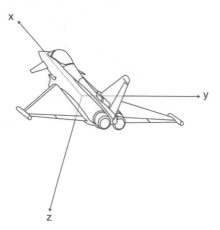

Figure 2.1 Platform axis system.

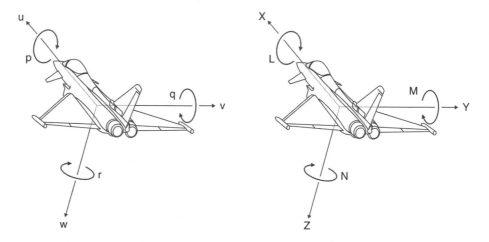

Figure 2.2 Dynamic parameters: linear and angular velocities (left); forces and moments (right).

This is often referred to as 'Flat Earth' and is shown in Figure 2.3 as a system of xyz-axes usually pointing North, East and Down, respectively, with its origin on a local ground plane. For this reason, these are sometimes called 'NED' axes although it is probably better to designate them as *ground axes*. The slight disadvantage with this scheme is that altitudes take negative values by definition. An alternative is to adopt 'ENU' axes, defining the xyz-axes pointing East, North and Up, respectively; this gives positive altitudes and also has the advantage of consistency with the xy-plane in map projections. In both schemes, because of the short time of interest, Earth rotation is ignored and, for convenience, Earth gravity is aligned with the prime vertical (which is normal to the horizontal datum).

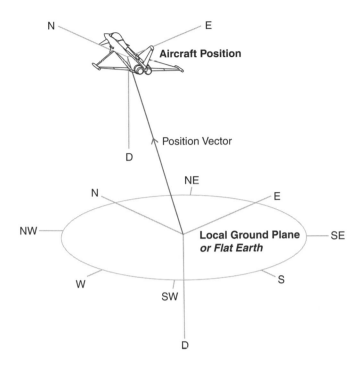

Figure 2.3 Local axis systems.

2.1.2.2 Navigational Axis System

A translation of the *ground axes* can be effected such that its origin coincides with the aircraft position, conventionally defined by its centre of mass. By definition, the resulting axis system is platform-centred with its horizontal axes parallel with the local horizontal datum. Thus, it is appropriate to introduce these as the *Navigational Axes*. Depending on the application, there are two possible implementations. The first, as shown in Figure 2.3, would be referred to the origin of 'Flat Earth'. The second would be referred to the instantaneous latitude and longitude defined on a curved Earth.

2.1.3   *Earth-Centred Axis Systems*

Positions are measured relative to surface geography using the ECEF frame that is developed in Section 3.1 for a spherical Earth representation and further developed in Section 3.6 for an ellipsoidal Earth representation. In both of these representations, the x-axis is aligned with the prime meridian that passes through Greenwich.

The ECEF Frame is the basis for converting between cartesian coordinates and aircraft position, defined by altitude, latitude and longitude, as shown in Figure 2.4 (cf. Figure 3.18). In aerospace applications, this is usually linked with the WGS84 ellipsoid and, for flights of short range and short duration, this is the global reference for selecting local axis systems.for excursions of long range or long duration it is no longer sufficient to use a local axis system because of the change of orientation that is associated with the rotation of the Earth as a function of time and the curvature of the Earth as a function of position. Thus, it is necessary to

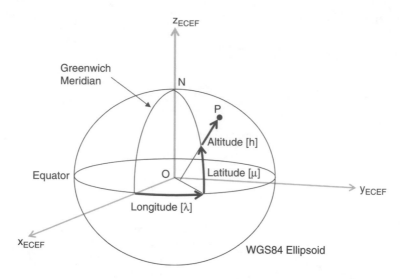

Figure 2.4 Position-fixing in the ECEF frame.

refer to the concept of the ECI Frame that is developed in Section 4.6.1 (cf. Figure 4.14). The relationship between ECI and ECEF frames appears in Figure 4.15.

In its commonly adopted form, the ECI frame has a fixed orientation in space, aligned with the equinox (i.e. x-axis) and mean equator (i.e. xy-plane) of Julian year 2000. This relates to the J2000 epoch, as discussed in Section 4.4. To a close approximation, the Earth rotates about the z-axis.

With reference to Section 4.6.2, the ECEF frame rotates with the Earth and its orientation is given by the right ascension of Greenwich (cf. (4.6)):

$$RA_G = 280.147 + \Omega D \tag{2.3}$$

where D is the number of Julian days after J2000 and Ω is the Earth rotation rate, which is defined for WGS84 in (3.58) and (4.6):

$$\Omega = 360.9856050 \text{ deg/day} \tag{2.4}$$

The ECEF position of an aircraft is derived from longitude, latitude and altitude. For spherical Earth models, the appropriate calculation is given in (3.2). More generally, consider an aircraft located at longitude (λ), geodetic latitude (μ) and altitude (h). Its coordinates (x,y,z) are given by (3.71):

$$\begin{aligned}
x &= (N + h) \cos \mu \cos \lambda \\
y &= (N + h) \cos \mu \sin \lambda \\
z &= [N(1 - e^2) + h] \sin \mu
\end{aligned} \tag{2.5}$$

where N is the length of the prime vertical, as given by (3.69):

$$N = \frac{a}{\sqrt{1 - e^2 \sin^2 \mu}} \tag{2.6}$$

In this context, a = equatorial radius (as defined by (3.60)) and e = eccentricity (i.e. *first* eccentricity as in (3.63)).

2.1.4   *Orientation*

Without question, the method for defining the orientation of an object in space is one of the biggest conceptually hurdles to be crossed by engineering students. In its most sophisticated form, this is formulated using *quaternions* and will be developed later in this chapter. In its simplest form, this is formulated as a sequence of so-called *elementary rotations*, which are rotations about a single axis of a given axis system. After any rotation, the axis system is re-orientated and the next rotation is performed from there.

The trigonometry is easy to construct; the big problem is to appreciate the equivalence of a *rotation* and a *projection* when applied to objects, with the consequences that this has for the sequencing of elementary rotations. In a rotation, the vector rotates and the axis system remains fixed; in a projection, the vector remains fixed and the axis system rotates. Although the mathematical principles are the same, it makes sense to distinguish elementary *rotations* from elementary *projections* and then to introduce a single method of specification.

2.1.4.1 Simple Rotations and Projections

Consider a given vector with coordinates expressed in a given axis system. Initially, the vector can be assigned unit length. If the vector were specified as a radius vector in the positive quadrant of xy-plane, making a fixed angle α with the x-axis, then this would be resolved using basis vectors **x** and **y**:

$$\mathbf{v} = \mathbf{x}\cos\alpha + \mathbf{y}\sin\alpha \qquad (2.7)$$

Applying a positive rotation through an angle θ, this would become:

$$\begin{aligned}\mathbf{v}' &= \mathbf{x}\cos(\alpha+\theta) + \mathbf{y}\sin(\alpha+\theta)\\ &= \mathbf{x}(\cos\alpha\cos\theta - \sin\alpha\sin\theta) + \mathbf{y}(\sin\alpha\cos\theta + \cos\alpha\sin\theta)\end{aligned} \qquad (2.8)$$

The change in xy-coordinates can be summarised, as follows:

$$\begin{pmatrix} \cos(\alpha+\theta) \\ \sin(\alpha+\theta) \end{pmatrix} = \begin{pmatrix} \cos\theta & -\sin\theta \\ \sin\theta & \cos\theta \end{pmatrix}\begin{pmatrix} \cos\alpha \\ \sin\alpha \end{pmatrix} \qquad (2.9)$$

This operation can be re-interpreted as shown in Figure 2.5, where the radius vector from the origin to a point P is fixed inside an axis system defined by basis vectors $\mathbf{x_1}$ and $\mathbf{y_1}$. By rotating

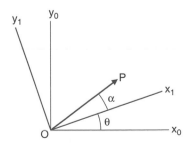

Figure 2.5 Elementary rotation in xy-plane.

the Ox_1y_1-frame the position vector will be rotated in the same way as expressed by (2.9). Adopting generalised coordinates, (2.7) can be re-written as:

$$\begin{aligned}\mathbf{v} &= \cos \alpha\, \mathbf{x}_1 + \sin \alpha\, \mathbf{y}_1 \\ &= x_1\, \mathbf{x}_1 + y_1\, \mathbf{y}_1\end{aligned} \tag{2.10}$$

Defining basis vectors $\mathbf{x_0}$ and $\mathbf{y_0}$ for the Ox_0y_0-frame, the new position vector (after the rotation has been applied) is:

$$\begin{aligned}\mathbf{v}' &= \cos (\alpha + \theta)\, \mathbf{x}_0 + \sin (\alpha + \theta)\, \mathbf{y}_0 \\ &= x_0\, \mathbf{x}_0 + y_0\, \mathbf{y}_0\end{aligned} \tag{2.11}$$

Thus, the rotation given by (2.9) can be written in the general form:

$$\begin{pmatrix} x_0 \\ y_0 \end{pmatrix} = \begin{pmatrix} \cos \theta & -\sin \theta \\ \sin \theta & \cos \theta \end{pmatrix} \begin{pmatrix} x_1 \\ y_1 \end{pmatrix} \tag{2.12}$$

This can be completely generalised in three dimensions by dropping the assumption of unit length and introducing a z-component. By construction, the rotation is a *positive* rotation about the z-axis (i.e. a clockwise rotation when viewed from the origin).[1] Because the z-component of a vector is not affected by a rotation about the z-axis, the full transformation can be expressed, as follows:

$$\textbf{z-Rotation:}\quad \begin{pmatrix} x_0 \\ y_0 \\ z_0 \end{pmatrix} = \begin{pmatrix} \cos \theta & -\sin \theta & 0 \\ \sin \theta & \cos \theta & 0 \\ 0 & 0 & 1 \end{pmatrix} \begin{pmatrix} x_1 \\ y_1 \\ z_1 \end{pmatrix} \tag{2.13}$$

Note that this transformation can be inverted by applying an elementary rotation through a negative angle. In this case, $\cos(-\theta) = \cos \theta$ and $\sin(-\theta) = -\sin \theta$ and, by inspection, it is seen that the matrix inverse is equal to the matrix transpose. This type of matrix is called *orthogonal* and the resulting transformations preserve right angles.

Now consider the corresponding method for projection. Effectively this takes the vector coordinates from one axis system and re-expresses them with respect to another axis system. To this end, Figure 2.6 shows the position vector of a point P relative to two

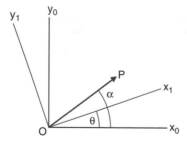

Figure 2.6 Elementary projection in xy-plane.

---

[1] Note that, if the z-axis were shown in Figure 2.5, it would point out of the page. Therefore a *positive* rotation appears as a counter-clockwise rotation.

axis systems, which are separated by an angle θ. On this occasion the vector is fixed with respect to the Ox_0y_0-frame (marked by a fixed angle α). By rotating the Ox_1y_1-frame the position vector will remain fixed but its coordinates (x_1,y_1) will change, as demonstrated below.

Again, define basis vectors $\mathbf{x}_0$ and $\mathbf{y}_0$ for Ox_0y_0 and $\mathbf{x}_1$ and $\mathbf{y}_1$ for Ox_1y_1. These basis vectors are interrelated by the following expressions:

$$\begin{aligned} \mathbf{x}_0 &= \mathbf{x}_1 \cos\theta - \mathbf{y}_1 \sin\theta \\ \mathbf{y}_0 &= \mathbf{x}_1 \sin\theta + \mathbf{y}_1 \cos\theta \end{aligned} \tag{2.14}$$

The position vector of point P relative to Ox_1y_1 is defined as:

$$\mathbf{r} = x_1\mathbf{x}_1 + y_1\mathbf{y}_1 \tag{2.15}$$

The corresponding position vector of point P relative to Ox_0y_0 (incorporating (2.11)) is then defined as:

$$\begin{aligned} \mathbf{r} &= x_0\mathbf{x}_0 + y_0\mathbf{y}_0 \\ &= x_0(\mathbf{x}_1 \cos\theta - \mathbf{y}_1 \sin\theta) + y_0(\mathbf{x}_1 \sin\theta + \mathbf{y}_1 \cos\theta) \\ &= (x_0 \cos\theta + y_0 \sin\theta)\mathbf{x}_1 + (x_0 \sin\theta - y_0 \cos\theta)\mathbf{y}_1 \end{aligned} \tag{2.16}$$

The rotation is obtained by equating coefficients of $\mathbf{x}_1$ and $\mathbf{y}_1$ from (2.15) and (2.16):

$$\begin{pmatrix} x_1 \\ y_1 \end{pmatrix} = \begin{pmatrix} \cos\theta & \sin\theta \\ -\sin\theta & \cos\theta \end{pmatrix} \begin{pmatrix} x_0 \\ y_0 \end{pmatrix} \tag{2.17}$$

As previously, the result obtained in the xy-plane can be expanded to include the z-axis. Thus the transformation in (2.17) can be rewritten, as follows:

$$\textbf{z-Projection:} \quad \begin{pmatrix} x_1 \\ y_1 \\ z_1 \end{pmatrix} = \begin{pmatrix} \cos\theta & \sin\theta & 0 \\ -\sin\theta & \cos\theta & 0 \\ 0 & 0 & 1 \end{pmatrix} \begin{pmatrix} x_0 \\ y_0 \\ z_0 \end{pmatrix} \tag{2.18}$$

Once again, the transformation can be inverted by applying an elementary projection through a negative opposite angle, which means that the transformation matrix is transposed.

Comparing (2.13) and (2.18), it should be clear that rotation and projection are equivalent mathematical processes. In fact, one process is just the inverse of the other. In addition, it should also be clear that the analysis performed in this section can easily be repeated for elementary projections and rotations about the y-axis and the x-axis. These observations will now be developed further.

2.1.4.2 Elementary Rotations

Having defined the trigonometry associated with elementary rotations and projections, it is useful to establish a standard notation for their subsequent application. There are a number of options, all equally valid, but the convention that is used throughout this book is, as follows:

$$R_x(\zeta) = \begin{pmatrix} 1 & 0 & 0 \\ 0 & \cos\zeta & \sin\zeta \\ 0 & -\sin\zeta & \cos\zeta \end{pmatrix} \tag{2.19}$$

$$R_y(\zeta) = \begin{pmatrix} \cos\zeta & 0 & -\sin\zeta \\ 0 & 1 & 0 \\ \sin\zeta & 0 & \cos\zeta \end{pmatrix} \tag{2.20}$$

$$R_z(\zeta) = \begin{pmatrix} \cos\zeta & \sin\zeta & 0 \\ -\sin\zeta & \cos\zeta & 0 \\ 0 & 0 & 1 \end{pmatrix} \tag{2.21}$$

Mathematically, these matrices implement what are strictly called 'elementary rotations', which is the opposite interpretation to that expounded up to now. Regardless of the differences in terminology, these matrices implement elementary rotations and projections, and therefore they have the following properties:

$$\begin{aligned} R_x(-\zeta) &= R_x^{-1}(\zeta) = R_x^T(\zeta) \\ R_y(-\zeta) &= R_y^{-1}(\zeta) = R_y^T(\zeta) \\ R_z(-\zeta) &= R_z^{-1}(\zeta) = R_z^T(\zeta) \end{aligned} \tag{2.22}$$

So, using this convention, an elementary projection relative to any given axis (assuming a positive angle ζ) would be applied using matrices $R_x(\zeta)$, $R_y(\zeta)$ or $R_z(\zeta)$, as appropriate. On the same basis, an elementary rotation would be applied using matrices $R_x(-\zeta)$, $R_y(-\zeta)$ or $R_z(-\zeta)$, as appropriate. Once the overall philosophy is understood and the initial confusion is resolved, it is quite natural to think in terms of mathematically defined 'elementary rotations' throughout and to adapt them for physical rotations or coordinate projections simply by changing the sign of the rotation angle. As ever, where sign changes are required, care is also required!

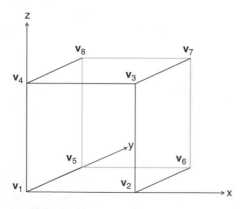

Figure 2.7 Vertex definitions for a cube.

By way of illustration, consider a cube shown in Figure 2.7, as defined by the following matrix of vertex coordinates:

$$\mathbf{V} = (\begin{matrix} \mathbf{v}_1 & \mathbf{v}_2 & \mathbf{v}_3 & \mathbf{v}_4 & \mathbf{v}_5 & \mathbf{v}_6 & \mathbf{v}_7 & \mathbf{v}_8 \end{matrix})$$

$$= \begin{pmatrix} 0 & 1 & 1 & 0 & 0 & 1 & 1 & 0 \\ 0 & 0 & 0 & 0 & 1 & 1 & 1 & 1 \\ 0 & 0 & 1 & 1 & 0 & 0 & 1 & 1 \end{pmatrix} \tag{2.23}$$

Assuming that the xy-plane is originally horizontal (with the z-axis pointing vertically upwards), the entire object can be rotated through an angle of 30° or $\pi/6$ radians about the x-axis by applying the following transformation:

$$\mathbf{V}' = R_x(-\pi/6)\mathbf{V}$$

For instance, this is the type of calculation that would be required to define the spatial position of an aircraft fuel tank at a bank angle of 30°, starting from a convenient datum frame. Verification can be achieved by simple calculation. For example, the new position of vertex $\mathbf{v}_4$ is derived as:

$$\mathbf{v}'_4 = \begin{pmatrix} 1 & 0 & 0 \\ 0 & \cos(\pi/6) & -\sin(\pi/6) \\ 0 & \sin(\pi/6) & \cos(\pi/6) \end{pmatrix} \begin{pmatrix} 0 \\ 0 \\ 1 \end{pmatrix} = \begin{pmatrix} 0 \\ -\sin(\pi/6) \\ \cos(\pi/6) \end{pmatrix} = \begin{pmatrix} 0 \\ -1/2 \\ \sqrt{3}/2 \end{pmatrix}$$

By inspection of Figure 2.6, this result is consistent with the physical rotation.

A related problem is to find the direction of the gravity vector, which is important in establishing the orientation of the fuel surface in equilibrium. Relative to the original datum frame in this particular example, the gravity vector is defined as:

$$\mathbf{g} = -\begin{pmatrix} 0 \\ 0 \\ g \end{pmatrix}$$

where g is the gravitational acceleration. Knowing that the tank is orientated at a bank angle of 30°, the components of gravity are established as:

$$\mathbf{g}' = R_x(\pi/6)\mathbf{g}$$

When fully expanded, this becomes:

$$\mathbf{g}' = -\begin{pmatrix} 1 & 0 & 0 \\ 0 & \cos(\pi/6) & \sin(\pi/6) \\ 0 & -\sin(\pi/6) & \cos(\pi/6) \end{pmatrix} \begin{pmatrix} 0 \\ 0 \\ g \end{pmatrix} = -g\begin{pmatrix} 0 \\ \sin(\pi/6) \\ \cos(\pi/6) \end{pmatrix} = -g\begin{pmatrix} 0 \\ 1/2 \\ \sqrt{3}/2 \end{pmatrix}$$

Again, by inspection, this result is consistent with the physical rotation.

In all problems involving rotations, the trick is to recognise what is being rotated and to associate vectors with the correct reference frame. If done systematically, then it should be obvious whether an object is actually being realigned with a given reference frame or whether it is simply being measured in a given reference frame.

2.1.4.3 Composite Rotations

An object can be subjected to an arbitrary sequence of elementary rotations in order to establish its orientation. In three-dimensional space, any orientation can be achieved with three rotations, each performed about a different axis. In special cases, one or more of these rotations might be zero. The general method is to apply the first rotation about one axis and, using the new orientation of the axis system, apply the next rotation about another axis and again, using the new orientation of the axis system, apply the final rotation about the remaining axis.

It is possible to specify orientation using more than three elementary rotations. Trivially, rotations of 2π radians can be applied indefinitely without effecting any physical change and a single rotation can be arbitrarily subdivided and applied in steps. However, there are genuine cases where multiple rotations are required.

One example is mechanical assembly involving many parts, each with its own local axis system. Ultimately, all local axis systems need to be reconciled with the reference frame for the assembled product, accommodating both *rigid-body* transformations (applied at points where attachments exist between physical parts) and *flexible-body* transformations (applied across physical parts and, if appropriate, between attachment points).

Another example of multiple rotations is shown in Figure 2.8. This starts from ECI axes (subscript '0'), and proceeds to ECEF axes (subscript '1'), to a set of intermediate axes aligned with the local meridian (subscript '2') and finally to the navigational axes of the platform (subscript '3').

The generic issue here is to identify the sequence of elementary rotations that will implement the desired composite rotation. To this end, consider a vector that is defined in the ECI frame (v_0) and proceed through a sequence of projections, as indicated by the frame subscripts in Figure 2.7:

$$\begin{aligned} \mathbf{v}_1 &= R_z(RA_G)\mathbf{v}_0 \\ \mathbf{v}_2 &= R_z(\lambda)\mathbf{v}_1 \\ \mathbf{v}_3 &= R_y(-\mu-\pi/2)R_y(-\mu)\mathbf{v}_2 \end{aligned} \tag{2.24}$$

This uses the results defined in (2.19), (2.20) and (2.21). Note that, in this case, positive rotations are being performed about the z-axes and negative rotations are being performed about the y-axis.[2]

Also note that successive rotations are performed with respect to the latest axis system, as mentioned already. The projection is constructed as follows:

$$\mathbf{v}_3 = R_y(-\mu-\pi/2)R_z(\lambda)R_z(RA_G)\mathbf{v}_0 \tag{2.25}$$

Because of the particular circumstances that apply here, it is possible to simplify the rotational sequence:

$$\mathbf{v}_3 = R_y(-\mu-\pi/2)R_z(\lambda + RA_G)\mathbf{v}_0 \tag{2.26}$$

This is a composite rotation, which can be written as a transformation:

$$\mathbf{v}_3 = E\mathbf{v}_0 \tag{2.27}$$

The transformation (i.e. rotation) matrix is obtained directly from (2.26):

$$E = R_y(-\mu - \pi/2)R_z(\lambda + RA_G) \tag{2.28}$$

---

[2] Obviously a rotation of $\pi/2$ radians is a half-revolution and could be applied in a positive (clockwise) sense or a negative (counter-clockwise) sense. In this case, it has been treated as negative in order to follow on from the preceding rotation.

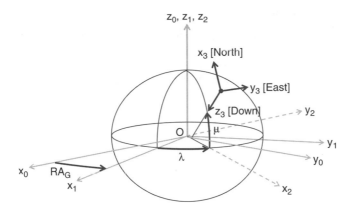

Figure 2.8 Relationship between ECI, ECEF and navigational axes

Having considered *projection*, the corresponding method for *rotation* must ensure that each step in the process is inverted and then the sequence re-assembled in a consistent manner. So, returning to the example shown in Figure 2.8, it is necessary to rearrange (2.24), as follows:

$$\begin{aligned}
\mathbf{v}_2 &= R_y(\mu + \pi/2)\mathbf{v}_3 \\
\mathbf{v}_1 &= R_z(-\lambda)\mathbf{v}_2 \\
\mathbf{v}_0 &= R_z(-RA_G)\mathbf{v}_1
\end{aligned} \tag{2.29}$$

Using the same simplification as before, the re-orientation is constructed as follows:

$$\mathbf{v}_0 = R_z(-RA_G)\,R_z(-\lambda)\,R_y(\mu + \pi/2)\mathbf{v}_3 \tag{2.30}$$

Given that the composite rotation matrix E has already been defined in 2.28, this can be written as a transformation:

$$\mathbf{v}_0 = E^T\mathbf{v}_3 \tag{2.31}$$

What this means is that, in order to perform a re-orientation, the elementary rotations that make up the original projection must each be inverted and then applied in the reverse sequence.

Note 2.1

Caution is needed here ... and a lot of it! This is a straightforward computational scheme but, from experience, nothing seems to cause so much confusion. Conceptually, it is difficult to grasp the practicalities of rotational sequences on first encounter but it is fundamental to many aspects of engineering geometry, which accounts for its frequent appearance in examinations. The big challenge is to recognise that placing an object in Frame A and then realigning it with Frame B is exactly the same as placing an object in Frame B and then projecting it into Frame A. It does not matter what 'Frame A' or 'Frame B' are, provided that their origins are coincident.

2.1.4.4 Direction Cosines

As part of any overview of coordinate geometry, it is always worth clarifying the concept of direction angles and direction cosines because these offer an alternative way of visualising the results of *projections* and *rotations*. Figure 2.9 shows the relative orientations of two axis systems, $Ox_0y_0z_0$ and $Oxyz$. *Direction angles* are drawn between all pairs of basis vectors, such that:

$$
\begin{aligned}
\mathbf{x}_0 &= \mathbf{x}\cos\alpha_1 + \mathbf{y}\cos\alpha_2 + \mathbf{z}\cos\alpha_3 \\
\mathbf{y}_0 &= \mathbf{x}\cos\beta_1 + \mathbf{y}\cos\beta_2 + \mathbf{z}\cos\beta_3 \\
\mathbf{z}_0 &= \mathbf{x}\cos\gamma_1 + \mathbf{y}\cos\gamma_2 + \mathbf{z}\cos\gamma_3
\end{aligned} \tag{2.32}
$$

$$
\begin{aligned}
\mathbf{x} &= \mathbf{x}_0\cos\alpha_1 + \mathbf{y}_0\cos\beta_1 + \mathbf{z}_0\cos\gamma_1 \\
\mathbf{y} &= \mathbf{x}_0\cos\alpha_2 + \mathbf{y}_0\cos\beta_2 + \mathbf{z}_0\cos\gamma_2 \\
\mathbf{z} &= \mathbf{x}_0\cos\alpha_3 + \mathbf{y}_0\cos\beta_3 + \mathbf{z}_0\cos\gamma_3
\end{aligned} \tag{2.33}
$$

Thus, the actual geometrical construction uses the cosine of direction angles, which gives a set of *direction cosines* (i.e. $\cos\alpha_1$, $\cos\beta_1$, and so on). Note that an equivalent formulation could be written that uses the dot product of each pair of basis vectors:

$$
\begin{aligned}
\mathbf{x}_0 &= \mathbf{x}(\mathbf{x}\bullet\mathbf{x}_0) + \mathbf{y}(\mathbf{y}\bullet\mathbf{x}_0) + \mathbf{z}(\mathbf{z}\bullet\mathbf{x}_0) & \mathbf{x} &= \mathbf{x}_0(\mathbf{x}_0\bullet\mathbf{x}) + \mathbf{y}_0(\mathbf{y}_0\bullet\mathbf{x}) + \mathbf{z}_0(\mathbf{z}_0\bullet\mathbf{x}) \\
\mathbf{y}_0 &= \mathbf{x}(\mathbf{x}\bullet\mathbf{y}_0) + \mathbf{y}(\mathbf{y}\bullet\mathbf{y}_0) + \mathbf{z}(\mathbf{z}\bullet\mathbf{y}_0) & \mathbf{y} &= \mathbf{x}_0(\mathbf{x}_0\bullet\mathbf{y}) + \mathbf{y}_0(\mathbf{y}_0\bullet\mathbf{y}) + \mathbf{z}_0(\mathbf{z}_0\bullet\mathbf{y}) \\
\mathbf{z}_0 &= \mathbf{x}(\mathbf{x}\bullet\mathbf{z}_0) + \mathbf{y}(\mathbf{y}\bullet\mathbf{z}_0) + \mathbf{z}(\mathbf{z}\bullet\mathbf{z}_0) & \mathbf{z} &= \mathbf{x}_0(\mathbf{x}_0\bullet\mathbf{z}) + \mathbf{y}_0(\mathbf{y}_0\bullet\mathbf{z}) + \mathbf{z}_0(\mathbf{z}_0\bullet\mathbf{z})
\end{aligned}
$$

An arbitrary position vector can be mapped on to the two axis systems, as follows:

$$
x\mathbf{x} + y\mathbf{y} + z\mathbf{z} = x_0\mathbf{x}_0 + y_0\mathbf{y}_0 + z_0\mathbf{z}_0
$$

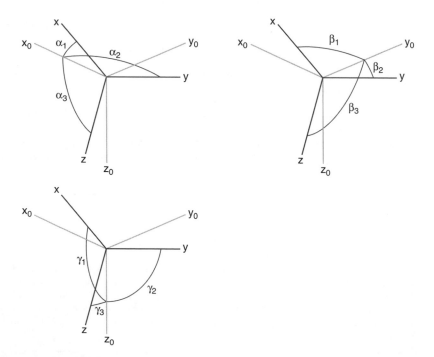

Figure 2.9 Direction angles.

Substitutions can be introduced from (2.32) or (2.33) and the vector components can then be equated to give a projection matrix and a rotation matrix, respectively:

$$
\begin{pmatrix} x \\ y \\ z \end{pmatrix} = \begin{pmatrix} \cos \alpha_1 & \cos \alpha_2 & \cos \alpha_3 \\ \cos \beta_1 & \cos \beta_2 & \cos \beta_3 \\ \cos \gamma_1 & \cos \gamma_2 & \cos \gamma_3 \end{pmatrix} \begin{pmatrix} x_0 \\ y_0 \\ z_0 \end{pmatrix}
\tag{2.34}
$$

$$
\begin{pmatrix} x_0 \\ y_0 \\ z_0 \end{pmatrix} = \begin{pmatrix} \cos \alpha_1 & \cos \beta_1 & \cos \gamma_1 \\ \cos \alpha_2 & \cos \beta_2 & \cos \gamma_2 \\ \cos \alpha_3 & \cos \beta_3 & \cos \gamma_3 \end{pmatrix} \begin{pmatrix} x \\ y \\ z \end{pmatrix}
\tag{2.35}
$$

Generically, a matrix of this type is called a *direction cosine matrix*. Often these matrices are written in shorthand form:

$$
\begin{pmatrix} x \\ y \\ z \end{pmatrix} = \begin{pmatrix} l_1 & l_2 & l_3 \\ m_1 & m_2 & m_3 \\ n_1 & n_2 & n_3 \end{pmatrix} \begin{pmatrix} x_0 \\ y_0 \\ z_0 \end{pmatrix}
\tag{2.36}
$$

$$
\begin{pmatrix} x_0 \\ y_0 \\ z_0 \end{pmatrix} = \begin{pmatrix} l_1 & m_1 & n_1 \\ l_2 & m_2 & n_2 \\ l_3 & m_3 & n_3 \end{pmatrix} \begin{pmatrix} x \\ y \\ z \end{pmatrix}
\tag{2.37}
$$

where

$$
\begin{aligned}
l_k &= \cos \alpha_k \\
m_k &= \cos \beta_k \\
n_k &= \cos \gamma_k
\end{aligned}
\tag{2.38}
$$

The fundamental constraints that apply to the x_0, y_0 and z_0 axes are, as follows:

$$
\begin{aligned}
\cos^2 \alpha_1 + \cos^2 \alpha_2 + \cos^2 \alpha_3 &= 1 \\
\cos^2 \beta_1 + \cos^2 \beta_2 + \cos^2 \beta_3 &= 1 \\
\cos^2 \gamma_1 + \cos^2 \gamma_2 + \cos^2 \gamma_3 &= 1
\end{aligned}
\tag{2.39}
$$

The equivalent constraints that apply to the x, y and z axes are, as follows:

$$
\begin{aligned}
\cos^2 \alpha_1 + \cos^2 \beta_1 + \cos^2 \gamma_1 &= 1 \\
\cos^2 \alpha_2 + \cos^2 \beta_2 + \cos^2 \gamma_2 &= 1 \\
\cos^2 \alpha_3 + \cos^2 \beta_3 + \cos^2 \gamma_3 &= 1
\end{aligned}
\tag{2.40}
$$

2.1.4.5 Euler Angles

The method of composite rotations is used specifically in order to define the orientation of a platform relative to its navigational reference frame (which is parallel with the ground reference frame). This is established by the sequence of rotations shown in Figure 2.10. It is

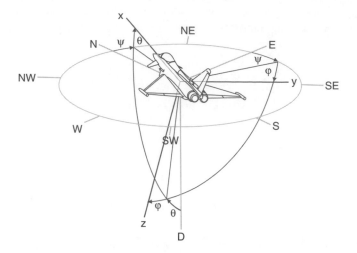

Figure 2.10 Frame rotation using Euler angles: azimuth (ψ) followed by pitch (θ) followed by roll (φ).

assumed that the aircraft is initially level with its xyz-axes aligned with North (**N**), East (**E**) and Down (**D**), respectively. The universal convention is to apply rotations in the following sequence[3]:

1. (ψ): *azimuth* angle clockwise about the **z**-axis;
2. (θ): *pitch* angle clockwise about the resulting **y**-axis;
3. (φ) or (ϕ): *roll* angle clockwise about the resulting **x**-axis.

Note that the roll angle is also known as the aircraft bank angle. In this context, these angles of rotation are called *Euler angles*. The following angular constraints apply:

$$-\pi/2 \leq \theta \leq \pi/2$$
$$-\pi < \varphi \leq \pi$$
$$0 \leq \psi < 2\pi$$

Euler angles give rise to the following projection from *navigational axes* (or *ground axes*) to *platform axes*:

$$\mathbf{A}_G = R_x(\varphi)R_y(\theta)R_z(\psi)$$

$$= \begin{pmatrix} 1 & 0 & 0 \\ 0 & \cos\varphi & -\sin\varphi \\ 0 & \sin\varphi & \cos\varphi \end{pmatrix} \begin{pmatrix} \cos\theta & 0 & \sin\theta \\ 0 & 1 & 0 \\ -\sin\theta & 0 & \cos\theta \end{pmatrix} \begin{pmatrix} \cos\psi & -\sin\psi & 0 \\ \sin\psi & \cos\psi & 0 \\ 0 & 0 & 1 \end{pmatrix} \quad (2.41)$$

---

[3] Mathematically, it does not matter what sequence is applied but, practically, it is crucial that everyone uses the same one. Also, the choice of frame subscripts is unimportant provided that they make good sense. Textbooks adopt different conventions and, as here, vary those conventions where it is convenient to do so.

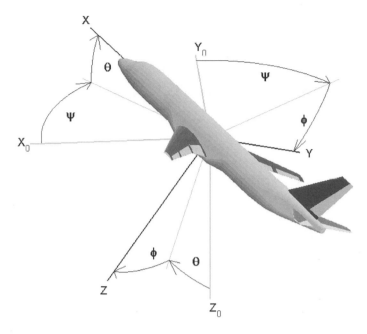

Figure 2.11 Visualisation of platform orientation.

This yields the familiar textbook expansion:

$$\mathbf{A}_G = \begin{pmatrix} \cos\theta\cos\psi & \cos\theta\sin\psi & -\sin\theta \\ \sin\varphi\sin\theta\cos\psi - \cos\varphi\sin\psi & \sin\varphi\sin\theta\sin\psi + \cos\varphi\cos\psi & \sin\varphi\cos\theta \\ \cos\varphi\sin\theta\cos\psi + \sin\varphi\sin\psi & \cos\varphi\sin\theta\sin\psi - \sin\varphi\cos\psi & \cos\varphi\cos\theta \end{pmatrix}$$

(2.42)

For clarity, Euler angles are again shown in Figure 2.11, in which the *platform axes* are marked as (x,y,z) while the *navigational axes* are marked as (x_0,y_0,z_0).

2.1.5   *Flight Axis System*

In atmospheric flight there is a requirement to account for the direction of the incident airflow. Starting with the platform axes, it is appropriate to derive a set of *flight axes*,[4] in which the x-axis is aligned with the velocity vector for any given flight condition. This needs to be done in a manner that is consistent with the definition of the airstream direction angles, namely the Angle of Attack (or AOA) (α) and Angle of Sideslip (or AOS) (β). This is illustrated in Figure 2.12.

The *first* step is to apply an elementary rotation that corresponds with the AOA. This is the pitch-up attitude of the platform with respect to the airflow and so the frame rotation must be applied about the y-axis through a negative angle ($-\alpha$). The *second* step

---

[4] Many books use the designation 'Flight Path Axes' or 'Wind Axes'.

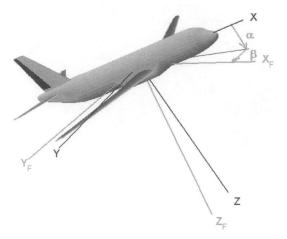

Figure 2.12 Flight axes.

is to apply an elementary rotation that corresponds with the AOS. This is a directional change that introduces a cross-wind velocity from starboard and so the subsequent frame rotation must be applied about the z-axis through a positive angle (β). These operations are combined as follows:

$$\mathbf{F}_A = R_z(\beta)R_y(-\alpha)$$

$$= \begin{pmatrix} \cos\beta & \sin\beta & 0 \\ -\sin\beta & \cos\beta & 0 \\ 0 & 0 & 1 \end{pmatrix} \begin{pmatrix} \cos\alpha & 0 & -\sin\alpha \\ 0 & 1 & 0 \\ \sin\alpha & 0 & \cos\alpha \end{pmatrix} \tag{2.43}$$

The complete projection from *platform axes to flight axes* is then given by:

$$\mathbf{F}_A = \begin{pmatrix} \cos\beta\cos\alpha & \sin\beta & -\cos\beta\sin\alpha \\ -\sin\beta\cos\alpha & \cos\beta & \sin\beta\sin\alpha \\ \sin\alpha & 0 & \cos\alpha \end{pmatrix} \tag{2.44}$$

By way of clarification, Figure 2.13 relates the airstream direction angles to velocity components (u,v,w) in platform axes. By convention, the velocity vector lies in the positive octant for positive α and β.

In the special case of zero sideslip (where $\beta = 0$), the Flight Path Axes are derived from a single rotation that corresponds with AOA. This is the flight condition that is prescribed in the analysis of aircraft stability, as will be developed in Volume 2, and so it is appropriate to refer to these axes as *stability axes* in this context.

Returning to the general case (where $\beta \neq 0$), the full sequence of elementary rotations that proceed from *ground axes* to *platform axes* and then to *flight axes* is shown in Figure 2.14. Based on (2.41) and (2.43), the associated projection is defined mathematically, as follows:

$$\mathbf{F}_G = \mathbf{F}_A.\mathbf{A}_G = R_z(\beta)R_y(-\alpha)R_x(\varphi)R_y(\theta)R_z(\psi) \tag{2.45}$$

---

[4] Many books use the designation 'Flight Path Axes' or 'Wind Axes'.

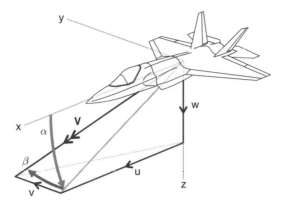

Figure 2.13 Airstream direction angles.

Figure 2.14 Ground axes, platform axes and flight axes.

As an alternative, it is possible to construct a direct projection from *ground axes* to *flight axes*, as shown in Figure 2.15. This is a combination of three elementary rotations, which establish the track angle (γ_3), the climb/dive angle (γ_2) and the bank angle (γ_1). Collectively, these can be referred to as *flight path angles*.

Following the method of elementary rotations, the required projection is:

$$\mathbf{F}_G = R_x(\gamma_1)R_y(\gamma_2)R_z(\gamma_3) \tag{2.46}$$

Reversing the process (as in Eshelby, 2000, Appendix A), the equivalent projection from *flight axes* to *ground axes* is achieved as:

$$\mathbf{G}_F = R_z(-\gamma_3)R_y(-\gamma_2)R_x(-\gamma_1) \tag{2.47}$$

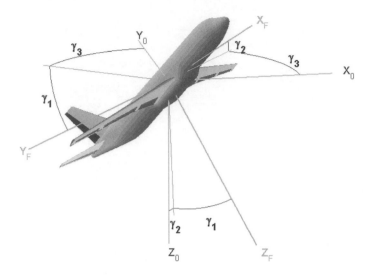

Figure 2.15 Flight path angles.

Thus, the circular projection starting and finishing with *ground axes* is:

$$\mathbf{G_F}.\mathbf{F_A}.\mathbf{A_G} = \mathbf{I} \tag{2.48}$$

where $\mathbf{I}$ is the 3-by-3 identity matrix.

It is interesting and useful to relate the Flight Path Angles ($\gamma_1, \gamma_2, \gamma_3$), the Euler Angles ($\psi, \theta, \phi$) and the airstream direction angles (α, β) (cf. Eshelby, 2000). First, the flight path angle (γ_2) is approximately:

$$\gamma_2 = \theta - \alpha \tag{2.49}$$

Secondly, the track angle (γ_2) is approximately:

$$\gamma_3 = \psi + \beta \tag{2.50}$$

This assumes small angles throughout. If the bank angles are zero then (2.49) becomes an exact equality. Note that the flight bank angle (γ_1) is not the same as the aircraft bank angle (ϕ) that was introduced in Section 2.1.4.5 (Euler Angles).

2.2   Changing Position and Orientation

Position is simply a translational displacement with components aligned with the axes of a given reference frame. This defines where an object is at a particular instant, as measured linearly along three independent (or orthogonal) directions. *Orientation* is a rotational displacement with components aligned with the axes of a given reference frame. This can be quantified in terms of Euler Angles (as discussed previously) or in terms of Quaternions (as will discussed later).

Movement relative to a flat Earth is purely a function of velocity components (u,v,w), as shown in Figure 2.2, and the projection matrix $\mathbf{A_G}$, as defined in (2.41) or (2.42). In this case,

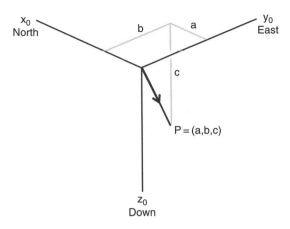

Figure 2.16 Position relative to ground axes.

aircraft position with respect to ground axes is defined by coordinates (x_0, y_0, z_0), with z_0 pointing downwards. Thus, the velocity over the ground is easily evaluated as:

$$\frac{d}{dt}\begin{pmatrix} x_0 \\ y_0 \\ z_0 \end{pmatrix} = \begin{pmatrix} \dot{x}_0 \\ \dot{y}_0 \\ \dot{z}_0 \end{pmatrix} = [\mathbf{A}_G]^T \begin{pmatrix} u \\ v \\ w \end{pmatrix} \qquad (2.51)$$

Note that altitude is *negative* ($-z_0$)! For clarification, the convention for position measurement is given in Figure 2.16.

Note 2.2

As a notational preference, time derivatives will normally be indicated using the standard operator (d/dt), e.g.

$$\frac{d}{dt}x \quad \text{or} \quad \frac{dx}{dt}.$$

Occasionally, for convenience and/or compactness, the alternative form will be used, with the variable being augmented with a dot directly above it, e.g.

$$\dot{x} \equiv \frac{dx}{dt}$$

Second-order derivatives will be represented in any of the following forms:

$$\ddot{x} \equiv \left(\frac{d}{dt}\right)^2 x \equiv \frac{d^2 x}{dt^2}$$

If required, higher-order derivatives (of order n) will be represented as follows:

$$x^{(n)} \equiv \left(\frac{d}{dt}\right)^n x \equiv \frac{d^n x}{dt^n}$$

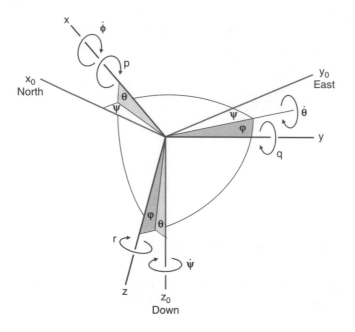

Figure 2.17 Rate of change in orientation.

Angular velocities p, q and r measure the rate at which the platform frame rotates. Thus the Euler angles that define the platform orientation must change over time. In order to calculate their rate of change, the individual rotations need to be transformed into platform axes, added together and then compared with the platform rates.

The relationship between angular velocities and Euler angles is depicted in Figure 2.17. Considering how projections are applied using elementary rotations, this can be transformed into the following mathematical prescription:

$$\begin{pmatrix} p \\ q \\ r \end{pmatrix} = \begin{pmatrix} \dot{\varphi} \\ 0 \\ 0 \end{pmatrix} + R_y(\theta) \begin{pmatrix} 0 \\ \dot{\theta} \\ 0 \end{pmatrix} + R_x(\varphi) R_y(\theta) \begin{pmatrix} 0 \\ 0 \\ \dot{\psi} \end{pmatrix} = \begin{pmatrix} \dot{\varphi} - \dot{\psi} \sin\theta \\ \dot{\theta} \cos\varphi + \dot{\psi} \cos\theta \sin\varphi \\ -\dot{\theta} \sin\varphi + \dot{\psi} \cos\theta \cos\varphi \end{pmatrix} \quad (2.52)$$

Thus, the Euler angle derivatives are defined about various axes of rotation but need to be harmonised with respect to platform axes. Rearranging this equation, the rate of frame rotation is found to be:

$$\frac{d}{dt} \begin{pmatrix} \varphi \\ \theta \\ \psi \end{pmatrix} \equiv \begin{pmatrix} \dot{\varphi} \\ \dot{\theta} \\ \dot{\psi} \end{pmatrix} = \begin{pmatrix} p + \bar{r}\, tan\theta \\ \bar{q} \\ \bar{r}\, sec\theta \end{pmatrix} \quad (2.53)$$

where

$$\bar{r} = q \sin\varphi + r \cos\varphi$$
$$\bar{q} = q \cos\varphi - r \sin\varphi$$

This formulation is perfectly adequate for most cases but, unfortunately, it breaks down as θ approaches ±90° (as tanθ becomes infinite) and it can induce significant numerical errors in rapid manoeuvres. These problems motivate the use of quaternions.

2.3   Rotating Axis Systems

2.3.1   *Inertial and Non-inertial Frames*

Inertial frames are crucial to the dynamics of vehicle motion because they define a fixed axis system within which the equations of motion can be defined. Newton's Newton's 2nd Law of Motion can then be used in order to derive true acceleration from a combination of force and inertia; hence the label *'inertial'*. This is fully justified since the axis system is *fixed* and, by definition, it does not accelerate or change its orientation. So, for flights with long range or long duration, the appropriate inertial frame is the Earth-Centred Inertial (ECI) frame.

This stipulation can be relaxed in special cases in which the axis system is not fixed *but* it does not accelerate and does not rotate to any appreciable extent over the time interval of interest. Accordingly, for flights of short range *and* short duration, it is sufficient to revert to a flat Earth reference frame. In this context, ground axes would constitute an inertial frame.

In all other situations, an axis system would be *non-inertial*. Acceleration and/or rotation of the reference frame would be great enough to invalidate dynamic calculations without specific compensation being applied. Note that, in a rotating axis system, the axis rotations will induce an apparent acceleration that has absolutely nothing to do with real forces or inertia. Looking at Figure 2.18, a fixed velocity vector **V** is resolved with respect to a rotating frame. Without loss of generality, the rotation is contained in the zx-plane. Over the sequence of time instants t_0, t_1 and t_2, the velocity components (u,w) are seen to change. Acceleration is defined as a change in velocity but, in this case, what is observed is not a real acceleration.

2.3.2   *Vector Differentiation*

Changes in vector quantities such as velocity and momentum are calculated in platform axes. If the frame is rotating then the calculation is being performed with respect to axis directions that are themselves changing.

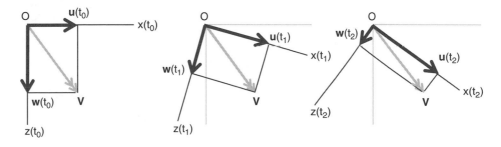

Figure 2.18 Virtual acceleration.

Conventionally, the velocity vector is defined in (2.1):

$$\mathbf{v} = u\mathbf{x} + v\mathbf{y} + w\mathbf{z}$$

Knowing that the rate of change can be calculated incrementally, the effect of axis system rotation can be constructed graphically as shown in Figure 2.19. The measured acceleration along the x-axis is $\partial u / \partial t$ but, because the x-axis has changed direction, this is a *partial* derivative. Over a time interval Δt, it can be seen that there are velocity changes perpendicular to the x-axis, which can be written as Δv_x and Δw_x. The lateral component can be derived from the forward velocity (u) and the yaw angle ($\Delta \psi$), such that:

$$\Delta v_x = u.\tan \Delta \psi$$

where

$$\Delta \psi = r.\Delta t$$

The average rate of change over the time interval is given by:

$$\frac{\Delta v_x}{\Delta t} = \frac{u.\tan \Delta \psi}{\Delta t} \approx \frac{u.\Delta \psi}{\Delta t} \approx r.u$$

Making Δt infinitesimally small so that true derivatives can be calculated, it is seen that the corresponding acceleration becomes:

$$\frac{dv_x}{dt} = r.u$$

Exactly the same reckoning can be applied to all other velocity components and the resulting process is summarised in Figure 2.19.

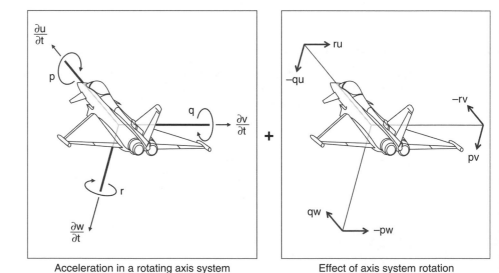

Acceleration in a rotating axis system   Effect of axis system rotation

Figure 2.19 Vector differentiation in a rotating axis system.

This process can be applied to any time-varying vector $\mathbf{r}(t)$. Assuming that its components are (x,y,z), the *total* derivative $d\mathbf{r}/dt$ (as calculated in a fixed axis system) is composed of two parts:

$$\frac{d}{dt}\begin{pmatrix} x \\ y \\ z \end{pmatrix} = \frac{\partial}{\partial t}\begin{pmatrix} x \\ y \\ z \end{pmatrix} + \begin{pmatrix} 0 & -r & q \\ r & 0 & -p \\ -q & p & 0 \end{pmatrix}\begin{pmatrix} x \\ y \\ z \end{pmatrix} \qquad (2.54)$$

The first term is the *partial* derivative of $\mathbf{r}$ with respect to time, calculated in the rotating axis system, and the second term compensates for the axis rotations.

The equation can be summarised as follows:

$$\frac{d}{dt}\mathbf{r} = \frac{\partial}{\partial t}\mathbf{r} + \mathbf{\Omega}\mathbf{r} \qquad (2.55)$$

where

$$\mathbf{\Omega} = \begin{pmatrix} 0 & -r & q \\ r & 0 & -p \\ -q & p & 0 \end{pmatrix} \qquad (2.56)$$

This is a compensation matrix that accounts for the instantaneous rotation of the platform axes. Thus, the term $\mathbf{\Omega}\mathbf{r}$ represents a set of fictitious accelerations that are simply a by-product of axis rotations.

An equivalent formulation is available as:

$$\frac{d}{dt}\mathbf{r} = \frac{\partial}{\partial t}\mathbf{r} + \mathbf{\omega} \times \mathbf{r} \qquad (2.57)$$

where the frame rotation is accounted by the cross-product of the angular velocity vector ($\mathbf{\omega}$) and the vector of interest ($\mathbf{r}$). The angular velocity is defined in (2.2).

2.3.3 *Poisson's Equation*

The above method of vector differentiation looks at the instantaneous rotation of an axis system in order to refer the calculation back to an inertial frame. An alternative method can be applied that considers the incremental change of vector defined within an inertial frame and then observed from within a rotating (noninertial) frame. The formal development is illustrated in Figure 2.20.

First, define a vector $\mathbf{r}$ in an inertial frame, with coordinates (x_0, y_0, z_0), that represents the position vector of point P. Over a small time interval Δt, an incremental change $\Delta\mathbf{r}$ gives a resultant vector $\mathbf{r}+\Delta\mathbf{r}$ that is directed to point Q. Applying the method of differential calculus, the total derivative is then defined as:

$$\frac{d\mathbf{r}}{dt} = \lim_{\Delta t \to 0}\left(\frac{\Delta\mathbf{r}}{\Delta t}\right)$$

Secondly, consider a rotated frame that is related to the inertial frame by a projection matrix $\mathbf{A}$. Knowing the original vector ($\mathbf{r}$) and its total derivative ($d\mathbf{r}/dt$), these can be re-expressed in the rotated frame as $\mathbf{A}\mathbf{r}$ and $\mathbf{A}(d\mathbf{r}/dt)$, respectively. Incidentally, a partial derivative ($\partial\mathbf{r}/\partial t$) would be re-expressed in the rotated frame as $\mathbf{A}(\partial r/\partial t)$.

Finally, consider a rotating frame within which the original vector ($\mathbf{r}$) is observed instantaneously as $\mathbf{A}(t)\mathbf{r}$ at time t, noting that the projection matrix is now time-dependent. Over a

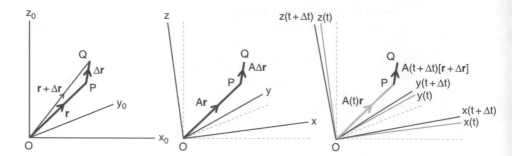

Figure 2.20 Alternative method for vector differentiation.

small time interval Δt, an incremental change Δr in the original vector would be observed from the rotating frame as $\mathbf{A}(t+\Delta t)(\mathbf{r}+\Delta \mathbf{r})$, i.e. the observer's frame has moved from $\mathbf{A}(t)$ to $\mathbf{A}(t+\Delta t)$. The new orientation can be defined as:

$$\mathbf{A}(t + \Delta t) = \mathbf{A}(t) + \frac{d\mathbf{A}(t)}{dt}\Delta t$$

Thus, the time derivative viewed from a rotating frame is constructed as follows:

$$\mathbf{A}(t)\frac{\partial \mathbf{r}}{\partial t} = \lim_{\Delta t \to 0} \left(\frac{\mathbf{A}(t + \Delta t)(\mathbf{r} + \Delta \mathbf{r}) - \mathbf{A}(t)\Delta \mathbf{r}}{\Delta t} \right)$$

$$= \lim_{\Delta t \to 0} \frac{1}{\Delta t} \left(\left(\mathbf{A}(t) + \frac{d\mathbf{A}(t)}{dt}\Delta t \right)(\mathbf{r} + \Delta \mathbf{r}) - \mathbf{A}(t)\Delta \mathbf{r} \right)$$

$$= \lim_{\Delta t \to 0} \frac{1}{\Delta t} \left(\mathbf{A}(t)\Delta \mathbf{r} + \frac{d\mathbf{A}(t)}{dt}\Delta t \ (\mathbf{r} + \Delta \mathbf{r}) \right)$$

$$= \lim_{\Delta t \to 0} \frac{1}{\Delta t} \left(\mathbf{A}(t)\Delta \mathbf{r} + \frac{d\mathbf{A}(t)}{dt}\mathbf{r}\Delta t \right)$$

Note that the vector $\mathbf{r}$ is expressed with respect to the original (inertial) frame but the derivative is expressed with respect to the rotating (noninertial) frame. This concept can be quite confusing but, recalling an earlier comment, a vector in one frame can be re-expressed in another by pre-multiplying by the relevant projection matrix, which is $\mathbf{A}(t)$ in this case.

This gives the partial derivative as:

$$\mathbf{A}(t)\left(\frac{\partial \mathbf{r}}{\partial t} \right) = \mathbf{A}(t)\left(\frac{d\mathbf{r}}{dt} \right) + \frac{d\mathbf{A}(t)}{dt}\mathbf{r} \tag{2.58}$$

Reverting back to the original method, the same quantity is derived as:

$$\mathbf{A}(t)\left(\frac{\partial \mathbf{r}}{\partial t} \right) = \mathbf{A}(t)\left(\frac{d\mathbf{r}}{dt} \right) - \Omega \ [\mathbf{A}(t)\mathbf{r}] \tag{2.59}$$

By simple comparison, Poisson's Equation is derived:

$$\frac{d\mathbf{A}(t)}{dt} = -\mathbf{\Omega}\mathbf{A}(t) \tag{2.60}$$

2.4   Quaternions

2.4.1   *Method of Construction*

Platform rotation is defined by an angular velocity vector $\boldsymbol{\omega} = p\mathbf{x}+q\mathbf{y}+r\mathbf{z}$, as defined in (2.2). This is interpreted in Figure 2.21, with an equivalent view that shows a single axis of rotation through the platform origin. Any orientation can be described by sequence of rotations in azimuth, pitch and roll; it can also be described by a single rotation about some axis in space, which will be called the *primary axis*. The orientation of this axis is defined by direction angles (α, β, γ) measured with respect to platform axes and depends on the orientation of the aircraft (as defined by Euler angles (ψ, θ, φ)). Note that, as the platform rotates, these direction angles will change.

The angle of rotation about the primary axis is defined as δ. Its rate of change is:

$$\frac{d\delta}{dt} = p.\cos\alpha + q.\cos\beta + r.\cos\gamma \tag{2.61}$$

This provides the basis for an alternative to Euler angles. The new scheme expresses aircraft orientation using four parameters, called *quaternions*. In this section, the method for their derivation is summarised.

Starting from ground axes (with basis vectors $\mathbf{x}_0, \mathbf{y}_0, \mathbf{z}_0$), a new axis system is defined with basis vectors $\mathbf{a}, \mathbf{b}, \mathbf{c}$. Vector $\mathbf{a}$ is aligned with the primary axis. Vectors $\mathbf{b}$ and $\mathbf{c}$ can be chosen in various ways but, for convenience, vector c is chosen to be orthogonal to $\mathbf{x}$ (and, of course, orthogonal to $\mathbf{a}$), as shown in Figure 2.22. Vector $\mathbf{b}$ is then derived in order to complete a right-handed axis system. Thus,

$$\mathbf{c} = \frac{\mathbf{a} \times \mathbf{x}_0}{\sin\alpha}$$
$$\mathbf{b} = \mathbf{c} \times \mathbf{a} \tag{2.62}$$

These vectors can be expanded in terms of the body axis components, to obtain

$$\mathbf{a} = \mathbf{x}_0 \cos\alpha + \mathbf{y}_0 \cos\beta + \mathbf{z}_0 \cos\gamma$$
$$\mathbf{b} = \frac{1}{\sin\alpha}\left[\mathbf{x}_0 \sin^2\alpha \; - \; (\mathbf{y}_0 \cos\beta - \mathbf{z}_0 \cos\gamma)\,\cos\alpha\right] \tag{2.63}$$
$$\mathbf{c} = \frac{1}{\sin\alpha}\left[\mathbf{y}_0 \cos\gamma \; - \; \mathbf{z}_0 \cos\beta\right]$$

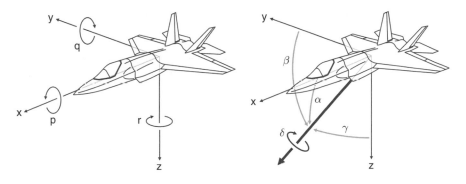

Figure 2.21 Platform rotational components.

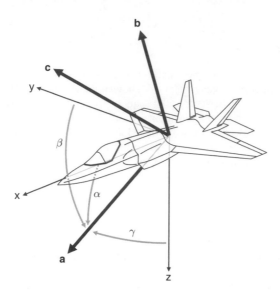

Figure 2.22 Axis system for analysing frame rotation.

Rotating the axis system about **a** through an angle δ will move the original x-axis to a new orientation defined by:

$$\mathbf{x} = \mathbf{a}\cos\alpha + (\mathbf{b}\cos\delta + \mathbf{c}\sin\delta)\sin\alpha \tag{2.64}$$

In component form, this becomes:

$$\begin{aligned}
\mathbf{x} = \ &\mathbf{x}_0[\cos^2\alpha + \sin^2\alpha\cos\delta] \\
&- \mathbf{y}_0[\cos\alpha\cos\beta(1-\cos\delta) + \cos\gamma\sin\delta] \\
&- \mathbf{z}_0\,[\cos\alpha\cos\gamma(1-\cos\delta) + \cos\beta\sin\delta]
\end{aligned} \tag{2.65}$$

Note that the primary axis is defined as:

$$\begin{aligned}
\mathbf{a} &= \mathbf{x}_0\cos\alpha + \mathbf{y}_0\cos\beta + \mathbf{z}_0\cos\gamma \\
&= \mathbf{x}\cos\alpha + \mathbf{y}\cos\beta + \mathbf{z}\cos\gamma
\end{aligned}$$

Using trigonometric formulae for half-angles, (2.65) is re-expressed as follows:

$$\begin{aligned}
\mathbf{x} = \ &\left[\cos^2\alpha\left(\cos^2\frac{\delta}{2} + \sin^2\frac{\delta}{2}\right) + (\cos^2\beta + \cos^2\gamma)\left(\cos^2\frac{\delta}{2} - \sin^2\frac{\delta}{2}\right)\right]\mathbf{x}_0 \\
&- \left[\cos\alpha\cos\beta\left(1 - \cos^2\frac{\delta}{2} + \sin^2\frac{\delta}{2}\right) + \left(2\cos\frac{\delta}{2}\sin\frac{\delta}{2}\right)\cos\gamma\right]\mathbf{y}_0 \\
&- \left[\cos\alpha\cos\gamma\left(1 - \cos^2\frac{\delta}{2} + \sin^2\frac{\delta}{2}\right) + \left(2\cos\frac{\delta}{2}\sin\frac{\delta}{2}\right)\cos\beta\right]\mathbf{z}_0
\end{aligned} \tag{2.66}$$

$$
\mathbf{x} = \left[\cos^2\frac{\delta}{2} + \left(\cos^2\alpha - \cos^2\beta - \cos^2\gamma \right) \sin^2\frac{\delta}{2} \right] \mathbf{x}_0
$$

$$
- 2 \left[\cos\alpha \cos\beta \sin^2\frac{\delta}{2} + \cos\gamma \cos\frac{\delta}{2} \sin\frac{\delta}{2} \right] \mathbf{y}_0 \tag{2.67}
$$

$$
- 2 \left[\cos\alpha \cos\gamma \sin^2\frac{\delta}{2} - \cos\beta \cos\frac{\delta}{2} \sin\frac{\delta}{2} \right] \mathbf{z}_0
$$

It is now possible to define the four *quaternions:*

$$
e_0 = \cos\left(\frac{\delta}{2} \right)
$$

$$
e_1 = \sin\left(\frac{\delta}{2} \right) \cos\alpha
$$

$$
e_2 = \sin\left(\frac{\delta}{2} \right) \cos\beta \tag{2.68}
$$

$$
e_3 = \sin\left(\frac{\delta}{2} \right) \cos\gamma
$$

Note, immediately, that these quantities are interrelated in the following way:

$$
e_0^2 + e_1^2 + e_2^2 + e_3^2 = 1 \tag{2.69}
$$

In addition, it should be noted that there are other possible definitions that involve sign changes and that would still satisfy (2.69). So the implementation is not unique but it is convenient to adopt straightforward definitions (and avoid minus signs).

This process can be repeated for the $\mathbf{y}$ and $\mathbf{z}$ vectors. Thus, the axis rotations are fully defined, as follows:

$$
\begin{aligned}
\mathbf{x} &= \left(e_0^2 + e_1^2 - e_2^2 - e_3^2 \right) \mathbf{x}_0 + 2(e_1 e_2 + e_3 e_0) \mathbf{y}_0 + 2(e_3 e_1 - e_2 e_0) \mathbf{z}_0 \\
\mathbf{y} &= 2(e_1 e_2 - e_3 e_0) \mathbf{x}_0 + \left(e_0^2 - e_1^2 + e_2^2 - e_3^2 \right) \mathbf{y}_0 + 2(e_2 e_3 + e_1 e_0) \mathbf{z}_0 \\
\mathbf{z} &= 2(e_3 e_1 + e_2 e_0) \mathbf{x}_0 + 2(e_2 e_3 - e_1 e_0) \mathbf{y}_0 + \left(e_0^2 - e_1^2 - e_2^2 + e_3^2 \right) \mathbf{z}_0
\end{aligned} \tag{2.70}
$$

The position vector of point P, relative to fixed axes, is defined as follows:

$$
\mathbf{r} = x_0 \mathbf{x}_0 + y_0 \mathbf{y}_0 + z_0 \mathbf{z}_0 \tag{2.71}
$$

Equivalently, this can be expressed as:

$$
\mathbf{r} = x\mathbf{x} + y\mathbf{y} + z\mathbf{z} \tag{2.72}
$$

Using the axis rotations given in (2.68), this can be re-written as:

$$
\begin{aligned}
\mathbf{r} &= \left[\left(e_0^2 + e_1^2 - e_2^2 - e_3^2 \right) x + 2\,(e_1 e_2 - e_3 e_0) y + 2\,(e_3 e_1 + e_2 e_0) z \right] \mathbf{x}_0 \\
&+ \left[2\,(e_1 e_2 + e_3 e_0) x + \left(e_0^2 - e_1^2 + e_2^2 - e_3^2 \right) y + 2\,(e_2 e_3 - e_1 e_0) z \right] \mathbf{y}_0 \\
&+ \left[2\,(e_3 e_1 - e_2 e_0) x + 2\,(e_2 e_3 + e_1 e_0) y + \left(e_0^2 - e_1^2 - e_2^2 + e_3^2 \right) z \right] \mathbf{z}_0
\end{aligned} \tag{2.73}
$$

Equating vector components in (2.69) and (2.73), the following transformation is obtained:

$$
\begin{pmatrix} x_0 \\ y_0 \\ z_0 \end{pmatrix} = \begin{pmatrix} e_0^2 + e_1^2 - e_2^2 - e_3^2 & 2(e_1e_2 - e_3e_0) & 2(e_3e_1 + e_2e_0) \\ 2(e_1e_2 + e_3e_0) & e_0^2 - e_1^2 + e_2^2 - e_3^2 & 2(e_2e_3 - e_1e_0) \\ 2(e_3e_1 - e_2e_0) & 2(e_2e_3 + e_1e_0) & e_0^2 - e_1^2 - e_2^2 + e_3^2 \end{pmatrix} \begin{pmatrix} x \\ y \\ z \end{pmatrix}
$$

Thus, the *projection* from ground axes to platform axes (cf. (2.42)) is given by:

$$
\begin{pmatrix} x \\ y \\ z \end{pmatrix} = \mathbf{A}_G \begin{pmatrix} x_0 \\ y_0 \\ z_0 \end{pmatrix}
$$

where

$$
\mathbf{A}_G = \begin{pmatrix} e_0^2 + e_1^2 - e_2^2 - e_3^2 & 2(e_1e_2 + e_3e_0) & 2(e_3e_1 - e_2e_0) \\ 2(e_1e_2 - e_3e_0) & e_0^2 - e_1^2 + e_2^2 - e_3^2 & 2(e_2e_3 + e_1e_0) \\ 2(e_3e_1 + e_2e_0) & 2(e_2e_3 - e_1e_0) & e_0^2 - e_1^2 - e_2^2 + e_3^2 \end{pmatrix} \tag{2.74}
$$

2.4.2   *Frame Rotation via Quaternions*

The components of rotation (p,q,r) are summarised in a vector (2.2):

$$
\boldsymbol{\omega} = p\mathbf{x} + q\mathbf{y} + r\mathbf{z}
$$

Also, for reference to Figure 2.22, the primary axis is given by:

$$
\mathbf{a} = \mathbf{x}\cos\alpha + \mathbf{y}\cos\beta + \mathbf{z}\cos\gamma
$$

Thus, the rate of rotation about the primary axis is easily derived as:

$$
\begin{aligned} \dot{\delta} &= \boldsymbol{\omega} \cdot \mathbf{a} \\ &= p\cos\alpha + q\cos\beta + r\cos\gamma \end{aligned} \tag{2.75}
$$

The quaternions (e_0, e_1, e_2, e_3) are defined in (2.68). Differentiating e_0, the rate of rotation appears explicitly:

$$
e_0 = \cos\frac{\delta}{2} \Rightarrow \dot{e}_0 = -\frac{1}{2}\sin\frac{\delta}{2}\dot{\delta}
$$

By a few simple substitutions, it is seen that the variation in this quaternion is a linear combination of the other quaternions.

$$
\dot{e}_0 = -\frac{1}{2}\sin\frac{\delta}{2}(p\cos\alpha + q\cos\beta + r\cos\gamma)
$$

$$
\dot{e}_0 = -\frac{1}{2}(pe_1 + qe_2 + re_3) \tag{2.76}
$$

The equivalent operations for e_1, e_2 and e_3, depend on Poisson's equation and the explicit form of the frame rotation matrix Ω. Thus,

$$\frac{d\mathbf{A_G}}{dt} = -\Omega\mathbf{A_G}$$

$$= -\begin{pmatrix} 0 & -r & q \\ r & 0 & -p \\ -q & p & 0 \end{pmatrix} \begin{pmatrix} e_0^2 + e_1^2 - e_2^2 - e_3^2 & 2(e_1e_2 + e_3e_0) & 2(e_3e_1 - e_2e_0) \\ 2(e_1e_2 - e_3e_0) & e_0^2 - e_1^2 + e_2^2 - e_3^2 & 2(e_2e_3 + e_1e_0) \\ 2(e_3e_1 + e_2e_0) & 2(e_2e_3 - e_1e_0) & e_0^2 - e_1^2 - e_2^2 + e_3^2 \end{pmatrix}$$

The diagonal elements of this equation are, as follows:

$$\begin{aligned}
e_0\dot{e}_0 + e_1\dot{e}_1 - e_2\dot{e}_2 - e_3\dot{e}_3 &= r(e_1e_2 - e_3e_0) - q(e_3e_1 - e_2e_0) \\
e_0\dot{e}_0 - e_1\dot{e}_1 + e_2\dot{e}_2 - e_3\dot{e}_3 &= p(e_2e_3 - e_1e_0) - r(e_1e_2 - e_3e_0) \\
e_0\dot{e}_0 - e_1\dot{e}_1 - e_2\dot{e}_2 + e_3\dot{e}_3 &= q(e_3e_1 - e_2e_0) - p(e_2e_3 - e_1e_0)
\end{aligned}$$

(2.77)

The fundamental constraint on quaternions is given by (2.62):

$$e_0^2 + e_1^2 + e_2^2 + e_3^2 = 1$$

The fundamental constraint on their rates of change is derived as:

$$e_0\dot{e}_0 + e_1\dot{e}_1 + e_2\dot{e}_2 + e_3\dot{e}_3 = 0 \qquad (2.78)$$

Thus, the relationships given by (2.70) can be re-written as:

$$\begin{aligned}
2(e_0\dot{e}_0 + e_1\dot{e}_1) &= r(e_1e_2 - e_3e_0) - q(e_3e_1 - e_2e_0) \\
2(e_0\dot{e}_0 + e_2\dot{e}_2) &= p(e_2e_3 - e_1e_0) - r(e_1e_2 - e_3e_0) \\
2(e_0\dot{e}_0 + e_3\dot{e}_3) &= q(e_3e_1 - e_2e_0) - p(e_2e_3 - e_1e_0)
\end{aligned}$$

Recalling (2.61), the dependency on the rate of change of e_0 can be removed easily, using the relationship:

$$-2e_0\dot{e}_0 = pe_1e_0 + qe_2e_0 + re_3e_0 \qquad (2.79)$$

The following equations are then obtained:

$$\begin{aligned}
2e_1\dot{e}_1 &= pe_1e_0 + re_1e_2 - qe_3e_1 \\
2e_2\dot{e}_2 &= qe_2e_0 + pe_2e_3 - re_1e_2 \\
2e_3\dot{e}_3 &= re_3e_0 + qe_3e_1 - pe_2e_3
\end{aligned}$$

(2.80)

Removing the common factor in each equation, the required quaternion derivatives are found. Combining these results with (2.76), the complete set of derivatives is:

$$\begin{aligned}
\dot{e}_0 &= -\frac{1}{2}(pe_1 + qe_2 + re_3) \\
\dot{e}_1 &= \frac{1}{2}(pe_0 + re_2 - qe_3) \\
\dot{e}_2 &= \frac{1}{2}(qe_0 + pe_3 - re_1) \\
\dot{e}_3 &= \frac{1}{2}(re_0 + qe_1 - pe_2)
\end{aligned}$$

(2.81)

This information can be summarised in matrix form, as follows:

$$
\begin{pmatrix} \dot{e}_0 \\ \dot{e}_1 \\ \dot{e}_2 \\ \dot{e}_3 \end{pmatrix} = \frac{1}{2} \begin{pmatrix} 0 & -p & -q & -r \\ p & 0 & r & -q \\ q & -r & 0 & p \\ r & q & -p & 0 \end{pmatrix} \begin{pmatrix} e_0 \\ e_1 \\ e_2 \\ e_3 \end{pmatrix}
\tag{2.82}
$$

This form has no explicit reference to derivatives of azimuth (ψ], pitch (θ] and roll (φ) and is free from their inherent limitations. However, this form does require an extra equation and its implementation is not without issues. In particular, the evaluation of these derivatives must conform to the constraint given in (2.78) and the resulting time update of the quaternion set must conform to the constraint given in (2.69).

For computation, (2.75) could usefully be re-expressed as:

$$
\dot{e} = \frac{1}{2} \Omega_e \; e
\tag{2.83}
$$

where

$$
\Omega_e = \begin{pmatrix} 0 & -p & -q & -r \\ p & 0 & r & -q \\ q & -r & 0 & p \\ r & q & -p & 0 \end{pmatrix}
\tag{2.84}
$$

2.4.3   *Relationship between Quaternions and Euler Angles*

The introduction to this parameterisation concludes with the method for translating between quaternions and measurable angles. Three key pieces of information are needed.

First, using quaternions (2.74), the projection from ground axes to platform axes is:

$$
\mathbf{A}_G = \begin{pmatrix} e_0^2 + e_1^2 - e_2^2 - e_3^2 & 2(e_1 e_2 + e_3 e_0) & 2(e_3 e_1 - e_2 e_0) \\ 2(e_1 e_2 - e_3 e_0) & e_0^2 - e_1^2 + e_2^2 - e_3^2 & 2(e_2 e_3 + e_1 e_0) \\ 2(e_3 e_1 + e_2 e_0) & 2(e_2 e_3 - e_1 e_0) & e_0^2 - e_1^2 - e_2^2 + e_3^2 \end{pmatrix}
$$

Secondly, the same projection expressed using Euler angles (2.42) is:

$$
\mathbf{A}_G = \begin{pmatrix} \cos\theta \cos\psi & \cos\theta \sin\psi & -\sin\theta \\ \sin\varphi \sin\theta \cos\psi - \cos\varphi \sin\psi & \sin\varphi \sin\theta \sin\psi + \cos\varphi \cos\psi & \sin\varphi \cos\theta \\ \cos\varphi \sin\theta \cos\psi + \sin\varphi \sin\psi & \cos\varphi \sin\theta \sin\psi - \sin\varphi \cos\psi & \cos\varphi \cos\theta \end{pmatrix}
$$

Finally, the constraint equation for quaternions (2.69) is:

$$
e_0^2 + e_1^2 + e_2^2 + e_3^2 = 1
$$

Taking the diagonal elements of the transformation matrix $\mathbf{A}_G$ in quaternion form, together with the constraint equation, the following relationships are derived:

$$
\begin{aligned}
A_{11} &= 2\left(e_0^2 + e_1^2\right) - 1 \\
A_{22} &= 2\left(e_0^2 + e_2^2\right) - 1 \\
A_{33} &= 2\left(e_0^2 + e_3^2\right) - 1
\end{aligned}
$$

where the symbol A_{ij} denotes the element of $\mathbf{A}_G$ at row-i and column-j.

These can be combined, as follows:

$$\mathbf{A}_{11} + \mathbf{A}_{22} + \mathbf{A}_{33} = 4e_0^2 - 1$$

Simple re-arrangement gives an expression for e_0, which can then be incorporated into the expressions for A_{11}, A_{22} and A_{33} via back-substitution. Thus, a set of quaternion expansions is available:

$$
\begin{aligned}
e_0^2 &= \frac{1}{4}(1 + A_{11} + A_{22} + A_{33}) \\
e_1^2 &= \frac{1}{4}(1 + A_{11} - A_{22} - A_{33}) \\
e_2^2 &= \frac{1}{4}(1 - A_{11} + A_{22} - A_{33}) \\
e_3^2 &= \frac{1}{4}(1 - A_{11} - A_{22} + A_{33})
\end{aligned}
\tag{2.85}
$$

Now the diagonal elements of the projection matrix $\mathbf{A}_G$ are also expressible as:

$$
\begin{aligned}
A_{11} &= \cos\theta\cos\psi \\
A_{22} &= \sin\varphi\sin\theta\sin\psi + \cos\varphi\cos\psi \\
A_{33} &= \cos\varphi\cos\theta
\end{aligned}
\tag{2.86}
$$

Applying these expressions to e_0, the following derivation is obtained:

$$
\begin{aligned}
4e_0^2 &= 1 + A_{11} + A_{22} + A_{33} \\
&= \left(\cos^2\frac{\psi}{2} + \sin^2\frac{\psi}{2}\right)\left(\cos^2\frac{\theta}{2} + \sin^2\frac{\theta}{2}\right)\left(\cos^2\frac{\varphi}{2} + \sin^2\frac{\varphi}{2}\right) \\
&\quad + \left(\cos^2\frac{\psi}{2} - \sin^2\frac{\psi}{2}\right)\left(\cos^2\frac{\theta}{2} - \sin^2\frac{\theta}{2}\right)\left(\cos^2\frac{\varphi}{2} + \sin^2\frac{\varphi}{2}\right) \\
&\quad + \left(\cos^2\frac{\psi}{2} - \sin^2\frac{\psi}{2}\right)\left(\cos^2\frac{\theta}{2} + \sin^2\frac{\theta}{2}\right)\left(\cos^2\frac{\varphi}{2} - \sin^2\frac{\varphi}{2}\right) \\
&\quad + \left(\cos^2\frac{\psi}{2} + \sin^2\frac{\psi}{2}\right)\left(\cos^2\frac{\theta}{2} - \sin^2\frac{\theta}{2}\right)\left(\cos^2\frac{\varphi}{2} - \sin^2\frac{\varphi}{2}\right) \\
&\quad + 8\cos^2\frac{\psi}{2}\sin^2\frac{\psi}{2}\cos^2\frac{\theta}{2}\sin^2\frac{\theta}{2}\cos^2\frac{\varphi}{2}\sin^2\frac{\varphi}{2} \\
&= 4\left(\cos\frac{\psi}{2}\cos\frac{\theta}{2}\cos\frac{\varphi}{2} + \sin\frac{\psi}{2}\sin\frac{\theta}{2}\sin\frac{\varphi}{2}\right)^2
\end{aligned}
$$

Taking positive square roots, an explicit parameterisation is obtained for e_0 in terms of the Euler angles. A similar deduction can be applied to the other quaternions. Thus the completion parameterisation is found to be:

$$
\begin{aligned}
e_0 &= \cos\frac{\psi}{2}\cos\frac{\theta}{2}\cos\frac{\varphi}{2} + \sin\frac{\psi}{2}\sin\frac{\theta}{2}\sin\frac{\varphi}{2} \\
e_1 &= \cos\frac{\psi}{2}\cos\frac{\theta}{2}\sin\frac{\varphi}{2} - \sin\frac{\psi}{2}\sin\frac{\theta}{2}\cos\frac{\varphi}{2} \\
e_2 &= \cos\frac{\psi}{2}\sin\frac{\theta}{2}\cos\frac{\varphi}{2} + \sin\frac{\psi}{2}\cos\frac{\theta}{2}\sin\frac{\varphi}{2} \\
e_3 &= \sin\frac{\psi}{2}\cos\frac{\theta}{2}\cos\frac{\varphi}{2} - \cos\frac{\psi}{2}\sin\frac{\theta}{2}\sin\frac{\varphi}{2}
\end{aligned}
\tag{2.87}
$$

These is the calculations that are necessary in order to initialise a simulation for a known orientation (ψ,θ,ϕ) at time $t = 0$. A set of inverse calculations are available by inspection of the projection matrix $\mathbf{A}_G$ (expressed using Euler angles) (2.42). Simple trigonometry dictates that:

$$\psi = \tan^{-1}\left(\frac{A_{12}}{A_{11}}\right)$$
$$\theta = \sin^{-1}(A_{13})$$
$$\varphi = \tan^{-1}\left(\frac{A_{23}}{A_{33}}\right)$$

(2.88)

Referring to the same matrix expressed using quaternions (2.74), the relevant matrix elements are:

$$A_{11} = e_0^2 + e_1^2 - e_2^2 - e_3^2$$
$$A_{12} = 2(e_1e_2 + e_3e_0)$$
$$A_{13} = 2(e_3e_1 - e_2e_0)$$
$$A_{23} = 2(e_2e_3 + e_1e_0)$$
$$A_{33} = e_0^2 - e_1^2 - e_2^2 + e_3^2$$

(2.89)

2.5 Line of Sight

Having established the mathematical principles involved in the spatial orientation, it is possible to establish lines of sight between an onboard observer and an offboard reference point. In essence, this is based on the range, bearing and elevation to a target but it has a more general application in the creation of outside-world visual scenes for use in flight simulation.

The geometrical scheme for visual referencing is shown in Figure 2.23. This shows an aircraft, with its position determined by the motion of the centre of gravity G. This is projected vertically downwards in order to give the current point G' on the ground, Note

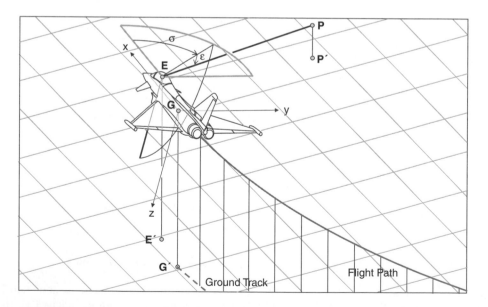

Figure 2.23 Observation from the pilot eye position.

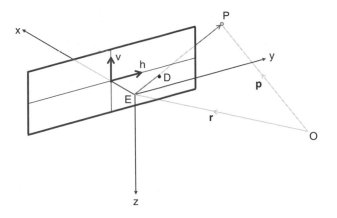

Figure 2.24 Projection on to a flat display surface.

that the flight path has associated with it a ground track, i.e. the locus of surface points that are overflown and are defined in xy-coordinates or lat-lon coordinates (as discussed in Chapter 2). The pilot sits forward of the centre of gravity and, for purposes of observation, the critical point is the so-called pilot eye position (shown as E), which is mid-way between left and right eyes, i.e. on the plane of symmetry of the head. This ignores binocular vision, in favour of a single vector that can be used for fixing relative position and for programming graphical displays. The observed point is P, which is shown above ground (as implied by the corresponding ground point P').

The line of sight is drawn from E to P. In this example, using E as the origin of the observational reference system (which is parallel with aircraft axis system), the relative positive vector is in the positive octant (i.e. its xyz-components are all positive). Thus, applying the standard convention for positive rotations, the orientation of vector EP can be defined by an azimuth angle (σ) and an elevation angle (ε). Positive azimuth is measured from the observer's x-axis around to starboard and, by virtue of the axis definitions, positive elevation is measured from the observer's xy-plane downwards.

Various display options can be invoked for this scenario, depending on the nature of the application and the sophistication of the available display screen equipment. The simplest is shown Figure 2.24. A flat display is placed orthogonal to the observer's x-axis, forward of the observer's position, and enables a field of view defined by horizontal and vertical offsets (measured as 'h' and 'v' coordinates on the display surface). As viewed from E, the target point is drawn as a display point D that lies on the true line of sight.

Relative to an origin at point O (somewhere!), trace vectors **p** and **r** out to the target point P and the eye position E, respectively. The line of sight is then determined as:

$$\mathbf{s} = \mathbf{p} - \mathbf{r} \tag{2.90}$$

This is projected on to the horizontal plane to create a vector w, where

$$\mathbf{w} = \mathbf{s} - (\mathbf{s} \bullet \hat{\mathbf{z}})\hat{\mathbf{z}} \tag{2.91}$$

All that is happening here is that the z-component of **s** is determined via a dot product between **s** and a unit vector along the z-axis and then is subtracted from **s**. Note that, by convention, a circumflex above the vector symbol denotes a vector of unit length. The resulting vector has the same xy-components as **s** but its z-component is zero, thereby demonstrating it to be a projection on to the xy-plane.

Azimuth and elevation angles (λ and θ, respectively) can be derived for the line of sight, as follows:

$$\lambda = \tan^{-1}\left(\frac{\hat{\mathbf{w}} \bullet \hat{\mathbf{y}}}{\hat{\mathbf{w}} \bullet \hat{\mathbf{x}}}\right) \tag{2.92}$$

$$\theta = \tan^{-1}\left(\frac{\hat{\mathbf{s}} \bullet \hat{\mathbf{z}}}{\hat{\mathbf{s}} \bullet \hat{\mathbf{w}}}\right) \tag{2.93}$$

Again, recognise that this relies on unit vectors. Basis vectors are predefined for x, y and z axes; the vectors **s** and **w** that are associated with the target have to normalised in this context (i.e. direction remains unchanged but length is scaled to one).

Now, the flat display projection is simple. If the display from the observer to the surface is R then the display coordinates are, as follows:

$$h = R \tan \lambda \tag{2.94}$$

$$v = R \tan \theta \ \sqrt{1 + \tan^2\lambda} \tag{2.95}$$

The inherent limitation is exposed by the tangent functions, such that wide-angle scenes would degrade outside the central region of the display. A simple solution would be to use multiple displays, as shown in Figure 2.25. In general, points that lie outside the display area would be ignored although care is always needed in order to ensure that objects that are drawn by interpolation are not inadvertently truncated as a result. An additional problem occurs with respect to target points behind the observer, depending on the implementation. The display coordinates might lie within the display area but they represent a reverse projection along the relevant sight-line; thus, the display would show spurious points that derive from objects behind the observer but projected in front of the observer. This is solved by ignoring all points with negative x-components (relative to the observer's frame of reference). Just to avoid any computational problems arising from division by zero or numerical errors close to the yz-plane, points with small positive components might also be ignored.

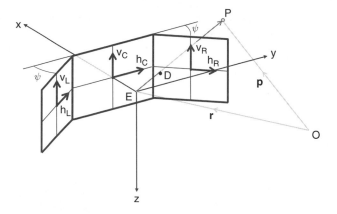

Figure 2.25 Projection on to multiple flat display surfaces.

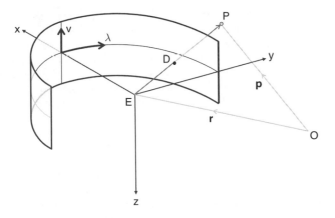

Figure 2.26 Projection on to a cylindrical display surface.

Having obtained the azimuth and elevation angles given by (2.85) and (2.86), plus the vertical displacement given by (2.88), two other display surfaces can be considered. Figure 2.26 shows a cylindrical display of a type often used for all-round immersive visualisation, with points located by azimuth and vertical offset. Figure 2.27 shows a spherical display, of a type often used for flight simulators housed in domes, with points located by azimuth and elevation. Whereas flat surfaces would normally be pixelated displays (or occasionally back-projected), the curved surfaces would be illuminated from a projection system that is nominally installed above the observer with a slight rearward offset. Clearly, in that configuration, a further analysis is required in order to establish the projection angles for any given target point. The principle is straightforward; in addition to azimuth and elevation angles from point E to point D (which are on the sight-line

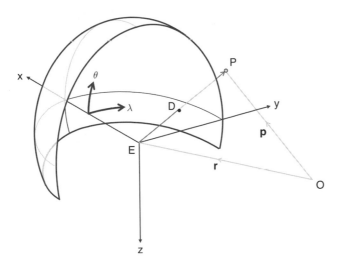

Figure 2.27 Projection on to a spherical display surface.

to point P), azimuth and elevation angles are needed from the display head to point D, duly compensated for any optical aberrations inherent in projection equipment.

The choice of display solution will depend on the type of technical investigation being undertaken. Smaller and flatter imagery is appropriate for engineering design and assessment while larger and more elaborate displays are appropriate for high-fidelity flight simulation. In particular, domes are very effective for wide-angle views of the outside world and many offer a high-resolution visual field directly ahead of the aircraft or steered via a pilot head/eye tracker, with a low-resolution visual field for peripheral vision (which can be blurred to simulate high-speed low-level flight). The overall effect is compelling and is often called 'the suspension of disbelief'. In other words, a simulated outside world appears very 'real', given the right context. Slightly different issues arise in low-speed low-level flight, especially associated with vertical take-off and landing and transitions between hover and forward flight.

At high speed, the focus of pilot attention is looking straight ahead in the direction of travel, occasionally scanning to coordinate the next manoeuvre. Essentially, below 50 knots, the focus shifts to assessing motion cues relative to the ground, with an increasingly interest in the view sidewards and downwards. Display resolution becomes very important, even down to the texture of the ground surface and the granularity of visual features, such as blades of grass and trees blowing in the wind. Ground handling is then conducted with as much of an all-round field of regard as is possible for a given aircraft type, based on the size and distribution of windows. Figure 2.28 gives a typical example of cockpit visibility for a commercial aircraft, showing limits of azimuth and elevation for twin flight crew based on nominal eye positions.

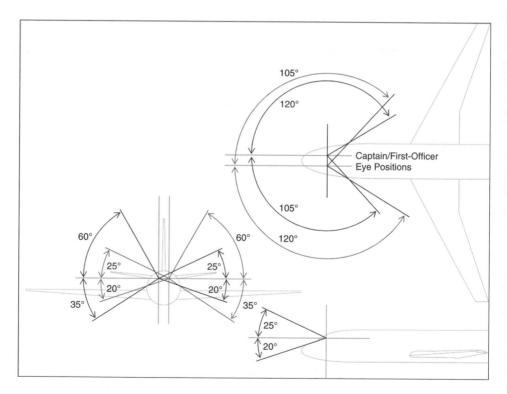

Figure 2.28 Cockpit visibility for a typical commercial aircraft.

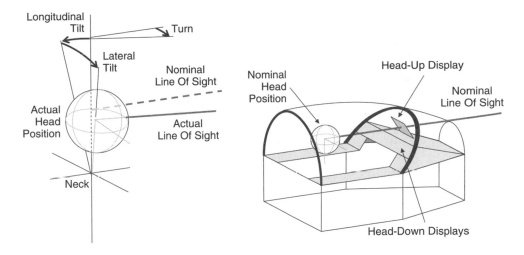

Figure 2.29 Projection on to a spherical display surface.

By way of adding a little more realism, Figure 2.29 illustrates the difference between *nominal* and *actual* head positions and their associated lines of sight. The trigonometry is the same in both cases; the distinction is that the eye position (denoted in earlier figures as point E) can move around. This is not a major issue for general observation of objects in the outside world. However, for objects observed through a Head-Up Display (HUD) that are aligned with projected symbology, this is potentially a serious matter. Geometrical symbols are designed to be viewed with both eyes close to the nominal slight-line and there is a notional eye-box (roughly 75 mm high and 200 mm wide) within which collimated images on the HUD are clearly visible to the observer. Near the boundary of that box, especially if one eye is outside, the geometrical perspective is altered. This would occur in the event of excessive tilting of the head.

As an aside, some military projects are using distributed electro-optic sensors in order to construct a full spherical view-field. This offers the opportunity to augment the visual scene with digital images of the outside world. Potentially this would include perspectives that otherwise would not be viewable, e.g. underneath the aircraft, and greatly enhance situational awareness in poor visibility. The fundamental problem is that digital images have to be acquired, processed and projected, which implies a latency between image capture and image display. This could result in misalignment of images relative to the actual visual scene, especially if viewed through a Helmet-Mounted Display (HMD), with the consequent risk of confusion, eye-strain or nausea. Solutions are also available for commercial aircraft that use a combination of electro-optic and radar inputs in order to create an all-weather visual enhancement (akin to the quality of night vision systems). If projected on to a HUD then much the same issues need to be addressed.

The spatial orientation of an aircraft is an essential part of the pilot-vehicle interface. As part of the primary flight information (which is safety-critical), it will appear on a HUD in many military aircraft and in increasing number of new civil aircraft. Therefore, it is worth considering the relevant symbology as an overlay to the outside-world visual scene, continuing and expanding the discussion on display geometry. Figure 2.30 depicts the particular geometry that defines instantaneous pitch and roll of an aircraft in flight. In this view, an aircraft is flying due North and entering a climb; its x-axis points on to a circle around the aircraft that is drawn in the vertical plane. The pitch angle is defined between the x-axis and the horizontal ground

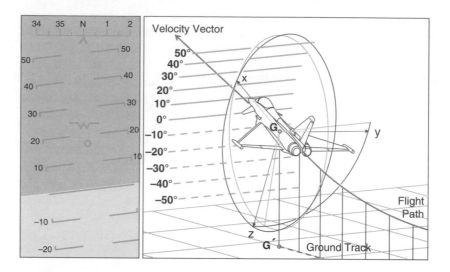

Figure 2.30 Basic head-up display symbology.

plane. The aircraft rolls about the x-axis and the yz-axes are shown in such a way as to imply a small bank angle to starboard. In addition, the velocity vector is shown. In effect, this simply re-expresses the Euler Angles and Airstream Direction Angles, as shown variously in Figures 2.10 through to 2.14.

Pitch attitudes are incorporated within the aircraft motion diagram, showing an artificial horizon at 0° and pitch angles in increments of 10° above and below. These are horizontal lines (or bars) which, by definition, are perpendicular to the vertical plane. Typical HUD symbology is shown on the left of Figure 2.30, against a background that suggests the division between land and sky. The HUD centre-line is aligned with the aircraft centre-line and, since the aircraft in this example is banking to starboard, the artificial horizon and the horizontal pitch bars are inclined to port. The current pitch attitude is indicated by the relationship between the pitch bars and the aircraft symbol that defines the orientation of the x-axis. This can appear in several forms; here, it is a 'W' with 'wings' while elsewhere it might be drawn as a circle with 'wings' and a 'fin'. In this display, a circle is used to indicate the direction of the velocity vector and, above the pitch bars, is a compass segment and a marker that shows the current heading (which happens to be North).

Chapter 3

Geospatial Reference Model

3.1 Spherical Earth

In its simplest representation, the Earth can be approximated as a sphere, rotating about a fixed axis that intersects the surface at the north and south poles. The plane that lies midway between the poles, orthogonal to the axis of rotation, bisects the earth into north and south hemispheres. On the surface, the line at which the hemispheres meet is the *equator*.

Positions on the surface are defined by *latitude* (measured as an angle north or south of the equator) and *longitude* (measured as an angle east or west of the international datum that passes through Greenwich). Lines of latitude are called *parallels* (i.e. parallel with the equator); lines of longitude are called *meridians*, which are great circles that pass through the poles. Parallels and meridians intersect each other at right angles, to create a grid pattern that is referred to as a graticule. This is shown in Figure 3.1, with an overlay of the world coastline.

The shortest distance between any two points is measured along a path that is called a *great circle*. This is defined by the intersection of the surface of a sphere and a plane passing through the centre. The resulting circle has the largest diameter possible and, accordingly, any other circle is called a *small circle*. In general, the property of minimum distance defines a *geodesic* (or a geodetic line) across a surface of arbitrary shape. The term 'geodesy' is used to refer to earth measurement on a large scale and, in particular, surveying with allowance for the Earth's curvature. This will be considered in some detail later in this chapter (in Section 3.7).

The area swept out between two meridians is a *lune*; the swept area between two parallels is a *zone*. The intersection of a lune and a zone is a quadrangle, which (in common with its plane equivalent) is a two-dimensional shape with four 'straight' sides that meet at right angles. This is shown in Figure 3.2. The determination of surface area enclosed between longitudes (λ_1, λ_2) and latitudes (ϕ_1, ϕ_2) is straightforward, constructed as a linear strip of length $(\lambda_2 - \lambda_1)\cos\phi$ and width $d\phi$ (positioned at an arbitrary latitude ϕ) and then integrated. Thus, the area (A) of the quadrangle is:

$$A = \int_{\phi_1}^{\phi_2} (\lambda_2 - \lambda_1) \, \cos\phi \, d\phi = (\lambda_2 - \lambda_1)(\sin\phi_2 - \sin\phi_1) \tag{3.1}$$

This can be expressed as a fraction of the total surface area by dividing by 4π.

Computational Modelling and Simulation of Aircraft and the Environment D.J. Diston
© 2009 John Wiley & Sons, Ltd

Figure 3.1 Spherical Earth.

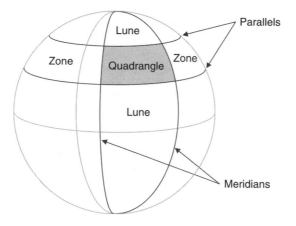

Figure 3.2 Spherical quadrangle.

Example 3.1

The area enclosed between longitudes (0°, 30°) and latitudes (15°, 45°) is as follows:

$$A = \frac{\pi}{6} \left(\sin \frac{\pi}{4} - \sin \frac{\pi}{12} \right) = 0.234723$$

measured in steradians. This represents 1.86787 % of the Earth surface and, assuming a radius of 6371km, the actual physical area is found to be 9527341km^2. This can be verified using the MATLAB Mapping Toolbox, as follows:

```
>>A = areaquad(15,0,45,30,6371)
   A = 9.5273e+006
```

Note that, in nautical miles,[1] the radius is 3440 nm and the area is 2777728 nm^2.

The origin of physical coordinates is located at the centre of the Earth (i.e. the point of intersection between the equatorial plane and the axis of rotation). The original of surface coordinates, otherwise known as *lat-lon*, is the intersection between the equator (0° latitude) and the Greenwich meridian (0° longitude). In order to provide a spatial reference frame, project a line out from the earth centre through the lat-lon origin and call this 'x'; then, rotate that line around the equator by 90° and call this 'y'; finally, align 'z' with the north pole. By convention, this is the Earth-Centred Earth-Fixed (ECEF) reference frame, shown in Figure 3.3. Note that an earth-fixed frame rotates with the earth.

Relative to the ECEF frame, the xyz-coordinates of an object can be determined from its longitude (λ), its geocentric latitude (ϕ) and its altitude (h), as follows:

$$\begin{aligned} x &= (R + h)\cos\phi\cos\lambda \\ y &= (R + h)\cos\phi\sin\lambda \\ z &- (R + h)\sin\phi \end{aligned} \qquad (3.2)$$

where R is the radius of the Earth.

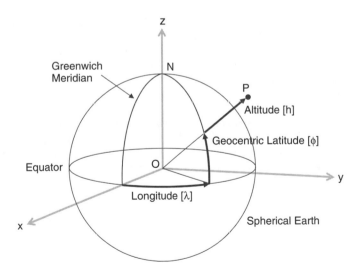

Figure 3.3 ECEF reference frame for a spherical Earth.

---

[1] The International Nautical Mile is defined as 1 nm = 1852 m.

The inverse calculation of (λ, ϕ, h) from (x,y,z) is conducted simply, as follows:

$$h = r - R$$
$$\phi = \tan^{-1}\left(\frac{z}{w}\right) \tag{3.3}$$
$$\lambda = \tan^{-1}\left(\frac{y}{x}\right)$$

where

$$r = \sqrt{x^2 + y^2 + z^2}$$
$$w = \sqrt{x^2 + y^2} \tag{3.4}$$

3.2   Spherical Trigonometry

Point-to-point navigation on a flat surface is trivial; it is based on the well-known rules of *plane trigonometry*. Consider the paths between three points A, B and C, drawn on a flat surface, as shown in Figure 3.4. These create a triangle ABC, with sides of length a, b and c. Recognising that this triangle can be constructed from two right-angled triangles, various applications of Pythagoras' theorem yield the following results:

$$\text{Sine Rule:} \quad \frac{\sin A}{a} = \frac{\sin B}{b} = \frac{\sin C}{c} \tag{3.5}$$

$$\text{Cosine Rule:} \quad a^2 = b^2 + c^2 - 2bc \, \cos A \tag{3.6}$$

Therein, a standard convention has been used such that the angle subtended at a particular vertex is given the same symbol as the vertex label.

The path from a current location to a destination is simply calculated as a range and heading, from the relevant xy-coordinates. For practical navigation in the real world, this is acceptable for short-range operations on what can be considered to be a 'Flat Earth'. Achieving the same capability over a spherical Earth is a completely different matter and requires the application of *spherical trigonometry*.

The basic problem is illustrated in Figure 3.5, showing three locations A, B and C, joined by paths that are constructed using great circles. These paths form a spherical triangle. In common with its plane counterpart, this has three sides of minimum length and three corners with interior angles: in contradistinction, the interior angles do not add up to 180° and the sides are measured as angles.

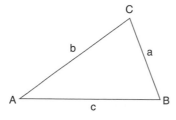

Figure 3.4 Plane triangle.

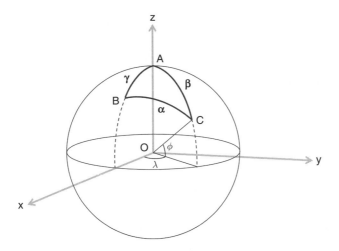

Figure 3.5 Spherical triangle.

Using the axis system shown, point A is at the North Pole and point C is defined by latitude and longitude. So the coordinates of C are as follows:

$$x = \cos\phi\cos\lambda$$
$$y = \cos\phi\sin\lambda \quad\quad\quad\quad (3.7)$$
$$z = \sin\phi$$

Applying the standard convention in geometry of using the same symbol to designate a vertex of a triangle and the interior angle at that vertex, it is obvious that $A = \lambda$. Also, noting that the sides of a spherical triangle are measured as angles (and that they are constructed using great circles), it is seen that $\phi = 90° - \beta$. Using these substitutions, the coordinates of C can be recast as:

$$x = \sin\beta\cos A$$
$$y = \sin\beta\sin A \quad\quad\quad\quad (3.8)$$
$$z = \cos\beta$$

The same problem is shown in Figure 3.6, referred to an axis system that has been rotated through an angle γ about the original y-axis. Thus, point B has become the new North Pole and the coordinates of point C are defined by a new latitude and a new longitude:

$$x' = \cos\phi'\cos\lambda'$$
$$y' = \cos\phi'\sin\lambda' \quad\quad\quad\quad (3.9)$$
$$z' = \sin\phi'$$

Applying similar reasoning as previously, it should be clear that $B = 180° - \lambda'$ and $\phi' = 90° - \alpha$. Making these substitutions, the coordinates become:

$$x' = -\sin\alpha\cos B$$
$$y' = \sin\alpha\sin B \quad\quad\quad\quad (3.10)$$
$$z' = \cos\alpha$$

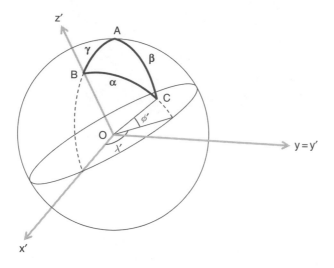

Figure 3.6 Equivalent spherical triangle.

The relationship between coordinates in the two axis systems can be established as an elementary rotation:

$$x' = x\cos\gamma - z\sin\gamma$$
$$y' = y$$
$$z' = x\sin\gamma + z\cos\gamma$$

(3.11)

Combining (3.8), (3.10) and (3.11), the following derivations can be made:

$$-\sin\alpha\cos B = (\sin\beta\cos A)\cos\gamma - (\cos\beta)\sin\gamma$$
$$\sin\alpha\sin B = \sin\beta\sin A$$
$$\cos\alpha = (\sin\beta\cos A)\sin\gamma + (\cos\beta)\cos\gamma$$

The first relationship is of no particular interest but the second and third relationships give rise to the fundamental rules of spherical trigonometry:

$$\text{Sine Rule:} \frac{\sin A}{\sin\alpha} = \frac{\sin B}{\sin\beta} = \frac{\sin C}{\sin\gamma}$$

(3.12)

$$\text{Cosine Rule:} \cos\alpha = \cos\beta\cos\gamma + \sin\beta\sin\gamma\cos A$$

(3.13)

Note that the cosine rule can be generalised for angles B and C and their respective opposite sides, β and γ. These can be used in a navigational context as shown in Figure 3.7 and as described in Example 3.2. For two points with lat-lon coordinates (ϕ_1,λ_1) and (ϕ_2,λ_2), respectively, the great-circle distance (α) is given by:

$$\alpha = \cos^{-1}[\cos\theta_1\cos\theta_2 + \sin\theta_1\sin\theta_2\cos(\lambda_2 - \lambda_1)]$$

(3.14)

where it is appropriate to use co-latitudes $\theta_1 = 90° - \phi_1$ and $\theta_2 = 90° - \phi_2$. For small distances (which is the most common calculation), this formula incurs significant inaccuracies. In order

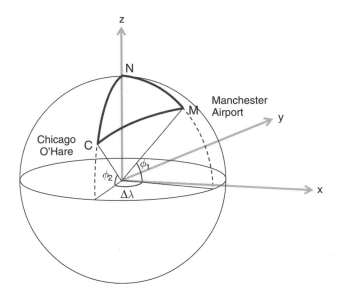

Figure 3.7 Navigation example.

to avoid this problem, an alternative method is available based on *haversines*, which will be developed in the next section.

Example 3.2

In order to find the approximate distance between Manchester Airport and Chicago O'Hare, construct a spherical triangle NMC, where N is the North Pole, M is Manchester and C is Chicago. Let the sides of this triangle be PM, PC and MC.

The airport locations are as follows:

Manchester Airport (53.3536°N 2.2747 °W)
Chicago O'Hare (41.9794°N 87.9044 °W)

The respective co-latitudes are:

Manchester Airport: NM = 36.6464°
Chicago O'Hare: NC = 48.0206°

The longitude difference is N = 85.6297° (subtended at the North Pole).
In this application, the cosine rule is:

$$\cos MC = \cos NM \cos NC + \sin NM \sin NC \cos N$$

This is evaluated as follows:

$$\cos MC = \cos 36.4646° \cos 48.0206° + \sin 36.4646° \sin 48.0206° \cos 85.6297°$$
$$= 0.570464$$

Therefore,

$$MC = 55.2174°$$

(continued)

Based on an approximate Earth radius of 6371 km, the distance along this great-circle path is *3315 nautical miles* (recalling that a nautical mile is 1852 m). Also, the angle between the North Pole and Chicago, subtended at Manchester, can also be found by applying the cosine rule:

$$\sin NC = \cos MC \cos NM + \sin MC \sin NM \cos M$$
$$\Rightarrow \quad \cos M = 0.430742$$
$$\Rightarrow \quad M = 64.4853°$$

Therefore, the flight must depart from Manchester on heading 295.5147° (i.e. just above East-North-East on the compass), which shows that great circle navigation may initially head north in order to reach a southerly destination. Note that the heading will vary continuously between Manchester and Chicago.

Standard trigonometric functions are sine, cosine and tangent, defined for any angle (θ). Equivalents functions defined for the complementary angle ($90°-\theta$) are cosecant, secant and cotangent, respectively. These are illustrated in Figure 3.8, relative to a unit circle, together with other less familiar functions.

The *versine* (or 'versed sine') is defined as:

$$\text{versin } \theta = 1 - \cos\theta \tag{3.15}$$

The versine of the complementary angle is the *coversine* (or 'coversed sine'):

$$\text{coversin}\theta = \text{versin}\left(\frac{\pi}{2} - \theta\right) = 1 - \sin\theta \tag{3.16}$$

The *exsecant* (or "external secant") is defined as:

$$\text{exsec}\theta = \sec\theta - 1 \tag{3.17}$$

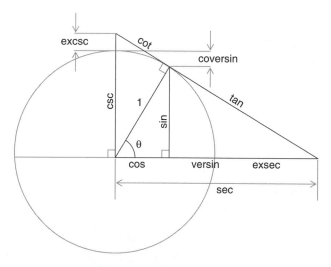

Figure 3.8 Trigonometric functions.

The exesecant of the complementary angle is the *excosecant*:

$$\text{excsc } \theta = \text{exsec}\left(\frac{\pi}{2} - \theta\right) = \csc \theta - 1 \tag{3.18}$$

The spherical cosine rule (3.11) can be re-expressed in the following way:

$$\begin{aligned}\cos \alpha &= \cos \beta \cos \gamma + \sin \beta \sin \gamma \cos A \\ &= \cos (\beta - \gamma) - \sin \beta \sin \gamma (1 - \cos C)\end{aligned}$$

In navigation, it is usual to apply a half-versine or *haversine*:

$$\text{hav}\theta = \frac{1}{2}(1 - \cos \theta) \tag{3.19}$$

Thus, the cosine rule can be re-developed as follows:

$$1 - 2\text{hav}\alpha = 1 - 2\text{hav}(\beta - \gamma) - 2 \sin \beta \sin \gamma \text{hav}A$$

With simplification, the end result is the *haversine formula*:

$$\text{hav}\alpha = \text{hav}(\beta - \gamma) + \sin \beta \sin \gamma \text{hav}A \tag{3.20}$$

This is important because it overcomes the inherent limitations of the cosine rule when calculating small distances. The new method of calculation is much more accurate and appears in the form:

$$\alpha = \text{hav}^{-1}[\text{hav}(\theta_1 - \theta_2) + \sin \theta_1 \sin \theta_2 \text{hav}(\lambda_1 - \lambda_2)] \tag{3.21}$$

The *haversine* can also be defined as:

$$h = \text{hav}\theta = \sin^2 \frac{\theta}{2} \tag{3.22}$$

which means that the *inverse haversine* is:

$$\theta = \text{hav}^{-1}(h) = 2 \sin^{-1}\sqrt{h} \tag{3.23}$$

Accordingly, the distance formula (3.19) can be quoted in terms of standard trigonometric functions, as follows:

$$\alpha = 2\sin^{-1}\sqrt{\sin^2\left(\frac{\theta_1 - \theta_2}{2}\right) + \sin \theta_1 \sin \theta_2 \sin^2\left(\frac{\lambda_1 - \lambda_2}{2}\right)} \tag{3.24}$$

This formula too suffers from inaccuracies, but only at the extremes of distance for diametrically opposing (or *antipodal*) points. Navigation would rarely, if ever, be performed over such a distance in a single path but, for consistency, a more elaborate formula is available that completely solves the problem:

$$\alpha = \tan^{-1}\left(\frac{\sqrt{[\sin \theta_2 \sin(\lambda_1 - \lambda_2)]^2 + [\sin \theta_1 \cos \theta_2 - \cos \theta_1 \sin \theta_2 \cos(\lambda_1 - \lambda_2)]^2}}{\cos \theta_1 \cos \theta_2 + \sin \theta_1 \sin \theta_2 \cos(\lambda_1 - \lambda_2)}\right) \tag{3.25}$$

Note 3.1

The equivalent to the haversine formula in plane trigonometry would be obtained from the cosine rule (3.6) by substituting:

$$\cos A = \cos^2\frac{A}{2} - \sin^2\frac{A}{2} = 1 - 2\ \sin^2\frac{A}{2}$$

Thus,

$$a^2 = b^2 + c^2 - 2bc\ \left(1 - 2\sin^2\frac{A}{2}\right)$$

With simple manipulation, this reduces to a convenient form, as follows:

$$a^2 = (b - c)^2 + 4bc\ \sin^2\frac{A}{2}$$

As in the case of the haversine formula, this is important because it overcomes the inherent limitations of the cosine rule when calculating small distances.

3.3   Great Circle Navigation

As discussed already, if two points on the surface of a sphere were joined by part of great circle then the path traversed between those points would have minimum length (assuming of course that a traveller does not go the long way round the Earth!).

The geometry is established as shown in Figure 3.9, with a position vector **r** tracing out a great circle path that is orthogonal to an axis vector (i.e. the centre-line vector) **c**. Without loss of generality, the path will be traversed in a clockwise direction around the axis vector and will

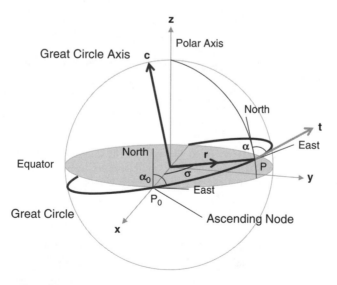

Figure 3.9 Great circle navigation on spherical Earth.

cross the equator at a point P_0 that lies on the x-axis of the ECEF reference frame. That point will be called the *ascending node* (which is a term coined from orbital mechanics[2]) because an object on that path will cross from the southern hemisphere to the northern hemisphere at that point. The heading of the great circle path at the ascending node is defined to be α_0, noting that course headings at any point are measured clockwise in azimuth from North. Distances between points can be found using spherical trigonometry, as discussed in Section 3.2. The question to be addressed here is how to plot a great circle course and how to determine the heading at any given point on that course.

From first principles, the position vector of a point P on a sphere of unit radius can be parameterised by the heading (α_0) measured at the ascending node (which effectively defines the inclination of the circular path) and a progression angle (σ) measured from the ascending node:

$$\mathbf{r} = \cos\sigma\mathbf{x} + \sin\sigma\sin\alpha_0\mathbf{y} + \sin\sigma\cos\alpha_0\mathbf{z} \qquad (3.26)$$

Recalling (3.2), it can be parameterised by longitude difference (λ) and latitude (θ):

$$\mathbf{r} = \cos\phi\cos\Delta\lambda\mathbf{x} + \cos\phi\sin\Delta\lambda\mathbf{y} + \sin\phi\mathbf{z} \qquad (3.27)$$

Note that longitude *difference* is the longitude measured from the ascending node, i.e. $\Delta\lambda = \lambda - \lambda_0$ where $\lambda =$ longitude. Thus the following equivalences are established:

$$
\begin{aligned}
\cos\Delta\lambda\cos\phi &= \cos\sigma \\
\sin\Delta\lambda\cos\phi &= \sin\sigma\sin\alpha_0 \\
\sin\phi &= \sin\sigma\cos\alpha_0
\end{aligned}
\qquad (3.28)
$$

Note that the longitude difference is often quoted in technical papers as:

$$\tan\Delta\lambda = \tan\sigma\sin\alpha_0 \qquad (3.29)$$

Normally it would be evaluated as $\Delta\lambda = \tan^{-1}(\sin\Delta\lambda/\cos\Delta\lambda)$.

Note 3.2

Knowing the tangent of an angle, the angle itself can be computed using the standard arc-tangent function atan (giving results in the range $(-\pi/2, \pi/2)$) or the four-quadrant arc-tangent function atan2 (for general results). In almost all programming languages, atan2(s,c) is equivalent to $\tan^{-1}$(s/c); one significant exception is Microsoft Excel® in which this calculation is specified as ATAN2(c,s). Note that atan2 gives a numerical result in the range $(-\pi, \pi)$.

Longitude difference can also be described by a point on the equator:

$$\mathbf{r}_0 = \frac{1}{r_0}(\cos\sigma\mathbf{x} + \sin\sigma\sin\alpha_0\mathbf{y}) \qquad (3.30)$$

where

$$r_0 = \sqrt{1 - \sin^2\sigma\cos^2\alpha_0} \qquad (3.31)$$

---

[2] Orbital Mechanics is discussed in Chapter 3.

Note that latitude can be derived from plane trigonometry:

$$\begin{aligned} \sin\theta &= \sin\sigma\cos\alpha_0 \\ \cos\theta &= \sqrt{1 - \sin^2\sigma\cos^2\alpha_0} \end{aligned} \tag{3.32}$$

where $\cos\theta = r_0 = \mathbf{r}\bullet\mathbf{r}_0$ (which is the dot product of $\mathbf{r}$ and $\mathbf{r}_0$).

The axis vector of the great circle is defined by:

$$\mathbf{c} = -\cos\alpha_0\mathbf{y} + \sin\alpha_0\mathbf{z} \tag{3.33}$$

which means that the tangent vector (i.e. the direction of travel) at a given point can be calculated as the cross product of $\mathbf{c}$ with $\mathbf{r}$, as follows:

$$\mathbf{t} = \mathbf{c}\times\mathbf{r} = -\sin\sigma\mathbf{x} - \cos\sigma\sin\alpha_0\mathbf{y} + \cos\sigma\cos\alpha_0\mathbf{z} \tag{3.34}$$

The directions of North and East are required at the current point on the great circle. Vector $\mathbf{r}_0$ (with length r_0) gives the equatorial projection of the current position and vector $\mathbf{z}$ defines the polar axis; therefore, local East is defined by:

$$\begin{aligned} \mathbf{e} &= \mathbf{z}\times\mathbf{r}_0 \\ &= \frac{1}{r_0}(-\sin\sigma\sin\alpha_0\mathbf{x} + \cos\sigma\mathbf{y}) \end{aligned} \tag{3.35}$$

Having found East, local North is then defined by:

$$\begin{aligned} \mathbf{n} &= \mathbf{r}\times\mathbf{e} \\ &= \frac{\sin\sigma\cos\alpha_0}{r_0}(-\cos\sigma\mathbf{x} - \sin\sigma\sin\alpha_0\mathbf{y}) + \frac{(1 - \sin^2\sigma\cos^2\alpha_0)}{r_0}\mathbf{z} \end{aligned} \tag{3.36}$$

From here, applying the cosine rule, the heading on the great circle is derived as:

$$\sin\alpha = \mathbf{t}\bullet\mathbf{e} = \frac{1}{r_0}\sin\alpha_0 \tag{3.37}$$

$$\cos\alpha = \mathbf{t}\bullet\mathbf{n} = \frac{1}{r_0}\cos\sigma\cos\alpha_0 \tag{3.38}$$

Recalling (3.31) and (3.32), the fundamental relationships that govern the variation of heading along a great circle are:

$$\cos\phi\sin\alpha = \sin\alpha_0 \tag{3.39}$$

$$\cos\phi\cos\alpha = \cos\sigma\cos\alpha_0 \tag{3.40}$$

The absence of any 'loss of generality' (as claimed at the start of this derivation) is justified by the absence of any dependency on where the ascending node lies on the equator. For convenience, it was initially located on the x-axis of the ECEF frame.

For practical navigation, define two points P_1 and P_2 on the surface of a sphere of unit radius, with respective position vectors:

$$\begin{aligned} \mathbf{r}_1 &= \cos\phi_1\cos\lambda_1\mathbf{x} + \cos\phi_1\sin\lambda_1\mathbf{y} + \sin\phi_1\mathbf{z} \\ \mathbf{r}_2 &= \cos\phi_2\cos\lambda_2\mathbf{x} + \cos\phi_2\sin\lambda_2\mathbf{y} + \sin\phi_2\mathbf{z} \end{aligned} \tag{3.41}$$

If the navigation starts at P_1 and ends at P_2 then the axis vector of the great circle is defined by:

$$\mathbf{c} = \frac{\mathbf{r}_1 \times \mathbf{r}_2}{\|\mathbf{r}_1 \times \mathbf{r}_2\|} \tag{3.42}$$

The heading (α_0) measured at the ascending node is determined as:

$$\sin \alpha_0 = \mathbf{c} \bullet \mathbf{z}_0 \tag{3.43}$$

where

$$\mathbf{z}_0 = (\mathbf{n} \bullet \mathbf{z})\mathbf{z}$$

Note that the dot product in this case, rather than equating to the cosine of an angle, actually equates to the sine of the complementary angle. Also note that a new vector $\mathbf{z}_0$ is required in order to keep the sign conventions consistent with the concept of an ascending node.

The unit vector pointing out towards the ascending node ($\mathbf{a}$) is calculated as:

$$\mathbf{a} = \frac{\mathbf{z} \times \mathbf{c}}{\cos \alpha_0} \tag{3.44}$$

Practical computation would utilise vector components of $\mathbf{a}$:

$$\mathbf{a} = a_x\mathbf{x} + a_y\mathbf{y} + a_z\mathbf{z} \tag{3.45}$$

The longitude of the ascending node (λ_0) is then given by:

$$\tan \lambda_0 = \frac{a_y}{a_x} \tag{3.46}$$

For interest, an explicit formula can be found by algebraic expansion:

$$\tan \lambda_0 = \frac{\cos \phi_1 \sin \phi_2 \sin \lambda_1 - \sin \phi_1 \cos \phi_2 \sin \lambda_2}{\cos \phi_1 \sin \phi_2 \cos \lambda_1 - \sin \phi_1 \cos \phi_2 \cos \lambda_2} \tag{3.47}$$

The progression angle for each terminal point of the navigation (σ_1, σ_2) obeys the cosine rule for plane trigonometry. Of course, if an angle σ is being determined from $\cos\sigma$, the solution is not unique. It is also necessary to find $\sin\sigma$. The complete solution for σ_1 is constructed as follows:

$$\mathbf{a}_1 = \mathbf{a} \times \mathbf{r}_1 \tag{3.48}$$

$$\sin \sigma_1 = \|\mathbf{a}_1\| \operatorname{sgn}(\mathbf{c} \bullet \mathbf{a}_1) \tag{3.49}$$

$$\cos \sigma_1 = \mathbf{a} \bullet \mathbf{r}_1 \tag{3.50}$$

$$\sigma_1 = \tan^{-1}\left(\frac{\sin \sigma_1}{\cos \sigma_1}\right) \tag{3.51}$$

The solution for σ_2 is constructed in the same way and then the great circle path is defined by all values of σ from σ_1 to σ_2. Well, not quite! As before, these angles would be evaluated using the four-quadrant arc-tangent function atan2, which yields results in the range $-\pi < \sigma \leq \pi$ (or $-180° < \sigma \leq 180°$). As with any circular path, the shortest route might cross over the discontinuity at $\sigma = \pi$ (or $\sigma = 180°$). for instance, the distance between $\sigma_1 = 179°$ and $\sigma_2 = -179°$ should calculated as $2°$ not $358°$. The computational fix for this problem is, if the distance appears to be greater than $180°$ (i.e. more than halfway round the sphere), add $360°$ to whichever of the terminal points that has a negative angle and then perform the calculation again. This generates angles that are outside the range $-180° < \sigma \le 180°$ but, computationally, that does not matter.

This method of great circle navigation is implemented in Algorithm 3.1; this is a MATLAB function that accepts the longitude and latitude of two positions (expressed in degrees, minutes and seconds) and returns a computed course from the first point to the second point, plus the distance traversed. As discussed, the course takes the short way round and, for convenience, progresses in nominal one-degree steps. As a way of handling courses that cross over the 'date line' at longitude $180°$, this code looks for large jumps in longitude (of more than 6 radians) and inserts NaN. This avoids problems when plotting graphs. Line-plotting in MATLAB can handle vectors of NaN-separated data,[3] which conveniently splits a line into segments and, in this case, conveniently splits navigational paths as they cross the date line. The resulting graph is then free from unsightly lines jumping from one side of the plot to the other (caused by using a continuous line to represent discontinuous data). The replacement of a few data elements by NaN is no problem here as the purpose of this is to illustrate the principles of calculation.

The context of this algorithm is illustrated by the following MATLAB commands:

```
>> MAN.lon = -2.2747 ; MAN.lat = 53.3536 ;
>> ORD.lon = -87.9044 ; ORD.lat = 41.9878 ;
>> LAX.lon = -118.4081 ; LAX.lat = 33.9425 ;
>> SYD.lon = 151.1772 ; SYD.lat = -33.9461 ;
>> HND.lon = 139.7794 ; HND.lat = 35.5522 ;
>> c1 = navigateGreatCircle(MAN,ORD) ;
>> c2 = navigateGreatCircle(ORD,LAX) ;
>> c3 = navigateGreatCircle(LAX,SYD) ;
>> c4 = navigateGreatCircle(SYD,HND) ;
>> c5 = navigateGreatCircle(HND,MAN) ;
```

Data structures are established for the longitude and latitude at five airports around the world, using data taken from Table 3.1. Then calculations are performed for a circumnavigation, proceeding westwards from Manchester to Chicago, Los Angeles, Sydney, Tokyo and finally back to Manchester. The actual paths (together with headings) followed are shown in Figure 3.10, with an overlay of the world coastline.[4] Note that the heading continually changes along a great circle path. For information, the total distance travelled is 38296 km or 20678 nautical miles (based on a mean Earth radius of 6371 km), with the distances for each leg given in Table 3.2.

---

[3] NaN is 'Not a Number'; it is treated as a special constant that is the result of the mathematical operation 0/0. (Likewise Inf or 'Infinity' is a special constant that results when dividing any nonzero number by zero.) NaN-separated data appear in MATLAB as vectors of numbers that contain occasional instances of NaN; when these vectors are used for data inputs for graph plotting (using functions plot2, plot3, etc.), the NaN entries are interpreted as delimiters between line segments. In this context, the NaN entries must occur in the same positions in each data input.

[4] This uses a data file 'coast.mat', which is supplied with the MATLAB Mapping Toolbox.

Table 3.1 Locations of selected international airports (data extracted from The Global Airport Database[5] and not validated for navigational use).

ICAO Code	IATA Code	Airport	LAT	LON	LAT	LON
EGCC	MAN	MANCHESTER	N53:21:13	W002:16:29	N53.3536°	W002.2747°
KORD	ORD	CHICAGO	N41:58:46	W087:54:16	N41.9878°	W087.9044°
KLAX	LAX	LOS ANGELES	N33:56:33	W118:24:29	N33.9425°	W118.4081°
YSSY	SYD	SYDNEY	S33:56:46	E151:10:38	S33.9461°	E151.1772°
RJTT	HND	TOKYO	N35:33:08	E139:46:46	N35.5522°	E139.7794°

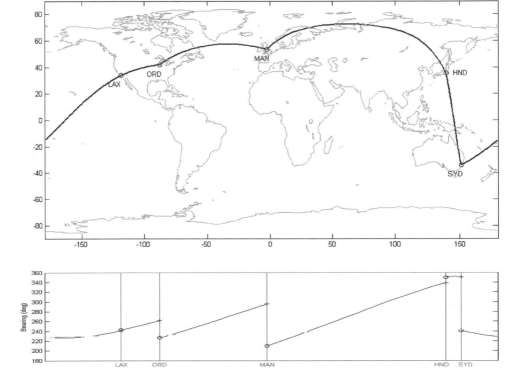

Figure 3.10 Westward circumnavigation of the Earth via great circles.

Table 3.2 Great circle distances between selected international airports.

From		To		km	nm
MAN	MANCHESTER	ORD	CHICAGO	6140	3315
ORD	CHICAGO	LAX	LOS ANGELES	2802	1513
LAX	LOS ANGELES	SYD	SYDNEY	12061	6512
SYD	SYDNEY	HND	TOKYO	7818	4221
HND	TOKYO	MAN	MANCHESTER	9475	5116

---

[5] www.partow.net/miscellaneous/airportdatabase/index.html.

```
function course = navigateGreatCircle(p1,p2)
unitVector = @(vector) vector/norm(vector) ;
position = @(lon,lat) (cos(lat)*cos(lon); cos(lat)*sin(lon);
sin(lat)) ;
radians = @(degrees) degrees*pi/180 ;
degrees = @(radians) radians*180/pi ;
findAngle = @(sine,cosine) atan2(sine,cosine) ;

z = (0; 0; 1) ;
r1 = position(radians(p1.lon),radians(p1.lat)) ;
r2 = position(radians(p2.lon),radians(p2.lat)) ;
c = unitVector(cross(r1,r2));
a = unitVector(cross(z,c)) ;
b = cross(c,a) ;

a1 = cross(a,r1) ;
Sigma1 = findAngle(norm(a1)*sign(dot(a1,c)),dot(a,r1)) ;

a2 = cross(a,r2);
Sigma2 = findAngle(norm(a2)*sign(dot(a2,c)),dot(a,r2)) ;
Sigma2 = atan2(sinSigma2,cosSigma2) ;

if abs(Sigma1-Sigma2) > pi
 if Sigma1 < 0
 Sigma1 = 2*pi+Sigma1 ;
 else
 Sigma2 = 2*pi+Sigma2 ;
 end
end
oneDegree = pi/180 ;
nIncrements = floor(abs(Sigma1-Sigma2)/oneDegree) ;
Sigma = linspace(Sigma1,Sigma2,nIncrements) ;
r = a*cos(Sigma) + b*sin(Sigma) ;
gcLon = atan2(r(2,:),r(1,:)) ;
gcLat = atan2(r(3,:),sqrt(r(1,:).^2+r(2,:).^2)) ;

Alpha0 = asin(dot(c,z)) ;
Alpha = findAngle(sin(Alpha0)./cos(gcLat),cos(Alpha0).*cos(Sigma)./
  cos(gcLat)) ;
longitudeDifference = abs(diff(gcLon)) ;
jump = find(longitudeDifference > 6) ;
gcLon(jump) = NaN ;
gcLat(jump) = NaN ;
Alpha(jump) = NaN ;

course.lon = degrees(gcLon) ;
course.lat = degrees(gcLat) ;
course.az = degrees(mod(Alpha,2*pi)) ;
course.km = abs(Sigma1-Sigma2)*6371 ;
```

Algorithm 3.1 'navigateGreatCircle'.

3.4   Rhumb Line Navigation

Navigation by great circles results in the shortest distance between origin and destination; however, most practical navigation is performed using *rhumb lines*, which are lines of constant heading. Without delving too far into the applicable mathematics, it is worth summarising the principles of navigating by this method and the differences vis-à-vis great circles.

For this purpose, a different way of expressing latitude is needed. For a spherical surface, the so-called *isometric latitude* is defined as follows:

$$\psi = \ln\left[\tan\left(\frac{\pi}{4} + \frac{\phi}{2}\right)\right] \tag{3.52}$$

where 'ln' denotes a natural logarithm. Using simply trigonometry

$$\tan^2\left(\frac{\pi}{4} + \frac{\phi}{2}\right) = \left(\frac{\cos\frac{\phi}{2} + \sin\frac{\phi}{2}}{\cos\frac{\phi}{2} - \sin\frac{\phi}{2}}\right)^2 = \frac{1 + 2\sin\frac{\phi}{2}\cos\frac{\phi}{2}}{1 - 2\sin\frac{\phi}{2}\cos\frac{\phi}{2}} = \frac{1 + \sin\phi}{1 - \sin\phi}$$

Thus, (3.52) is equivalent to:

$$\psi = \frac{1}{2}\ln\left[\frac{1 + \sin\phi}{1 - \sin\phi}\right] \tag{3.53}$$

For completeness, the conversion from isometric latitude to geocentric latitude on a sphere is achieved as:

$$\phi = 2\tan^{-1}\left(\varepsilon^{\psi}\right) - \frac{\pi}{2} \tag{3.54}$$

where ε is the base of natural logarithms ($\varepsilon \approx 2.71828182845904$). For the other widely used definition, the conversion is:

$$\phi = \sin^{-1}\left(\frac{1 - e^{2\psi}}{1 - e^{2\psi}}\right) \tag{3.55}$$

Note 3.3

Other definitions of latitude can be found (e.g. mathworld.wolfram.com). These are definitely more obscure and less frequently quoted; for instance:

$$\psi = \sinh^{-1}(\tan\phi)$$

$$\psi = \ln(\tan\phi + \sec\phi) \equiv gd^{-1}(\phi)$$

where 'gd' represents the little-known Gudermannian function. The corresponding inverse relationships are:

$$\phi = \tan^{-1}(\sinh\psi)$$

$$\theta = gd\ (\psi)$$

The remarkable property of isometric latitude (Ψ) is that it enables the simplest of calculations for the rhumb line between two points P_1 and P_2. The heading is an azimuth angle (α) (measured clockwise from North):

$$\alpha = \tan^{-1}\left(\frac{\lambda_2 - \lambda_1}{\Psi_2 - \Psi_1}\right) \tag{3.56}$$

where (λ_1, Ψ_1) and (λ_2, Ψ_2) specify the longitude and isometric latitude of points P_1 and P_2, respectively. Equally remarkable is the simplicity of the distance calculation for a rhumb line on the surface of a sphere, namely:

$$s = \frac{M_2 - M_1}{\cos \alpha} \tag{3.57}$$

where M_1 and M_2 are the meridian distances from the equator to P_1 and P_2, respectively. Note that the definition of meridian distance on a sphere is $M = R\phi$, where R is the radius of the sphere and ϕ is the (geocentric) latitude. This formula can be applied for $\cos\alpha > 10^{-6}$ without numerical problems; otherwise the rhumb line distance approximates to $s \approx R\cos\phi \, |\lambda_2 - \lambda_1|$. The only remaining issue is the discontinuity in longitude at $\pm \pi$ (or $\pm 180°$). Thus, values of longitude may have to be incremented by at $\pm 2\pi$ (or $\pm 360°$) in order for the calculation to be valid.

For illustration, the same circumnavigation is considered as in Section 3.3. The results of the rhumb line calculations are summarised in Table 3.3, giving heading and distance for each leg, plus the excess distance travelled relative to the equivalent great circle route. The total distance is 21702 nautical miles (vs 20678 nautical miles). The comparison is shown graphically in Figure 3.11, with the curved paths being the great circle routes and the straight paths being the rhumb lines. With the exception of higher latitudes, excess distances are surprisingly small, which justifies the practice of flying along rhumb lines between way-points.

As an observation on Figure 3.11, the use of isometric latitude has resulted in an obvious distortion in the vertical scale. In fact, all maps suffer distortion simply because a curved surface cannot be projected on to a flat surface without sacrificing length, angles or areas to some extent, or some combination of all three. In this case, the map projection is attributed to Gerardus Mercator (1512–94) and is called the *Mercator Projection*.

Table 3.3 Rhumb line distances between selected international airports.

From		To		Rhumb Line			Excess	
				Heading	km	nm	km	nm
MAN	MANCHESTER	ORD	CHICAGO	258.79	6500	3510	360	194
ORD	CHICAGO	LAX	LOS ANGELES	251.47	2815	1520	13	7
LAX	LOS ANGELES	SYD	SYDNEY	231.37	12093	6530	32	17
SYD	SYDNEY	HND	TOKYO	351.27	7818	4221	0	0
HND	TOKYO	MAN	MANCHESTER	280.40	10965	5921	1490	805

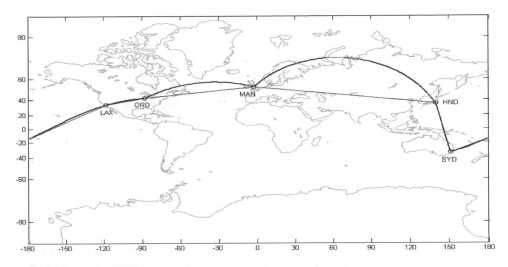

Figure 3.11 Comparison of great circles and rhumb lines.

3.5   Reference Ellipsoids

3.5.1   *World Geodetic System (WGS84)*

As mentioned already, the Earth is not exactly spherical; it can be approximated by various reference ellipsoids, each with axial symmetry and each with different dimensions. The standard Earth model that is adopted for aerospace applications is the World Geodetic System 1984 (WGS84). Its main parameters are as follows:

$$\text{Earth rotation rate} \quad \omega = 7.292115 \times 10^{-5} \text{rad.s}^{-1} \quad (3.58)$$

$$\text{flattening parameter} \quad f = 1/298.257223563 \quad (3.59)$$

$$\text{equatorial radius} \quad a = 6378137 \text{ m} \quad (3.60)$$

$$\text{polar radius} \quad c = 6356752.3 \text{ m} \quad (3.61)$$

Note that the interrelationship between equatorial radius (a) and polar radius (c) is determined in this instance by the ellipticity (or flattening parameter):

$$f = \frac{a - c}{a} \quad (3.62)$$

Another way of summarising the dimensional ratio (or shape) of an ellipse is to use *eccentricity*, which can be defined in two forms:

$$\text{First Eccentricity}: \quad e = \sqrt{\frac{a^2 - c^2}{a^2}} \quad (3.63)$$

$$\text{Second Eccentricity}: \quad \varepsilon = \sqrt{\frac{a^2 - c^2}{c^2}} \quad (3.64)$$

Using WGS84 dimensions, $e = 0.081819218$ and $\varepsilon = 0.082094465$.

3.5.2   *Geoid Approximations*

With reference to gravitational modelling in Chapter 4, an equipotential surface can be calculated that represents a best-fit to the global Mean Sea Level (MSL). This is called the **geoid** (cf. Section 5.2.10) and, conceptually, this extends around the Earth such that any water surface would correspond with the local geoid heights, assuming a hypothetical network of tunnels to provide interconnection with the sea. In this context, tidal effects are not included. Because the geoid shape is due to the mass distribution of the Earth, the effects of physical peaks and troughs are smoothed out and the resulting variation in geoid height is within ± 100 metres.

The relationship between a reference ellipsoid, the geoid and the local topography is illustrated in Figure 3.12. These are all surface definitions but vary in shape and position. The inclusion of a geoid model serves to illustrate two issues. First, the true geoid is approximated by any numerical computation. Secondly, the geoid is a global 'best fit' and does not necessarily define the MSL at any given place or time. Thus, for mapping and surveying, it is appropriate to use different ellipsoids in defining the horizontal datum in various parts of the world. The underlying rationale is shown schematically in Figure 3.13, where North America and Europe have their own datum definitions that approximate the geoid profile in the respective regions. In effect, the difference between the geoid and the relevant ellipsoid can be ignored for practical purposes.

A sample list of reference ellipsoids is presented in Table 3.4, using data from the WGS84 definition document (NIMA, 2004). The associated parameters are the equatorial radius (a) and the flattening parameter (or ellipticity) (f), as discussed in the previous section. For consistency with published sources, these are also presented as increments relative to the WGS84 ellipsoid (i.e. Δa and Δf), such that:

$$\Delta a = a_{WGS84} - a_{ellipsoid}$$
$$\Delta f = f_{WGS84} - f_{ellipsoid}$$

Each of these is used in a particular geographical locale in order to provide a horizontal datum for the purpose of surveying and mapmaking. Examples of applicability are given in Table 3.5, together with the coordinates for the centre of the WGS ellipsoid in the local frame.

Although geoid heights are not particularly great, the concept of a 'mean sea level' is relevant to aerospace applications, as a datum both for atmospheric properties and for terrain elevation. This implies overlaying geoid elevation data into a reference ellipsoid and

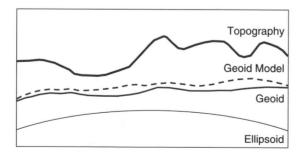

Figure 3.12 Surface definitions.

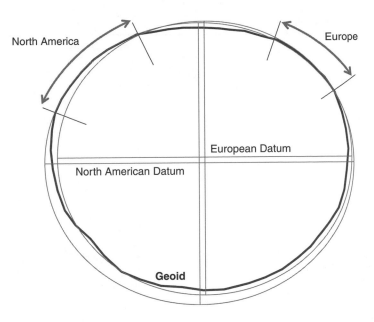

Figure 3.13 Reference ellipsoids.

Table 3.4 Reference ellipsoids.

Name	Code	a	1/f	Δa	Δf.10⁴
Airy 1830	AA	6377563.396	299.3249646	+573.604	+0.11960023
Australian National	AN	6378160.000	298.25	−23	−0.00081204
Bessel 1841					
Ethiopia, Indonesia, Japan, Korea	BR	6377397.155	299.1528128	+739.845	+0.10037483
Namibia	BN	6377483.865	299.1528128	+653.135	+0.10037483
Clarke 1866	CC	6378206.400	294.9786982	−69.400	−0.37264639
Clarke 1880	CD	6378249.145	293.465	−112.145	−0.54750714
Everest					
Brunei, East Malaysia	EB	6377298.556	300.8017	+838.444	+0.28361368
India 1830	EA	6377276.345	300.8017	+860.655	+0.28361368
India 1956	EC	6377301.243	300.8017	+835.757	+0.28361368
Pakistan	EF	6377309.613	300.8017	+827.387	+0.28361368
West Malaysia, Singapore 1948	EE	6377304.063	300.8017	+832.937	+0.28361368
West Malaysia 1969	ED	6377295.664	300.8017	+841.336	+0.28361368
Geodetic Reference System 1980	RF	6378137.000	298.257222101	0	−0.00000016
Helmert	HE	6378200.000	298.3	−63	+0.00480795
Hough 1960	HO	6378270.000	297.0	−133	−0.14192702
Indonesian 1974	ID	6378160.000	298.247	−23	−0.00114930
International 1924	IN	6378388.000	297.0	−251	−0.14192702
Krassovsky	KA	6378245.000	298.3	−108	+0.00480795
Modified Airy	AM	6377340.189	299.3249646	+796.811	+0.11960023
Modified Fischer 1960	FA	6378155.000	298.3	−18	+0.00480795
South American 1969	SA	6378160.000	298.25	−23	−0.00081204
WGS72	WD	6378135.000	298.26	+2	+0.00031211
WGS84	WE	6378137.000	298.257223563	0	0

Table 3.5 Datum definitions.

Datum	Code	Ellipsoid	Δx	Δy	Δz
CAPE	CAP	CD			
South Africa			−136	−108	−292
CARTHAGE	CGE	CD			
Tunisia			−263	+6	+431
EUROPEAN 1950	EUR	IN			
Egypt	EUR-F		−130	−117	−151
Iran	EUR-H		−117	−132	−189
Tunisia	EUR-T		−112	−77	−145
INDIAN					
Bangladesh	IND-B	EA	+282	+726	+254
India and Nepal	IND-I	EC	+295	+736	+257
TOKYO	TOY	BR			
Japan	TOY-A		−148	+507	+685
South Korea	TOY-B		−146	+507	+687
AUSTRALIAN GEODETIC 1984 (AGD84)	AUG	AN			
Australia and Tasmania			+134	−48	+149
EUROPEAN 1950	EUR	IN			
Mean Solution	EUR-M		−87	−98	−121
British Isles	EUR-K		−86	−96	−120
IRELAND 1965	IRL	AM			
Ireland			+506	−122	+611
ORDNANCE SURVEY OF GREAT BRITATIN 1936 (OSGB36)	OGB	AA			
Mean Solution	OGB-M		+375	−111	+431
England	OGB-A		+371	−111	+434
Scotland and Shetland Islands	OGB-C		+384	−111	+425
Wales	OGB-D		+370	−108	+434
NORTH AMERICAN 1983 (NAD83)	NAR	RF			
Contiguous United States (CONUS)	NAR-C		0	0	0
Hawaii	NAR-H		+1	+1	−1
SOUTH AMERICAN 1969	SAN	SA			
Mean Solution	SAN-M		−57	+1	−41
S-42 (PULKOVA 1942)	SPK	KA			
Hungary	SPK-A		+28	−121	−77
Poland	SPK-B		+23	−124	−82
Kazakhstan	SPK-E		+15	−130	−84

then using this as the foundation for digital elevation data (i.e. topography and bathymetry). As mentioned in Chapter 1, accurate simulations in coastal regions require a shoreline definition, such as the NGA World Vector Shoreline (or **WVS**). This implies a further overlay on top of the terrain representation and, to add complication, wind and tides will cause this to vary.

For interest the general form of the geoid surface can be obtained from data supplied with the MATLAB Mapping Toolbox via Algorithm 3.2. The result is shown in Figure 3.14, with colour-coded geoid heights and an overlay of the world coastline (as supplied in a MATLAB data files 'geoid.mat' and 'coast.mat', respectively). The relevant theoretical foundation is presented in Section 4.2, with specific mention of the geoid in Section 5.2.10.

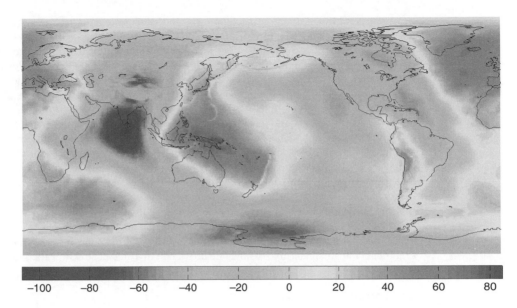

Figure 3.14 EGM96 geoid heights (in metres).

```
function plotGeoid
egm96 = load('geoid.mat') ; % geoid defined for longitude 0 to 360
coast = load('coast.mat') ; % coastline defined for longitude -180 to 180
coast = shift(coast) ; % align coastline data with geoid

figure
geoshow(geoid, geoidrefvec, 'DisplayType','texturemap') ;
set(gca,'XLim',(0,360),'YLim',(-90,90),'Visible','off','NextPlot',
'add');

plot(coast.long,coast.lat,'k') ;

hcb = colorbar('horiz') ;
set(hcb,'XTick',(-100:20:80)) ;

function newCoastline = shift(coastline)
% Shift coast data by 180 degrees longitude
% Note: coastline is defined by NaN-separated vectors

long = (coast.long; NaN; coast.long+360) ;
lat = (coast.lat; NaN; coast.lat) ;

tooSmall = long<0 ;
tooBig = long>360 ;

long(tooSmall|tooBig) = NaN ;
lat(tooSmall|tooBig) = NaN ;

newCoast.long = long ;
newCoast.lat = lat ;
```

Algorithm 3.2 'plotGeoid'.

As a general observation, issues often arise in the reuse of data, as here; often the format and/or content will not be entirely appropriate for the intended application and some rearrangement will be required. In this case, in order to be compatible with the geoid data, the coastline data must span longitudes from 0° eastwards through 360°; as supplied, the actual span is from 180 °W to 180 °E. Because line-plotting in MATLAB can handle vectors of NaN-separated data, a simple method is to create an oversized data set by wrapping two sets of coastline data across four hemispheres (extending from 180 °W to 540 °E). Values that correspond with the first and last hemispheres can then be trimmed by replacing those values by NaN. This may not be mathematically elegant but it is a useful programming trick.

The difference between spherical Earth and ellipsoidal Earth representations would be imperceptible to the human eye; similarly the offsets between WGS84 and other ellipsoids (depending on the choice of horizontal datum) might seem trivial. Relevant to the scale of the Earth, geoid height variations would be even smaller. So why bother going to this much detail? The answer is that the implications for mapping and precision navigation are significant. Given that tracking tasks can require 1-metre accuracy, reference models of the Earth need to be an order of magnitude more accurate for simulation purposes. As an example, it is reasonable to expect that an aircraft should be able to navigate its way on to a given runway in the simulated world.

3.6   Coordinate Systems

3.6.1   *Geocentric and Geodetic Latitude*

Finding positions on the surface of an ellipsoid is less straightforward than for a sphere. The problem is best introduced by taking a two-dimensional slice through the Earth model that coincides with the zx-plane, as shown in Figure 3.15. Thus, define a point P_0 on the ellipse, with coordinates (x_0, z_0). The geocentric latitude is ϕ and the so-called *geodetic* latitude[6] is μ. The geodetic latitude defines the orientation of the local normal at point P_0. The equation for an ellipse is:

$$\left(\frac{x}{a}\right)^2 + \left(\frac{z}{c}\right)^2 = 1 \tag{3.65}$$

The tangent line at P has gradient:

$$\frac{dz}{dx} = -\left(\frac{c}{a}\right)^2 \frac{x_0}{z_0} \tag{3.66}$$

The corresponding normal has gradient:

$$\tan \mu = \left(\frac{a}{c}\right)^2 \frac{z_0}{x_0} \tag{3.67}$$

By inspection of Figure 3.15, it should be recognised that

$$\tan \phi = \frac{z_0}{x_0} \tag{3.68}$$

---

[6] Graphical User Interface.

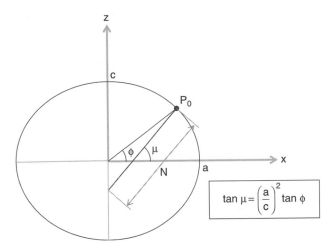

Figure 3.15 Elliptical geometry.

This leads to the relationship between geocentric and geodetic latitude, i.e.

$$\tan \mu = \left(\frac{a}{c}\right)^2 \tan \phi \qquad (3.69)$$

3.6.2   ***Parametric or Reduced Latitude***

An alternative method of constructing elliptical geometry is shown in Figure 3.16. The uses a circle, which is called the *auxiliary circle*, with radius equal to the semi-major axis of the ellipse (which is the equatorial radius of the Earth in this context). At any point on the circumference of the circle, the equivalent point on the ellipse is produced by scaling the z-coordinate by the ratio of the semi-minor axis to the semi-major axis (i.e. the ratio of polar radius to equatorial radius).

Accordingly, define a point P on the circle with coordinates:

$$\begin{aligned} x &= a \cos \theta \\ z &= a \sin \theta \end{aligned} \qquad (3.70)$$

The angle θ is the *parametric* latitude and it is also referred to as the *reduced* latitude. The equivalent point on the ellipse is below P in a direction that is parallel with the z-axis. It is denoted by P_0 and has coordinates:

$$\begin{aligned} x_0 &= a \cos \theta \\ z_0 &= c \sin \theta \end{aligned} \qquad (3.71)$$

By direct substitution and simple manipulation thereafter, it is easily shown that these coordinates satisfy (3.65).

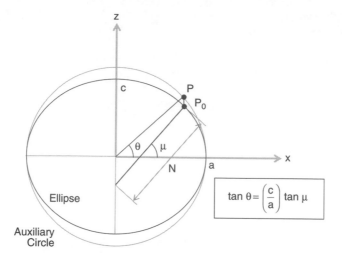

Figure 3.16 Equivalent elliptical geometry.

By construction, it can be seen that

$$x = x_0$$

$$\frac{z_0}{z} = \frac{c}{a}$$

By definition, it can be seen that

$$\tan \phi = \frac{z_0}{x_0}$$

$$\tan \theta = \frac{z}{x_0}$$

Thus, the relationship between *geocentric* latitude and *parametric* latitude is established as:

$$\tan \theta = \left(\frac{a}{c}\right) \tan \phi \qquad (3.72)$$

Also, with reference to (3.69), the relationship between *geodetic* latitude and *parametric* latitude is readily derived as:

$$\tan \mu = \left(\frac{a}{c}\right) \tan \theta \qquad (3.73)$$

(**Warning**: *Avoid confusion with earlier notation in which θ denoted co-latitude.*)

It is interesting to note that, using this formulation, the radial distance (r) from the centre of the Earth to a point on the surface is given by:

$$r^2 = a^2 \cos^2\theta + c^2 \sin^2\theta$$

$$= \frac{a^2 + c^2 \tan^2\theta}{1 + \tan^2\theta}$$

For practical purposes it is more useful to re-express this relationship is terms of geodetic latitude, using (3.73):

$$r = \sqrt{\frac{a^4 + c^4 \tan^2\mu}{a^2 + c^2 \tan^2\mu}} \tag{3.74}$$

Recognising that the tangent function is singular for angles of $\pm\pi/2$ and numerically inconvenient for angles close to those extremes, an alternative and well-conditioned formula is, as follows:

$$r = \sqrt{\frac{a^4 \cos^2\mu + c^4 \sin^2\mu}{a^2 \cos^2\mu + c^2 \sin^2\mu}} \tag{3.75}$$

3.6.3 *Cartesian Coordinates*

In this context, the normal that passes through a given point P_0 on the ellipse is called the *prime vertical*. When projected downwards into the ellipse, the prime vertical intersects the z-axis at a point vertically below point P_0. The distance from that point to P_0 is often denoted by the symbol N. Thus, the x-coordinate of P_0 is:

$$x_0 = N\cos\mu$$

Noting (3.68) and (3.69), the z-coordinate of P_0 is:

$$z_0 = x_0 \tan\phi = N\cos\mu \left(\frac{c}{a}\right)^2 \tan\mu = N\left(\frac{c}{a}\right)^2 \sin\mu$$

From the definition of first eccentricity (3.31), the coordinates of P_0 can be written as follows:

$$\begin{aligned} x_0 &= N\cos\mu \\ z_0 &= N(1-e^2)\sin\mu \end{aligned} \tag{3.76}$$

These coordinates necessarily satisfy the equation for the ellipse (3.62), such that

$$\frac{N^2 \cos^2\mu}{a^2} + \frac{N^2 (1-e^2)^2 \sin^2\mu}{a^2 (1-e^2)} = 1$$

$$N^2 \left(1 - \sin^2\mu\right) + N^2 \left(1 - e^2\right)\sin^2\mu = a^2$$

$$N^2 \left(1 - e^2 \sin^2\mu\right) = a^2$$

$$\Rightarrow \quad N = \frac{a}{\sqrt{1-e^2 \sin^2\mu}} \tag{3.77}$$

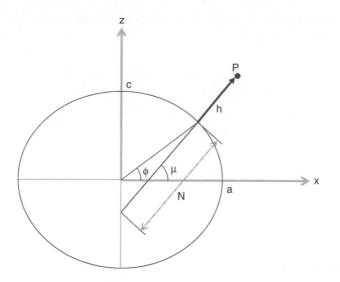

Figure 3.17 Generalised elliptical geometry.

From here it is a simple matter to derive the coordinates of a point P vertically above P_0, defined by geodetic latitude (μ) and altitude (h) as shown in Figure 3.17. By inspection, it is seen that the coordinates of P are:

$$x = (N + h) \cos \mu$$
$$z = [N (1 - e^2) + h] \sin \mu \tag{3.78}$$

Having obtained these basic relationships, it is a straightforward task to obtain generalised expressions for xyz-coordinates relative to the ECEF frame shown in Figure 3.18. For a

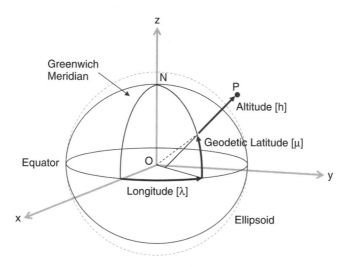

Figure 3.18 ECEF reference frame for an ellipsoidal Earth.

position defined by longitude (λ), geodetic latitude (μ) and altitude (h), these are derived as follows:

$$x = (N + h) \cos \mu \cos \lambda$$
$$y = (N + h) \cos \mu \sin \lambda \qquad (3.79)$$
$$z = [N(1 - e^2) + h] \sin \mu$$

3.6.4   *Approximate Cartesian Coordinates*

A simplified derivation is available based on the radial distance (r) from the centre of the Earth. In that case, altitude (h) is measured along a radial direction and is approximately equal to actual altitude (along the prime vertical) because the eccentricity of the Earth is very small. Thus, the xyz-coordinates become:

$$x = (r + h) \cos \phi \cos \lambda$$
$$y = (r + h) \cos \phi \sin \lambda \qquad (3.80)$$
$$z = (r + h) \sin \phi$$

This is the approach that might well be applied for standard models of atmospheric properties (cf. Chapter 5). Recalling (3.73) and (3.69), these coordinates are functions of geodetic latitude by virtue of the following equivalences:

$$r = \sqrt{\frac{a^4 \cos^2 \mu + c^4 \sin^2 \mu}{a^2 \cos^2 \mu + c^2 \sin^2 \mu}}$$

$$\sin \phi = \frac{c^2 \sin \mu}{\sqrt{a^4 \cos^2 \mu + c^4 \sin^2 \mu}} \qquad (3.81)$$

$$\cos \phi = \frac{a^2 \cos \mu}{\sqrt{a^4 \cos^2 \mu + c^4 \sin^2 \mu}} \qquad (3.82)$$

Equations (3.78) and (3.79) are justified by the trigonometrical interpretation shown in Figure 3.19.

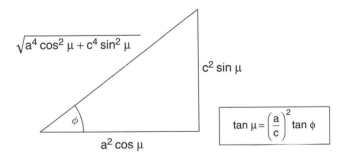

Figure 3.19 Relationship between geocentric and geodetic latitude.

3.6.5   *Latitude, Longitude and Altitude*

Section 3.6.3 considered the accurate calculation of cartesian coordinates (x,y,z) from longitude, latitude and altitude (λ,μ,h). The inverse calculation of (λ,μ,h) from (x,y,z) can be conducted using various methods, as summarised by Fok and Iz (2003).

One of the better known methods (Bowring, 1976) is initialised, as follows:

$$\tan \phi = \frac{z}{w} \tag{3.83}$$

$$\theta = \tan^{-1}\left[\left(\frac{a}{c}\right)\tan \phi\right] \tag{3.84}$$

where

$$w = \sqrt{x^2 + y^2} \tag{3.85}$$

and where (3.84) is a rearrangement of (3.70). It then proceeds as an iteration:

$$\mu = \tan^{-1}\left[\frac{z + (\varepsilon^2 c)\sin^3\theta}{w - (e^2 a)\cos^3\theta}\right] \tag{3.86}$$

$$\theta = \tan^{-1}\left[\left(\frac{c}{a}\right)\tan \mu\right] \tag{3.87}$$

where (3.87) is a rearrangement of (3.73). New values of μ are calculated repeatedly until convergence is achieved, i.e. the latest value of θ (from (3.84) on the first pass and from (3.87) otherwise) is used in order to calculate a new value of μ in (3.86) and vice versa. This process continues until the difference between successive values of μ is smaller that a predefined threshold (remembering to use the modulus of the difference!). The method is simple and gives fast convergence, which probably accounts for its popularity. In fact, for many applications, a single iteration may suffice.

The remaining calculation is for altitude, which can be arranged in the following convenient (and well-conditioned) form:

$$h = h_1 \cos^2\mu + h_2 \sin^2\mu \tag{3.88}$$

where

$$h_1 = \frac{w}{\cos \mu} - N \quad \text{and} \quad h_2 = \frac{z}{\sin \mu} - N(1 - e^2)$$

The parameter w is defined by (3.85).

3.7   **Navigation on an Ellipsoidal Earth**

3.7.1   *Differential Geometry*

A navigational course is defined by the change in latitude and longitude as a function of distance travelled. The key relationship between geodetic latitude and reduced latitude is shown clearly in Figure 3.16. This reflects the following equivalence between coordinates of any point of the perimeter of the ellipse, as defined by (3.71) and (3.76):

$$a \cos \theta = N \cos \mu$$
$$c \sin \theta = N(1 - e^2) \sin \mu \tag{3.89}$$

where the length of the prime vertical (N) is given by (3.77):

$$N = \frac{a}{\sqrt{1 - e^2 \sin^2 \mu}}$$

This is consistent with (3.73):

$$\tan \theta = \left(\frac{c}{a}\right) \tan \mu$$

and also gives rise to the following relationships:

$$\sin \theta = \frac{\sin \mu \sqrt{1 - e^2}}{\sqrt{1 - e^2 \sin^2 \mu}} \tag{3.90}$$

$$\cos \theta = \frac{\cos \mu}{\sqrt{1 - e^2 \sin^2 \mu}} \tag{3.91}$$

For a position defined by longitude (λ), geodetic latitude (μ) and altitude (h), the equivalent xyz-coordinates are defined by (3.79). Using parametric latitude (θ), as above, the coordinates of a point on the surface of an ellipsoid (h = 0) can be written as follows:

$$\begin{aligned} x &= a \, \cos \theta \cos \lambda \\ y &= a \, \cos \theta \sin \lambda \\ z &= c \, \sin \theta \end{aligned} \tag{3.92}$$

Now, taking a differential change in latitude and longitude, the corresponding differential change in position is given by:

$$\begin{aligned} dx &= -a \, \sin \theta \cos \lambda d\theta - a \cos \theta \sin \lambda d\lambda \\ dy &= -a \, \sin \theta \sin \lambda d\theta + a \cos \theta \cos \lambda d\lambda \\ dz &= c \, \cos \theta d\theta \end{aligned} \tag{3.93}$$

The total distance travelled (ds) is derived as follows:

$$ds^2 = dx^2 + dy^2 + dz^2$$

$$ds^2 = a^2 (\sin \theta \cos \lambda d\theta + \cos \theta \sin \lambda d\lambda)^2 + a^2 (\sin \theta \sin \lambda d\theta - \cos \theta \cos \lambda d\lambda)^2 + c^2 (\cos \theta d\theta)^2$$

$$ds^2 = a^2 \left(\sin^2 \theta + \frac{c^2}{a^2} \cos \theta \right) d\theta^2 + a^2 \cos^2 \theta d\lambda^2$$

Recalling the definition of ellipticity (3.61), this becomes:

$$ds^2 = a^2 \left(1 - e^2 \cos^2\theta\right)d\theta^2 + a^2 \left(\cos^2\theta\right)d\lambda^2 \tag{3.94}$$

Note that, as this relates to distance travelled, it is actually a scale of measurement. Formally, 'ds' is called a *metric*.

From (3.73), the change in parametric latitude can be related to a change in geodetic latitude:

$$\frac{d\theta}{\cos^2\theta} = \left(\frac{c}{a}\right)\frac{d\mu}{\cos^2\mu}$$

From (3.90), it is seen that

$$\frac{\cos\theta}{\cos\mu} = \frac{1}{\sqrt{1-e^2\sin^2\mu}}$$

Therefore,

$$d\theta = \left(\frac{c}{a}\right)\left(\frac{1}{\sqrt{1-e^2\sin^2\mu}}\right)^2 d\mu$$

$$d\theta^2 = \frac{1-e^2}{\left(1-e^2\sin^2\mu\right)^2}d\mu^2 \tag{3.95}$$

Recalling the definition of first eccentricity (3.63), which can be recast in the form:

$$e^2 = 1 - \left(\frac{c}{a}\right)^2 \quad \text{or} \quad 1 - e^2 = \left(\frac{c}{a}\right)^2$$

Also from (3.76), it can easily be shown that:

$$1 - e^2\cos^2\theta = \frac{1-e^2}{1-e^2\sin^2\mu} \tag{3.96}$$

Thus, the metric for the ellipsoid can be rewritten as follows:

$$ds^2 = a^2\left(\frac{1-e^2}{1-e^2\sin^2\mu}\right)\frac{1-e^2}{\left(1-e^2\sin^2\mu\right)^2}d\mu^2 + a^2\left(\frac{\cos\mu}{\sqrt{1-e^2\sin^2\mu}}\right)^2 d\lambda^2$$

$$ds^2 = \frac{a^2(1-e^2)^2}{\left(1-e^2\sin^2\mu\right)^3}d\mu^2 + \frac{a^2\cos^2\mu}{1-e^2\sin^2\mu}d\lambda^2 \tag{3.97}$$

This can be written in the general form:

$$ds^2 = A(d\mu)^2 + B(d\lambda)^2$$

and then differentiated with respect to latitude (μ) and longitude (λ):

$$2(ds)\frac{ds}{d\mu} = 2A(d\mu) \Rightarrow \frac{ds}{d\mu} = A\frac{d\mu}{ds} \Rightarrow \frac{ds}{d\mu} = \sqrt{A}$$
$$2(ds)\frac{ds}{d\lambda} = 2B(d\lambda) \Rightarrow \frac{ds}{d\lambda} = B\frac{d\lambda}{ds} \Rightarrow \frac{ds}{d\lambda} = \sqrt{B}$$

(3.98)

Applying these formulae to (3.97), the variations in latitude and longitude are expressed as:

$$R_M \equiv \frac{ds}{d\mu} = \frac{a(1 - e^2)}{\left(1 - e^2 \sin^2\mu\right)^{3/2}}$$

(3.99)

$$R_P \equiv \frac{ds}{d\lambda} = \frac{a \cos\mu}{\left(1 - e^2 \sin^2\mu\right)^{1/2}}$$

(3.100)

where R_M is the radius of curvature along the meridian at longitude λ and R_P is the so-called radius of curvature in the local plane of latitude (i.e. the *parallel* at latitude μ). Thus, the metric given in (3.96) can be written in the compact form:

$$ds^2 = (R_M)^2 d\mu^2 + (R_P)^2 d\lambda^2$$

(3.101)

With reference to Figure 3.16, noting the significance of $\cos\mu$ in resolving measurements along the prime vertical into any plane parallel with the equator, the radii of curvature can be written as they often appear in texts:

$$R_M = \frac{a\,(1 - e^2)}{\left(1 - e^2 \sin^2\mu\right)^{3/2}}$$

(3.102)

$$R_N = \frac{a}{\left(1 - e^2 \sin^2\mu\right)^{1/2}}$$

(3.103)

where R_M is the meridian radius of curvature (as before) and R_N is the so-called *transverse* radius of curvature. Note that R_N is identical to N, which the length of the prime vertical, as defined in (3.77). Also, from (3.98) and (3.103), it is clear that

$$R_P = R_N \cos\mu$$

(3.104)

Given that $R_N = N$, together with that coordinate equivalence in (3.89), this can be re-written as follows:

$$R_P = a \cos\theta$$

(3.105)

The metric given in (3.100) can be expressed equivalently in terms of longitude and parametric latitude. From (3.94), the variation of parametric latitude with respect to geodetic latitude is:

$$\frac{d\theta}{d\mu} = \frac{(1 - e^2)^{1/2}}{1 - e^2 \sin^2\mu}$$

Recalling (3.95), the equivalence between parametric and geodetic latitudes is:

$$1 - e^2 \cos^2\theta = \frac{1 - e^2}{1 - e^2 \sin^2\mu}$$

Thus, (3.101) can be manipulated as follows:

$$R_M = \frac{a \, (1 - e^2)^{1/2}}{(1 - e^2 \sin^2\mu)^{1/2}} \frac{(1 - e^2)^{1/2}}{(1 - e^2 \sin^2\mu)}$$

$$R_M = a\sqrt{1 - e^2 \cos^2\theta} \, \frac{d\theta}{d\mu} \tag{3.106}$$

Combining these two results, i.e. (3.105) and (3.106), the metric in (3.100) can be re-expressed metric as:

$$ds^2 = a^2 \left[(1 - e^2 \cos^2\theta) d\theta^2 + \cos^2\theta \, d\lambda^2 \right] \tag{3.107}$$

As a final observation in this section, the ratio of variations in latitude and longitude associated with a distance travelled (s) gives the instantaneous course heading (α), measured as an azimuth angle from North (increasing clockwise):

$$\tan\alpha = \left(\frac{ds}{d\lambda} \right) \Big/ \left(\frac{ds}{d\mu} \right)$$

This can be evaluated in the form:

$$\tan\alpha = \cos\mu \frac{1 - e^2 \sin^2\mu}{1 - e^2} \left(\frac{d\lambda}{d\mu} \right) \tag{3.108}$$

Alternatively, in terms of parametric latitude, this can be expressed as:

$$\tan\alpha = \frac{\cos\theta}{\sqrt{1 - e^2 \cos^2\theta}} \left(\frac{d\lambda}{d\theta} \right) \tag{3.109}$$

In passing, note that this equation is consistent with (3.107), applying Pythagoras' theorem to a right-angled triangle. By this reasoning, the trigonometry can be completed:

$$\sin\alpha = a \cos\theta \left(\frac{d\lambda}{ds} \right) \tag{3.110}$$

$$\cos\alpha = a\sqrt{1 - e^2 \cos^2\theta} \left(\frac{d\theta}{ds} \right) \tag{3.111}$$

3.7.2   Geodesics

As stated already, a geodesic line defines the shortest path between two points on a smooth surface (i.e. a surface that is continuous and differentiable). On a flat surface, geodesics are straight lines; on spherical surfaces, geodesics are great circles. Two useful and remarkable facts are that (i) a geodesic is the path followed by a particle sliding smoothly across a surface at constant speed and (ii) the radius of curvature of a geodesic is locally aligned with the normal vector to that surface.

For a particle of mass m and velocity v, the angular momentum is calculated by resolving the linear momentum along the local *tangent* of latitude, i.e. mv.sinα, and multiplying by the radius of the local *parallel* of latitude, i.e. a.cosθ or N.cosμ (depending on whether parametric or geodetic latitude is used). Under the conservation of angular momentum, the following relationship is true:

$$\cos\theta\sin\alpha = k \tag{3.112}$$

where k is a constant. The variation in distance travelled along a geodesic path can be derived from (3.111) as follows:

$$\frac{ds}{d\theta} = a\frac{\sqrt{1 - e^2\cos^2\theta}}{\cos\alpha} \tag{3.113}$$

where

$$\cos\theta\cos\alpha = \sqrt{\cos^2\theta - k^2} \tag{3.114}$$

Performing the necessary substitution, the variation becomes:

$$\frac{ds}{d\theta} = a\cos\theta\sqrt{\frac{1 - e^2\cos^2\theta}{\cos^2\theta - k^2}} \tag{3.115}$$

This constant (k) can be established from (3.112) simply by setting a particular value of latitude. For some point on the equator (i.e. $\theta = 0$), the constant can be written as:

$$k = \sin\alpha_0 \tag{3.116}$$

where α_0 is a constant angle, which means that (3.112) can be rewritten as

$$\cos\theta\sin\alpha = \sin\alpha_0 \tag{3.117}$$

Noting the equivalence of *geocentric* latitude on an ellipsoid and *parametric* latitude on a sphere, this is identical to Equation (3.39). This leads to the conclusion that a geodesic drawn in the surface of an ellipsoid is equivalent to a great circle drawn on the surface of the auxiliary (or circumscribed) sphere. This means that the geodesic trajectory is a function of the progression angle (σ) around a great circle (cf. Figure 3.9).

Recalling Section 3.3, the key relationships for great circles (using parametric latitude) are given by (3.28), (3.29), (3.32), (3.39) and (3.40):

$$\left\{\begin{array}{l} \cos\theta\cos\Delta\lambda = \cos\sigma \\ \cos\theta\sin\Delta\lambda = \sin\sigma\sin\alpha_0 \\ \tan\Delta\lambda = \tan\sigma\sin\alpha_0 \end{array}\right\} \tag{3.118}$$

$$\left\{\begin{array}{l} \sin\theta = \sin\sigma\cos\alpha_0 \\ \cos\theta = \sqrt{1 - \sin^2\sigma\cos^2\alpha_0} \end{array}\right\} \tag{3.119}$$

$$\left\{\begin{array}{l} \cos\theta\sin\alpha = \sin\alpha_0 \\ \cos\theta\cos\alpha = \cos\sigma\cos\alpha_0 \\ \tan\alpha = \cos\sigma\tan\alpha_0 \end{array}\right\} \tag{3.120}$$

From (3.119), it is seen that:

$$\frac{d\theta}{d\sigma} = \frac{\cos \sigma \cos \alpha_0}{\cos \theta} \tag{3.121}$$

From (3.109), the variation of longitude with parametric latitude is:

$$\frac{d\lambda}{d\theta} = \tan \alpha \frac{\sqrt{1 - e^2 \cos^2 \theta}}{\cos \theta} \tag{3.122}$$

From (3.115) and (3.116), the variation of distance with parametric latitude is:

$$\frac{ds}{d\theta} = a \cos \theta \sqrt{\frac{1 - e^2 \cos^2 \theta}{\cos^2 \theta - \sin^2 \alpha_0}} \tag{3.123}$$

where, in order to be consistent with (3.113), the sign of the square root must be equal to the sign of $\cos \alpha$.

Other useful derivations from (3.119) are:

$$\cos^2 \theta - \sin^2 \alpha_0 = \cos^2 \sigma \cos^2 \alpha_0$$
$$1 - e^2 \cos^2 \theta = (1 - e^2)\left(1 + E^2 \sin^2 \sigma\right) \tag{3.124}$$

where

$$E^2 = \frac{e^2}{1 - e^2} \cos \alpha_0 \tag{3.125}$$

Referring back to (3.101) and (3.106), the following correspondence is seen:

$$a\sqrt{1 - e^2 \cos^2 \theta} \, d\theta = \frac{a(1 - e^2)}{\left(1 - e^2 \sin^2 \mu\right)^{3/2}} \, d\mu \tag{3.126}$$

This collection of equations from (3.118) to (3.126), together with (3.73), provides the basis for the remaining derivations in this section.

Note 3.4

Unfortunately this subject comprises many different theoretical elements and it can seem that the task at hand is buried under a lot of mathematics. So, to recap, the task here is to determine a geodesic path (which is a line of minimum length) between two points on the Earth's surface, which is approximated by an ellipsoid. In particular, where does the path go and how long is it?

By definition, the longitude and geodetic latitude of each point is known. Using (3.76), the equivalent parametric latitude can be found. The next step is to go back to Section 3.3 and determine the great circle path around the auxiliary sphere. That will sort out where the ascending node lies on the equator, what the heading is at the equator (α_0) and what the progression angles are for the two points, via (3.51). The remaining step is to obtain formulas for distance and longitude along a geodesic path, based on a spherical great circle. The great circle provides a starting point for calculation but the actual end-result is affected by the eccentricity of the ellipsoid.

3.7.3   *Geodesic Trajectory*

The variation in longitude with respect to the progression angle (σ) is derived from (3.120), (3.122) and (3.123), as follows:

$$\frac{d\lambda}{d\sigma} = \left(\frac{d\lambda}{d\theta}\right)\left(\frac{d\theta}{d\sigma}\right)$$

$$= \left(\tan\alpha\,\frac{\sqrt{1 - e^2\cos^2\theta}}{\cos\theta}\right)\left(\frac{\cos\sigma\cos\alpha_0}{\cos\theta}\right)$$

$$= \left(\tan\alpha\,\frac{\sqrt{1 - e^2\cos^2\theta}}{\cos\theta}\right)\left(\frac{\cos\theta\cos\alpha}{\cos\theta}\right)$$

$$= \frac{\sin\alpha}{\cos\theta}\sqrt{1 - e^2\cos^2\theta}$$

$$= \frac{\sin\alpha_0}{\cos^2\theta}\sqrt{1 - e^2\cos^2\theta}$$

Applying relationships between parametric latitude (θ) and progression angle (σ) given in (3.121) and (3.124), this equation becomes:

$$\frac{d\lambda}{d\sigma} = \frac{\sin\alpha_0}{1 - \sin^2\sigma\,\cos^2\alpha_0}\sqrt{(1 - e^2)\,(1 + E^2\sin^2\sigma)} \tag{3.127}$$

where E is given by (3.125). Thus, the longitude of a point on the geodesic path starting from P_1 is derived by integration:

$$\lambda = \lambda_1 + \sin\alpha_0\sqrt{1 - e^2}\int_{\sigma_1}^{\sigma}\frac{\sqrt{1 + E^2\sin^2\sigma}}{1 - \sin^2\sigma\cos^2\alpha_0}\,d\sigma \tag{3.128}$$

where λ_1 is the longitude of P_1, σ_1 is the progression angle of P_1, and σ is the progression angle of any point on the geodesic path between P_1 and P_2.

As is popular in texts on geodesy and mapping, it is possible to apply a Taylor expansion to the integrand in (3.127) and then perform the integration on the resulting power series, the longitude difference of the geodesic path is:

$$\Delta\lambda = \tan^{-1}(\sin\alpha_0\tan(\sigma - \sigma_1)) - e^2\sin\alpha_0\sum_{n=0}^{\infty}L_{2n}e^{2n}\sin 2n\sigma \tag{3.129}$$

where the first few coefficients L_{2n} are, as follows:

$$L_0 = \frac{1}{2}\left(1 + \frac{1 + \sin^2\alpha_0}{8}e^2 + \frac{3 + 2\sin^2\alpha_0 + 3\sin^4\alpha_0}{64}e^4 + \frac{5\,(5 + 3\sin^2\alpha_0 + 3\sin^4\alpha_0 + 5\sin^6\alpha_0)}{1024}e^6\right)$$

$$L_2 = \frac{1}{32}\left(1 + \frac{1 + \sin^2\alpha_0}{2}e^2 + \frac{15\,(5 + 6\sin^2\alpha_0 + 5\sin^4\alpha_0)}{256}e^4\right)\cos^2\alpha_0$$

$$L_4 = \frac{1}{512}\left(1 + \frac{5\,(1 + \sin^2\alpha_0)}{16}e^2\right)\cos^4\alpha_0$$

$$L_6 = \frac{5}{24576}\cos^6\alpha_0$$

Parametric latitude is related to progression angle via (3.119):

$$\sin \theta = \sin \sigma \cos \alpha_0$$
$$\cos \theta = \sqrt{1 - \sin^2 \sigma \cos^2 \alpha_0} \tag{3.130}$$

Geodetic latitude is then determined as:

$$\mu = \tan^{-1}\left[\left(\frac{a}{c}\right)\frac{\sin \theta}{\cos \theta}\right] \tag{3.131}$$

3.7.4   *Geodesic Length*

The variation in distance with respect to the progression angle (σ) is derived from (3.121), (3.123) and (3.125), as follows:

$$\frac{ds}{d\sigma} = \left(\frac{ds}{d\theta}\right)\left(\frac{d\theta}{d\sigma}\right)$$
$$= \left(a\frac{\cos \theta}{\cos \sigma \cos \alpha_0}\sqrt{(1 - e^2)\,(1 + E^2 \sin^2 \sigma)}\right)\left(\frac{\cos \sigma \cos \alpha_0}{\cos \theta}\right)$$
$$= \left(a\sqrt{1 - e^2}\right)\sqrt{1 + E^2 \sin^2 \sigma}$$

Noting the definition of first eccentricity (3.63), the rationalised expression is:

$$\frac{ds}{d\sigma} = c\,\sqrt{1 + E^2 \sin^2 \sigma} \tag{3.132}$$

Thus, the length along a geodesic path between two points P_1 and P_2 is:

$$s = c\int_{\sigma_1}^{\sigma_2}\sqrt{1 + E^2 \sin^2 \sigma}d\sigma \tag{3.133}$$

where σ_1 and σ_2 are the progression angles of P_1 and P_2 respectively.

Applying a Taylor expansion to the integrand in (3.133) and then performing the integration on the resulting power series, the length of the geodesic path is:

$$s(\sigma) = c(S_0\sigma - S_2\,\sin 2\sigma - S_4\,\sin 4\sigma - S_6\,\sin 4\sigma - \ldots) \tag{3.134}$$

where

$$S_0 \approx 1 + \frac{1}{4}E^2 - \frac{3}{64}E^4 + \frac{5}{256}E^6$$

$$S_2 \approx \frac{1}{8}\left(E^2 - \frac{1}{4}E^4 + \frac{15}{128}E^6\right)$$

$$S_4 \approx \frac{1}{256}\left(E^4 - \frac{3}{4}E^6\right)$$

$$S_6 \approx \frac{1}{3072}E^6$$

Thus, the distance between to points P_1 and P_2 is given by:

$$\Delta s = s(\sigma_2) - s(\sigma_1)$$

In its fully expanded form, this involves subtraction of 'sines' of double angles. A well-known method is given in Note 3.4, which reduces numerical errors associated with small differences. Thus the distance formula can be expressed, as follows:

$$\Delta s = c \left[A_0 \Delta\sigma - 2A_2 (\cos 2\sigma_m \sin \Delta\sigma) - 2A_4 (\cos 4\sigma_m \sin 2\Delta\sigma) - \ldots \right] \qquad (3.135)$$

where $\Delta\sigma = \sigma_2 - \sigma_1$ and $\sigma_m = (\sigma_2 + \sigma_1)/2$.

Note 3.5

$$
\begin{aligned}
2\cos(\alpha+\beta)\sin(\alpha-\beta) &= 2\ (\cos\alpha\cos\beta - \sin\alpha\sin\beta)\ (\sin\alpha\cos\beta - \cos\alpha\sin\beta) \\
&= 2\sin\alpha\cos\alpha\ (\sin^2\beta + \cos^2\beta) - 2\sin\beta\cos\beta\ (\sin^2\alpha + \cos^2\alpha) \\
&= 2\sin\alpha\cos\alpha - 2\sin\beta\cos\beta \\
&= \sin 2\alpha - \sin 2\beta
\end{aligned}
$$

Note 3.6

Numerous algorithms exist for determining geodesic paths and enthusiasts are recommended to study the original papers of researchers such as Rainsford (1955), Sodano (1965), Bowring (1969) and Vincenty (1975). Those who follow this up will see the extent to which the basic method developed in Sections 3.7.3 and 3.7.4 can be refined in order to enable compact and efficient algorithms. However, be aware that the algebra becomes intricate and care is required in its interpretation.

3.7.5   *Meridian Distances*

By definition, a meridian is related to a great circle with $\alpha = \alpha_0 = 0$. With reference to (3.125), $E^2 = e^2/(1 - e^2)$, and therefore the distance calculation (3.133) simplifies to the following:

$$s = c \int_{\sigma_1}^{\sigma_2} \sqrt{1 + \frac{e^2}{1 - e^2} \sin^2\sigma}\ d\sigma = a \int_{\sigma_1}^{\sigma_2} \sqrt{1 - e^2 \cos^2\sigma}\ d\sigma$$

Moreover, (3.120) simplifies to the following:

$$\cos\theta = \cos\sigma$$

The distance calculation in terms of parametric latitude becomes:

$$s = a \int_{\theta_1}^{\theta_2} \sqrt{1 - e^2 \cos^2\theta}\, d\theta \qquad (3.136)$$

Applying (3.126), the distance calculation in terms of geodetic latitude becomes:

$$s = a \left(1 - e^2\right) \int_{\mu_1}^{\mu_2} \frac{1}{\left(1 - e^2 \sin^2 \mu\right)^{3/2}} \, d\mu \tag{3.137}$$

Specifically, setting $\mu_1 = 0$ (i.e. the equator) and $\mu_2 = \mu$ (i.e. any other latitude), the calculation produces the distances between the equator and a point at latitude μ. This is expressed as:

$$M = a \left(1 - e^2\right) \int_0^{\mu} \frac{d\mu}{\left(1 - e^2 \sin^2 \mu\right)^{3/2}} \tag{3.138}$$

This equation cannot be integrated exactly, for reasons that are of no great importance here. However, it can be integrated via a series expansion:

$$M = a \left(M_0 \mu - M_2 \sin 2\mu + M_4 \sin 4\mu - M_6 \sin 6\mu + \ldots\right) \tag{3.139}$$

where

$$M_0 \approx 1 - \frac{1}{4} e^2 - \frac{3}{64} e^4 - \frac{5}{256} e^6$$

$$M_2 \approx \frac{3}{8} \left(e^2 + \frac{1}{4} e^4 + \frac{15}{128} e^6\right)$$

$$M_4 \approx \frac{15}{256} \left(e^4 + \frac{1}{4} e^6\right)$$

$$M_6 \approx \frac{35}{3072} e^6$$

With reference to Snyder (1987) (summarising the work of Adams (1949)), this is the basis for the definition of *rectifying* latitude:

$$\omega = \frac{\pi}{2} \left(\frac{M(\mu)}{M(\pi/2)}\right) \tag{3.140}$$

This maps geodetic latitudes on to a sphere with correct distances along meridians. $M(\pi/2)$ is the distance from the equator to the pole and is evaluated from the formula for M by setting $\mu = \pi/2$ (which means that all the sine terms evaluate to zero). For the WGS84 ellipsoid, the polar distance is calculated as 10001.949km.s

An alternative formulation relates the rectifying latitude and the geodetic latitude:

$$\omega = \mu - W_2 \sin 2\mu + W_4 \sin 4\mu - W_6 \sin 6\mu + W_8 \sin 8\mu - \ldots \tag{3.141}$$

where

$$W_2 \approx \frac{3}{2} \left(E_1 - \frac{3}{8} E_1^3\right)$$

$$W_4 \approx \frac{15}{16} \left(E_1^2 - \frac{1}{2} E_1^4\right)$$

$$W_6 \approx \frac{35}{48} E_1^3$$

$$W_8 \approx \frac{315}{512} E_1^4$$

and where

$$E_1 = \frac{1 - \sqrt{1 - e^2}}{1 + \sqrt{1 - e^2}} \tag{3.142}$$

For completeness, the inverse relationship is quoted as follows:

$$\mu = \omega + U_2 \sin 2\omega + U_4 \sin 4\omega + U_6 \sin 6\omega + U_8 \sin 8\omega + \dots \tag{3.143}$$

where

$$U_2 \approx \frac{3}{2} E_1 - \frac{27}{32} E_1{}^3$$

$$U_4 \approx \frac{21}{16} E_1{}^2 - \frac{55}{32} E_1{}^4$$

$$U_6 \approx \frac{151}{96} E_1{}^3$$

$$U_8 \approx \frac{1097}{512} E_1{}^4$$

3.7.6   *Rhumb Lines*

As stated in Section 3.4, *rhumb lines* are lines of constant heading. However, in this context, they are defined on an ellipsoidal surface rather than a spherical surface. Thus, in comparison with (3.52), the general definition of *isometric latitude* is as follows:

$$\psi = \ln \left[\tan\left(\frac{\pi}{4} + \frac{\mu}{2}\right) \left(\frac{1 - e \sin \mu}{1 + e \sin \mu}\right)^{\frac{e}{2}} \right] \tag{3.144}$$

where 'ln' denotes a natural logarithm. This is equivalent to:

$$\psi = \frac{1}{2} \ln \left[\left(\frac{1 + \sin \mu}{1 - \sin \mu}\right) \left(\frac{1 - e \sin \mu}{1 + e \sin \mu}\right)^{e} \right] \tag{3.145}$$

The conversion from isometric latitude to geodetic latitude can be performed iteratively, for $n \geq 1$, as follows:

$$\mu_n = 2 \tan^{-1} \left(\varepsilon^{\psi} \left(\frac{1 - e \sin \mu_{n-1}}{1 + e \sin \mu_{n-1}}\right)^{\frac{e}{2}} \right) - \frac{\pi}{2} \tag{3.146}$$

where ε is the base of natural logarithms ($\varepsilon \approx 2.71828182845904$). The equivalent spherical formula (3.54) is used to give the initial estimate:

$$\mu_0 = 2 \tan^{-1} \left(e^{\psi}\right) - \frac{\pi}{2} \tag{3.147}$$

The rhumb line determination then proceeds in the same way as for spherical navigation. The heading is an azimuth angle (α) (measured clockwise from North) is given by (3.56):

$$\alpha = \tan^{-1}\left(\frac{\lambda_2 - \lambda_1}{\Psi_2 - \Psi_1}\right) \tag{3.148}$$

where (λ_1, Ψ_1) and (λ_2, Ψ_2) specify the longitude and isometric latitude of points P_1 and P_2, respectively. The distance between P_1 and P_2 is given by (3.57):

$$s = \frac{M_2 - M_1}{\cos\alpha} \tag{3.149}$$

where M_1 and M_2 are the meridian distances from the equator to P_1 and P_2, respectively. Recall that the definition of meridian distance on an ellipsoid is discussed in Section 3.7.5.

3.8   Mapping

Mapping is the process of producing a representation of a curved two-dimensional surface. This enables the surface of the Earth (or any other body in space) to be calibrated so that any point can located with respect to a datum surface. In other words, this is a process of establishing the surface topography and locating significant features. In the preceding sections, a framework has been developed based on latitude, longitude and height and, in turn, this has been related to ECEF cartesian coordinates. Previously, for points above the surface, height is referred to as *altitude*. Where surface points are being considered, this is more appropriately designated as *elevation*; below the surface, this is designated as *depth*.

The framework in this section will be based on orthogonal coordinates so that a model of the Earth's surface (in part of in full) can be created inside a rectangular box. The simplest map construction would be an xyz-frame where 'x' is longitude, 'y' is latitude and 'z' is altitude. A projection on to the yz-plane, zx-plane or any other vertical plane would produce a profile of the surface topography along the specific azimuth direction. A projection on to the horizontal xy-plane would produce a conventional map (in plan form), upon which elevations are marked by peaks and troughs and a set of contours in-between. The big problems are associated with distortion and resolution.

In this case, *distortion* in the x-scale occurring as an x-increment on the map represents a distance that tends to zero as the y-ordinate moves away from the equator (i.e. proportional to $\cos\mu$). Geographical features become increasingly spaced out towards the extreme north and south of the map. The *resolution* of a map not only determines the smallest object that will be seen on the map but also the minimum spacing between objects that will create a recognisable map of a given area.

3.9   General Principles of Map Projection

Two-dimensional mapping is the process of producing a flat two-dimensional representation of a curved two-dimensional surface. This enables the surface of the Earth (in part or in full) to be projected on to a map (as conventionally understood) with latitude (μ) and longitude (λ) being projected on to map coordinates (x,y):

$$x = \mathbf{f}(\lambda, \mu)$$
$$y = \mathbf{g}(\lambda, \mu)$$

(3.150)

An ideal map would have no distortion and infinite resolution. This would mean that:

- distances and areas have the correct relative magnitude;
- azimuth angles are reproduced correctly at every point;
- latitude and longitude are shown correctly at every point;
- great circles are depicted as straight lines;
- all features of interest and importance are visible.

Setting aside the issue of finite resolution (which is a fundamental limitation based on absolute size), it is not possible to achieve the other *ideal* four attributes on the same map. In effect, a map is a flattened-out representation of a spherical or ellipsoidal surface. However, it is possible to achieve one or more attributes. Thus, the main types of map projection are categorised, as follows:

Equidistant	Distances are represented correctly from a central point to all other points on the map
Equal area	All areas are represented with the correct relative size (but not necessarily the correct shape)
Azimuthal	Directions are represented correctly from a central point to all other points on the map
Conformal	Angles of intersection are correctly represented between pairs of short lines and thus small areas have the correct shape

In terms of actual projections, the most basic concept is mapping of points on the surface of a sphere on to a tangent plane. This can be achieved geometrically by setting the centre of projection at the centre of the sphere, at the point that is diametrically opposite the tangent point or at a point that is infinitely distant from the tangent point. This provides the basis for *gnomonic*, *stereographic* and *orthographic* projections, respectively. These are shown schematically in Figure 3.20. In each case, point A is at the intersection of the sphere and the tangent plane. The line drawn orthogonal to the tangent plane at A extends to the centre of the sphere (point O), the oppositional point (point P) and hence to infinity. A generic point of the surface (designated as point B) would be projected as point B' on the tangent plane. Thus, that plane is the mapping surface. It can be located at any point on the surface that is being mapped, i.e. at any desired orientation.

The tangent point is the central point of the map. For all three projections, great circles that pass through the central point are shown as straight lines and, thus, azimuth angles are shown correctly at that point. So, these are all azimuthal projections. A number of distinctions can be highlighted, as follows:

1. In gnomonic and stereographic projections, scale increases with distance from the central point: in orthographic projection, it decreases such that the representation appears as it would in a long-range image.
2. In stereographic and orthographic projections, it is possible to map an entire hemisphere: in gnomonic projection, this is not possible.
3. In orthographic and gnomonic projections, shape is not preserved: in stereographic projection, circles on the sphere appear as circles on the map.

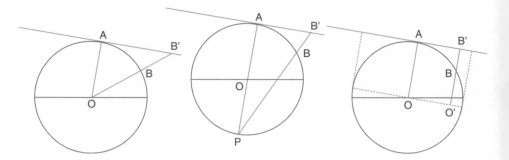

Figure 3.20 Gnomonic, stereographic and orthographic projections.

More elaborate projections can be developed by changing the projection surface from flat to curved. Commonly used methods employ a cone or a cylinder (as shown in Figures 3.21 and 3.22) at an appropriate orientation. In the simplest configuration, the surface is tangential to the sphere along a line called the *standard tangent*.

For a cone, this will always be a small circle; for a cylinder, this will always be a great circle. In many cases, conical projections will be made with respect to a surface with its apex on the north-south polar axis, such that the standard tangent becomes a *standard parallel*. In cylindrical projections, a north-south orientation for the cylinder means that the standard tangent is the equator. However, other orientations are often used; a horizontal orientation leads to a so-called *transverse* projection and anything else will lead to an *oblique* projection.

As an observation on conical and cylindrical projections, the scale is correct on the standard tangent but distortions become progressively worse as the distance increases away from the tangent. A common approach is to reduce the diameter of the cone or cylinder slightly, thereby intersecting the sphere along two small circles. The scale will be correct along each circle and the scale errors will be generally reduced over a wider area.

There are many, many map projections in existence, ranging from the frequently used planforms to some quite elegant and artistic depictions. Each projection has a pair of equations associated with it, generating xy-coordinates from latitude and longitude. Depending on the

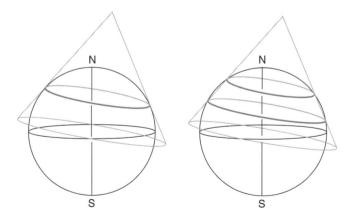

Figure 3.21 Conical projection surface showing standard tangent(s).

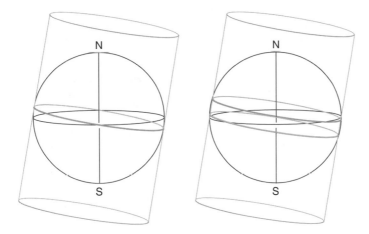

Figure 3.22 Cylindrical projection surface showing standard tangent(s).

parameters involved, the projections can exhibit considerable variation and produce bizarrely distorted representations of land mass.

Equations for a spherical Earth are relatively easy to follow if they are based on geometric projections, albeit the corresponding equations for an ellipsoidal Earth are necessarily more sophisticated. Equations that are based on mathematical projections are altogether more complicated and their origins appear arcane.

The MATLAB Mapping Toolbox offers an excellent overview of this entire topic, plus a large set of functions for creating and manipulating maps and map data. At the time of writing, it supports over sixty projections, as listed in Table 3.6. By way of introduction, a function called **viewmaps** is available, which allows a user to investigate the properties of about twenty projections. This is controlled via a GUI,[7] as shown in Figure 3.23.

Table 3.6 Map Projections available in the MATLAB Mapping toolbox.

Aitoff Projection	Kavraisky V Projection
Albers Equal-Area Conic Projection	Kavraisky VI Projection
Apianus II Projection	Lambert Azimuthal Equal-Area Projection
Balthasart Cylindrical Projection	Lambert Conformal Conic Projection
Behrmann Cylindrical Projection	Lambert Equal-Area Cylindrical Projection
Bolshoi Sovetskii Atlas Mira Projection	Loximuthal Projection
Bonne Projection	McBryde-Thomas Flat-Polar Parabolic Projection
Braun Perspective Cylindrical Projection	McBryde-Thomas Flat-Polar Quartic Projection
Breusing Harmonic Mean Projection	McBryde-Thomas Flat-Polar Sinusoidal Projection
Briesemeister Projection	Mercator Projection
Cassini Cylindrical Projection	Miller Cylindrical Projection
Central Cylindrical Projection	Mollweide Projection
Collignon Projection	Murdoch I Conic Projection

---

[7] A Coastline Extractor is available online from the NOAA National Geophysical Data Center and can be found at rimmer.ngdc.noaa.gov/mgg/coast/get_coast.html.

Table 3.6 (Continued).

Craster Parabolic Projection	Murdoch III Minimum Error Conic Projection
Eckert I Projection	Orthographic Projection
Eckert II Projection	Plate Carrée Projection
Eckert III Projection	Polyconic Projection
Eckert IV Projection	Putnins P5 Projection
Eckert V Projection	Quartic Authalic Projection
Eckert VI Projection	Robinson Projection
Equal-Area Cylindrical Projection	Sinusoidal Projection
Equidistant Azimuthal Projection	Stereographic Projection
Equidistant Conic Projection	Tissot Modified Sinusoidal Projection
Equidistant Cylindrical Projection	Transverse Mercator Projection
Fournier Projection	Trystan Edwards Cylindrical Projection
Gall Isographic Projection	Universal Polar Stereographic Projection
Gall Orthographic Projection	Universal Transverse Mercator Projection
Gall Stereographic Projection	Van der Grinten I Projection
Globe	Vertical Perspective Azimuthal Projection
Gnomonic Projection	Wagner IV Projection
Goode Homolosine Projection	Werner Projection
Hammer Projection	Wetch Cylindrical Projection
Hatano Asymmetrical Equal-Area	Wiechel Projection
Projection	Winkel I Projection

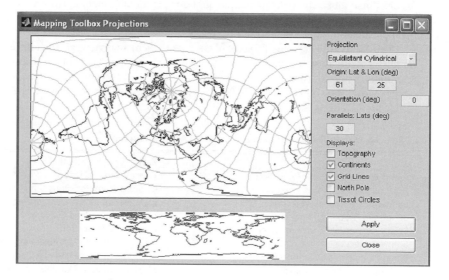

Figure 3.23 Graphical user interface for MATLAB 'viewmaps' function.

From this, a number of examples have been generated and these are depicted in Figures 3.24 to 3.26. As an observation, maps can have virtually any shape and orientation and can be centred on any location, depending on the preference of the mapmaker. For instance, the loximuthal projection in Figure 3.26 has a layout that is quite popular in Australia. Incidentally, Figure 3.1 (which was not generated using the Mapping Toolbox) is an orthographic projection centred on the Black Sea.

Figure 3.24 Polar stereographic projection.

3.10   Mercator Projection

The Mercator Projection is a cylindrical projection with the standard tangent at the equator. The x-ordinate wraps around the equator and therefore is simply a scaled measure of longitude. The y-ordinate is isometric latitude, as defined in (3.52) or (3.53). Thus, the equations of this projection for the *sphere* are:

$$x = R(\lambda - \lambda_0)$$
$$y = R \ln \left[\tan \left(\frac{\pi}{4} + \frac{\phi}{2} \right) \right] \tag{3.151}$$

where 'ln' denotes a natural logarithm and R is the radius of the sphere.

The equivalent projection for the *ellipsoid* is given by:

$$x = a(\lambda - \lambda_0)$$
$$y = a \ln \left[\tan \left(\frac{\pi}{4} + \frac{\mu}{2} \right) \left(\frac{1 - e \sin \mu}{1 + e \sin \mu} \right)^{\frac{e}{2}} \right] \tag{3.152}$$

where a is the equatorial radius. This applies the definition of isometric latitude from (3.144) or (3.145). Noting the earlier discussion (Section 3.7.6), the inverse projection will be based on an iterative method for converting from isometric latitude (Ψ) to geodetic latitude (μ), as in (3.146).

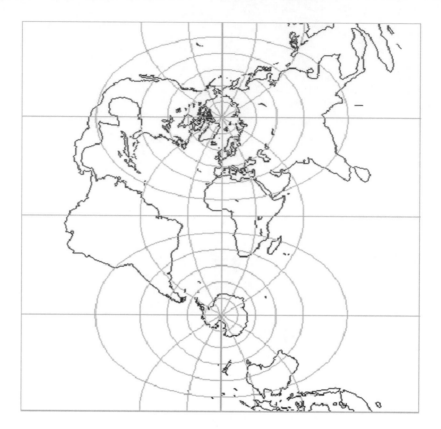

Figure 3.25 Transverse Mercator Projection.

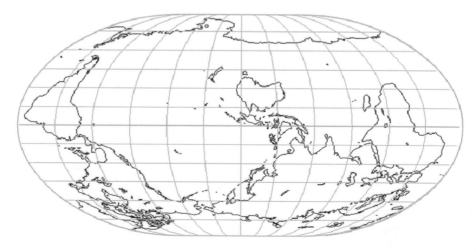

Figure 3.26 Loximuthal Projection.

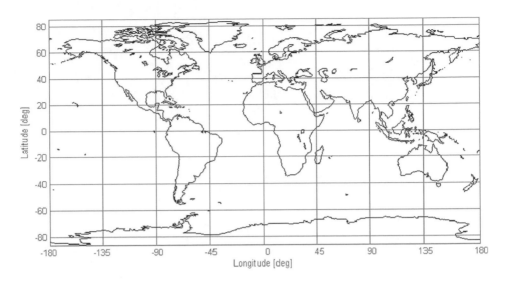

Figure 3.27 Lat-Lon Projection of the world.

A comparison between a standard lat-lon projection and a Mercator projection is given by Figures 3.27 and 3.28. Note that xy-coordinates in (3.151) and (3.152) have units of length but that large-area maps are usually calibrated with respect to latitude and longitude. The map in Figure 3.28 has a familiar appearance as it is used one of the more widely used representations of planet Earth. It is easy to draw and, importantly, easy to understand. In contrast, Figure 3.28 is distorted, with land masses compressed towards the equator and land masses stretched towards the poles. This is odd-looking but its inherent value lies in the use of isometric latitude, which allows for the determination of rhumb lines directly from the map (as discussed in Section 3.2.7).

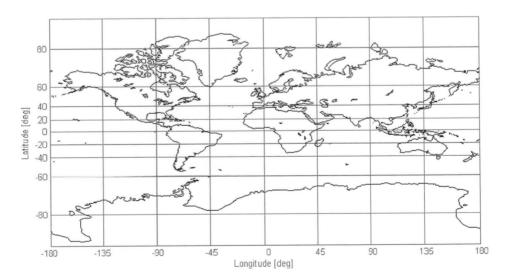

Figure 3.28 Mercator Projection of the world.

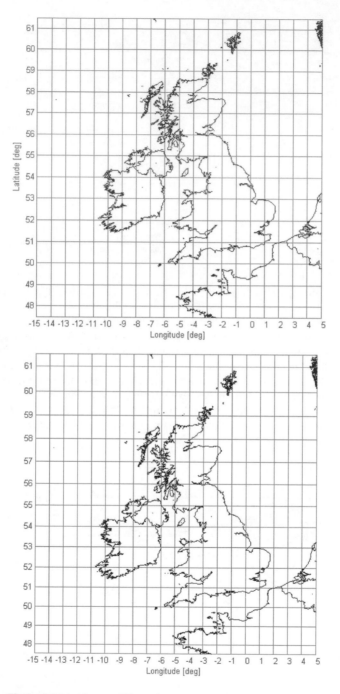

Figure 3.29 Lat-Lon and Mercator projections of the British Isles.

When applied to a small region, this distortion is hardly discernable except by close comparison. Figure 3.29 shows a segment of the World Vector Shoreline extracted[8] for the British Isles. This is overlaid on to a graticule that covers the quadrangle 47.5°N to 61.5°N and 15°W to 2°E. The separation of parallels definitely increases towards North but the impact of the overall map is slight. The problem is that the north-south scale is varying, which is inconvenient for a long, narrow country like the UK (albeit of small surface area). An alternative approach is discussed in the next section, founded on the Transverse Mercator projection.

3.11   Transverse Mercator Projection

3.11.1   *Forward Projection*

The Transverse Mercator Projection is a cylindrical projection with the standard tangent at a selected meridian. This will be the *central meridian* of the map. Not surprisingly, this is rather more complicated than the standard Mercator projection. The defining equations of this projection (as well as many others) are given by Snyder (1987). The projection for the *sphere* is, as follows:

$$x = Rk_0 \left[\frac{1}{2} \ln \left(\frac{1 + \cos\phi \sin\,\Delta\lambda}{1 - \cos\phi \sin\,\Delta\lambda} \right) \right]$$

$$y = Rk_0 \left[\tan^{-1} \left(\frac{\tan\phi}{\cos\Delta\lambda} \right) - \phi_0 \right]$$

(3.153)

where R is the radius of the sphere, k_0 is the scale along the central meridian, λ_0 is the longitude of the central meridian and $\Delta\lambda$ is the longitude difference:

$$\Delta\lambda = \lambda - \lambda_0$$ (3.154)

The map origin corresponds with (geocentric) latitude ϕ_0 and longitude λ_0.

The equivalent projection for the *ellipsoid* is, as follows:

$$x = R_N k_0 \left(X_1 \Delta\lambda + X_3 \Delta\lambda^3 + X_5 \Delta\lambda^5 \right)$$

$$y = R_N k_0 \left(Y_2 \Delta\lambda^2 + Y_4 \Delta\lambda^4 + Y_6 \Delta\lambda^6 \right) + k_0 \left(M - M_0 \right)$$

(3.155)

where

$$X_1 = \cos\mu$$

$$X_3 = \frac{1}{6} \left(1 - \tan^2\mu + \varepsilon^2 \cos^2\mu \right) \cos^3\mu$$

$$X_5 = \frac{1}{120} \left(5 - 18\tan^2\mu + \tan^4\mu + 72\varepsilon^2 \cos^2\mu - 58\varepsilon^2 \right) \cos^5\mu$$

$$Y_2 = \frac{1}{2} \tan\mu \cos^2\mu$$

[8] Great Britain constitutes the UK minus Northern Ireland and the Channel Islands.

$$Y_4 = \frac{1}{24}\left(5 - \tan^2\mu + 9\varepsilon^2\cos^2\mu + 4\varepsilon^4\cos^4\mu\right)\tan\mu\cos^4\mu$$

$$Y_6 = \frac{1}{720}\left(61 - 58\tan^2\mu + \tan^4\mu + 600\varepsilon^2\cos^2\mu - 330\varepsilon^2\right)\tan\mu\cos^6\mu$$

M is the meridian distance (defined by (3.139)) and M_0 is the corresponding distance from the equator to the reference latitude μ_0. The transverse radius of curvature is R_N (defined in (3.102)), which is equal to the length of the prime vertical (N) (defined by (3.77)). Also, ε is the second eccentricity (defined by (3.66)). In this case, the origin of map coordinates corresponds with (geodetic) latitude μ_0 and longitude λ_0. At the poles, i.e. $\mu = \pm\,\pi/2$, note that the projection simply becomes:

$$\begin{aligned} x &= 0 \\ y &= k_0(M - M_0) \end{aligned} \tag{3.156}$$

In order to shown the basic form of implementation, the projection formulae are placed in Algorithm 3.3 (named 'applyTranverseMercatorProjection'). Typically, cartographic codes are quite abstract because they are composed mainly of series expansions. However, it is important to appreciate the need for efficient evaluation of polynomials (shown as a nested calculation), for minimum use of nonlinear functions (e.g. sine and cosine evaluation) and readable layout (e.g. breaking a long formula into several steps via intermediate variables). It should be stressed that the optimal form for computation is usually not the optimal form for algebraic manipulation.

An example application is given in the following code fragment, which calculates the easting and northing of Manchester Airport (53.3536 °N − 2.2747 °W) relative to a map origin at 49°N 2°W, based on WGS84 and a scale factor $k_0 = 0.9996$. WGS84 ellipsoid parameters are programmed in Algorithm 3.4 (named 'set_WGS84'). The result obtained is −18_284 m East 484_188 m North, to the nearest metre.

```
>> (x,y) = applyTransverseMercatorProjection( ...
set_WGS84,0.9996,53.3536,-2.2747,49,-2)
x =
-1.828423139131702e+004
y =
4.841882092506726e+005
```

The scale factor along the central meridian is defined by k_0. The scale factor at the working point on the map projection is given by:

$$k = k_0\left(1 + K_2\Delta\lambda^2 + K_4\Delta\lambda^4 + K_6\Delta\lambda^6\right) \tag{3.157}$$

where

$$K_2 = \frac{1}{2}\left(1 + \varepsilon^2\cos^2\mu\right)\cos^2\mu$$

$$K_4 = \frac{1}{24}\left(5 - 4\tan^2\mu + 42\varepsilon^2\cos^2\mu + 13\varepsilon^4\cos^4\mu - 28\varepsilon^2\right)\cos^4\mu$$

$$K_6 = \frac{1}{720}\left(61 - 148\tan^2\mu + 16\tan^4\mu\right)\cos^6\mu$$

```
function (x,y) = applyTransverseMercatorProjection( ...
 ellipsoid,k0,latObject,lonObject,latOrigin,lonOrigin)

a = ellipsoid.a ;
e12 = ellipsoid.e1^2 ;
e22 = ellipsoid.e2^2 ;

S = sin(latObject) ;
C = cos(latObject) ;
S2 = S.^2 ;
C2 = C.^2 ;
T = S./C ;
T2 = T.^2 ;
T4 = T2.^2 ;
EC2 = e22.*C2 ;
EC4 = EC2.^2 ;

N = a./sqrt(1 - e12*S2) ;
A = (lonObject - lonOrigin).*cos(latObject) ;
A2 = A.^2 ;

m0 = 1 - (1/4 - (3/64 - (5/256)*e12)*e12)*e12 ;
m2 = (3/8 + (3/32 + (45/1024)*e12)*e12)*e12 ;
m4 = (15/256 + (45/1024)*e12)*e12*e12 ;
m6 = (35/3072)*e12*e12*e12 ;
M = a*( ...
 m0*latObject-m2*sin(2*latObject)+m4*sin(4*latObject)-
 m6*sin(6*latObject)) ;
M0 = a*( ...
 m0*latOrigin-m2*sin(2*latOrigin)+m4*sin(4*latOrigin)-
 m6*sin(6*latOrigin)) ;x1 = 1 ;
x1=1;
x3 = (1 - T2 + EC2)/6 ;
x5 = (5 - 18*T2 + T4 + 72*EC2 - 58*e22)/120 ;
y0 = M - M0 ;
y2 = 1/2 ;
y4 = (5 - T2 + 9*EC2 + 4*C2)/24 ;
y6 = (61 - 58*T2 + T4 + 600*EC2 - 330*e22)/720 ;

x = k0*N.*(x1 + (x3 + x5.*A2).*A2).*A ;
y = k0*(y0 + N.*T.*((y2 + (y4 + y6.*A2).*A2).*A2)) ;
```

Algorithm 3.3 'applyTransverseMercatorProjection'.

```
function wgs84 - setWGS84

wgs84.a = 6378137 ;
wgs84.f = 1/298.257223563 ;
wgs84.c = wgs84.a*(wgs84.f - 1) ;
wgs84.e1 = sqrt(1 - (wgs84.c/wgs84.a)^2) ;
wgs84.e2 = wgs84.e1^2/(1 - wgs84.e1^2) ;
```

Algorithm 3.4 'set_WGS84'.

3.11.2   *National Grid of Great Britain*

The Transverse Mercator projection is used as the basis for regional grid references, e.g. state plane coordinate systems in the United States. In the UK, the Ordnance Survey defined a grid system for Great Britain[9] called OSGB36, which is based on the re-triangulation performed between 1936 and 1953. As listed in Table 3.4, this uses the Airy 1830 ellipsoid (code AA), for which parameters are defined in Table 3.5. The projection has its *true* origin at 49°N 2°W. However, the coordinate is referenced to a *false* origin that is 100 km north and 400 km west of the true origin. The scale factor along the central meridian is $k_0 = 0.9996012717$. The variation of scale factor is plotted against latitude and longitude in Figure 3.30, with each vertical increment equating to a 1% variation.

Note 3.7

Northern Ireland is covered by the Irish Grid System. This uses the Modified Airy ellipsoid (code AM) and an origin at 53.5°N 8°W, with a false origin offset by 250 km south and 200 km west. In this case, the scale factor along the central meridian is $k0 = 1.000035$.

The map of the British Isles is drawn in Figure 3.31 using a Transverse Mercator projection with OSGB36 parameters. This is planar grid with orthogonal measurements taken from the true origin. Conventionally, these are called eastings and northings and are expressed as distances. To illustrate the effect of surface curvature, an overlay has been included for the quadrangle from 49 °N to 61 °N and 11 °W to 2 °E.

Grid coordinates are defined in the context of 500 km and 100 km squares, as shown in Figure 3.32. Grid zones are designated by a two-letter code; locations within the zone are then specified by a coordinate pair composed of an easting and a northing, which can take values between 0 and 100,000. Thus, a six-digit sequence would resolve a location to the nearest 1000 metres; eight digits would resolve to the nearest 100 metres and so on.

In order to highlight the significance of the reference ellipsoid, the Airy1830 parameters are programmed in Algorithm 3.5 (named 'set_Airy1830') and then the coordinates of Manchester Airport will be recalculated using the method that was applied to WGS84 in Section 3.11.1.

```
function Airy1830 = set_Airy1830

Airy1830.a = 6377563.396 ;
Airy1830.f = 1/299.32496946 ;
Airy1830.c = Airy1830.a*(Airy1830.f - 1) ;
Airy1830.e1 = sqrt(1 - (Airy1830.c/Airy1830.a)^2) ;
Airy1830.e2 = Airy1830.e1^2/(1 - Airy1830.e1^2) ;
```

Algorithm 3.5 'set_Airy1830'.

---

[9] Underscore characters have been inserted in order to make it easier to read large numbers.

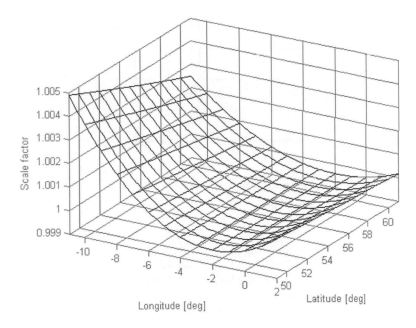

Figure 3.30 Variation of scale factor for OSGB36.

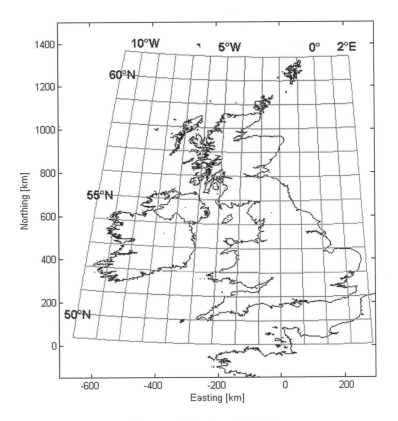

Figure 3.31 Transverse Mercator Projection of British Isles.

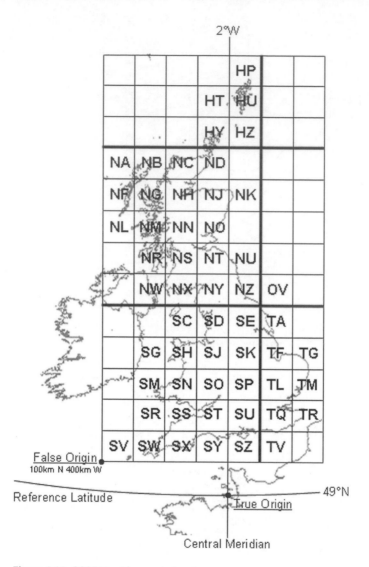

Figure 3.32 OSGB36 grid zone designators.

This is performed in the following code fragment, which calculates eastings and northings relative to a map origin at 49 °N 2 °W, based on Airy1830 and a scale factor $k_0 = 0.9996$.

```
>> (x,y) = applyTransverseMercatorProjection(...
set_Airy1830,0.9996,53.3536,-2.2747,49,-2)
x = -1.828244614289820e+004
y = 4.841457200135240e+005
```

The new coordinates are -18_282 m East 484_146 km North, which shows an offset of 2 m West 42 m South relative to WGS84. Although relatively small, this would be significant for, say, attempting to land on a runway. Note that this calculation is performed for a location that is close to the central meridian. In passing, the effect of scale factor can be established by resetting the value of k_0 from 0.9996 (used for the UTM grid in Section 3.8.6) to 0.9996012717, as appropriate to OSGB36. This shows a further shift of 616mm North and 23 mm West.

```
>> (x,y) = applyTransverseMercatorProjection(...
set_Airy1830,0.9996012717,53.3536,-2.2747,49,-2)
x =
-1.828246940198860e+004
y =
4.841463359480099e+005
```

3.11.3   *Universal Transverse Mercator (UTM) Grid*

The global application of the Transverse Mercator projection gives the so-called Universal Transverse Mercator (UTM) grid system, as shown in Figure 3.33. This is effective between latitudes 80 °S and 84 °N and is based on a partition of the surface into longitudinal segments that are six degrees wide. The partition boundaries occur at the prime meridian (running through Greenwich) and at multiples of 6 degrees to east and west. Each strip is partitioned into sections that span 8 degrees of latitude, from 80 °S to 72 °N, with the northern-most section spanning 12 degrees.

The nominal grid is then defined by 20 bands of latitude, labelled 'C' to 'X' (excluding 'I' and 'O') from south to north, and 60 segments of longitude, labelled '01' to '60' eastwards from 180°. This would give 1200 grid zones. However, there are five exceptions that cover the seas around Norway, where zone widths are multiplied by 0.5, 1.5 or 2.0. Zones are labelled by longitude followed by latitude. For instance, zone 30T occupies the quadrangle between longitudes 6 °W and 0° and latitudes 40 °N and 48 °N; it covers the west coast of France, part of the north coast of Spain and the Bay of Biscay.

Thus, the UTM grid has 1197 zones, recognising that 32X, 34X and 36X are not used (because 31X, 33X, 35X and 37X are wider than the standard zones). Also note that 31V is narrow and 32V is wide.

In this case, the Transverse Mercator projection is applied along the central meridian of each segment, with a true origin on the equator and a scale factor of $k_0 = 0.9996$. Based on WGS84 parameters, a six-degree segment is 667_917 metres wide at the equator. Also the distance along any meridian from equator to either pole is 10_001_949 metres. Using the true origin, locations to the west of the central meridian would have negative x-ordinates and locations south of the equator would have negative y-ordinates. In order to assign positive coordinates to all locations, all longitudinal segments are given false easting of 500_000 metres and the southern hemisphere is given a false northing of 10_000_000 metres.[10]

---

[10] www.partow.net/miscellaneous/airportdatabase/index.html.

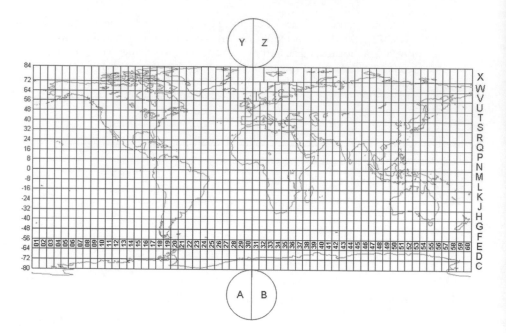

Figure 3.33 UTM grid for the world.

Remember that the UTM zones do not extend to the poles. In Figure 3.33, there are additional semi-circular zones shown around the poles, labelled as 'A', 'B', 'Y' and 'Z'. These are defined using the Universal Polar Stereographic (UPS) projection, which will be discussed in Section 3.13.

3.11.4   *Projection Geometry*

The map geometry associated with the Transverse Mercator projection is illustrated schematically in Figure 3.34, using the general layout of OSGB36 (as discussed in Section 3.11.2). This shows a given position or *working point* in relation to the true origin, false origin, latitude, longitude and grid north. The meridian of longitude points towards the North Pole. The convergence angle is the angle between grid north and the meridian projection. Northing and Easting are measured from the False Origin; the location of the true origin is measured by False Northing and False Easting (i.e. $-100\,\mathrm{km}$ North $400\,\mathrm{km}$ East in OSGB36). The central meridian is clearly marked and a perpendicular line is drawn that passes through the point. The point of intersection between central meridian and the perpendicular is called the *footprint latitude*.

Map coordinates are given as series expansions in (3.155):

$$
\begin{aligned}
x &= R_N k_0 \left(X_1 \Delta\lambda + X_3 \Delta\lambda^3 + X_5 \Delta\lambda^5 \right) \\
y &= R_N k_0 \left(Y_2 \Delta\lambda^2 + Y_4 \Delta\lambda^4 + Y_6 \Delta\lambda^6 \right) + k_0 \left(M - M_0 \right)
\end{aligned}
\tag{3.158}
$$

where $\Delta\lambda = \lambda - \lambda_0$ and the coefficients X_n and Y_n are specified in Section 3.8.4. With reference to Figure 3.35, easting (E) and northing (N) are obtained from x and y, respectively, by adding the false easting (FE) and false northing (FN), respectively.

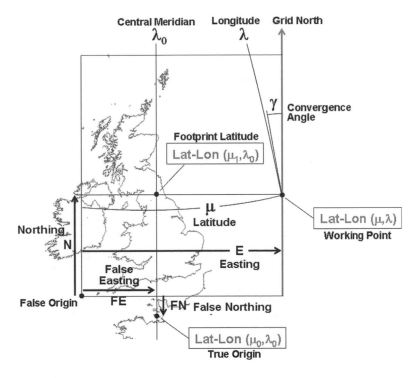

Figure 3.34 Map geometry of the Transverse Mercator Projection.

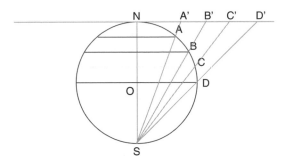

Figure 3.35 Geometry of the polar stereographic projection.

The scale factor at the working point on the map projection is given in (3.157):

$$k = k_0 \left(1 + K_2 \Delta\lambda^2 + K_4 \Delta\lambda^4 + K_6 \Delta\lambda^6\right)$$

where k_0 is the scale factor along the central meridian and the coefficients K_n are specified in Section 3.8.4.

With reference to the Technical Manual on Universal Grids, TM8358.2 (DMA, 1989), the convergence angle can be specified as follows:

$$\gamma = G_1 \Delta\lambda + G_3 \Delta\lambda^3 + G_5 \Delta\lambda^5 \tag{3.159}$$

where

$G_1 = \sin\mu$

$G_3 = \dfrac{1}{3}\left(1 + \varepsilon^2 \cos^2\mu + 2\varepsilon^4 \cos^4\mu\right)\tan\mu\cos\mu$

$G_5 = \dfrac{1}{15}\left(2 - \tan^2\mu + 15\varepsilon^2 \cos^2\mu\left(1 - \tan^2\mu\right) + 5\varepsilon^4 \cos^4\mu\left(7 - 10\tan^2\mu\right)\right)\tan\mu\cos^3\mu$

The footprint latitude is the latitude on the central meridian that has the same northing as that of the working point. Thus, it corresponds with a meridian distance (i.e. the distance along the central meridian from the equator) that is given by:

$$M = M_0 + \frac{y}{k_0} \tag{3.160}$$

By definition, xy is the y-ordinate measured with respect to the true origin and is the difference between the northing and the false northing. In the earlier discussion, M_0 was identified as the meridian distance from the equator to the true origin at the reference latitude μ_0. From here, the rectifying latitude is calculated (cf. (3.140)):

$$\omega_1 = \frac{\pi}{2}\left(\frac{M}{M_P}\right) \tag{3.161}$$

where M_P is the distance from the equator to the pole. Thence the footprint latitude can be calculated via (3.143):

$$\mu_1 = \omega_1 + U_2 \sin 2\omega_1 + U_4 \sin 4\omega_1 + U_6 \sin 6\omega_1 + U_8 \sin 8\omega_1 + \dots \tag{3.162}$$

where

$$U_2 \approx \frac{3}{2}E_1 - \frac{27}{32}E_1^3$$

$$U_4 \approx \frac{21}{16}E_1^2 - \frac{55}{32}E_1^4$$

$$U_6 \approx \frac{151}{96}E_1^3$$

$$U_8 \approx \frac{1097}{512}E_1^4$$

The parameter E_1 is defined by (3.142):

$$E_1 = \frac{1 - \sqrt{1 - e^2}}{1 + \sqrt{1 - e^2}}$$

The meridian radius of curvature (R_M) (cf. (3.102)) and the transverse radius of curvature (R_N) (cf. (3.103) or (3.77)), are evaluated at the footprint latitude:

$$R_{M1} = \frac{a(1 - e^2)}{\left(1 - e^2 \sin^2 \mu_1\right)^{3/2}} \tag{3.163}$$

$$R_{N1} = \frac{a}{\sqrt{1 - e^2 \sin^2 \mu_1}} \tag{3.164}$$

Also, note the following relationship:

$$R_{N1} = R_{M1}\left(1 + \varepsilon^2 \cos^2 \mu_1\right) \tag{3.165}$$

3.11.5   *Inverse Projection*

The *forward* projection was presented in Section 3.11.1. Using the mathematics developed in the previous section, it is possible to specify the *inverse* projection to generate lat-lon from xy-coordinates. The appropriate equations are as follows:

$$\lambda = \lambda_0 + N_1 D + N_3 D^3 + N_5 D^5$$
$$\mu = \mu_1 + \left(T_2 D^2 + T_4 D^4 + T_6 D^6\right) \frac{R_{N1}}{R_{M1}} \tag{3.166}$$

where

$$N_1 = \sec \mu_1$$

$$N_3 = \frac{\sec \mu_1}{6}\left(1 + 2\tan^2 \mu_1 + \varepsilon^2 \cos^2 \mu_1\right)$$

$$N_5 = \frac{\sec \mu_1}{120}\left(\begin{array}{l} 5 + 28\tan^2 \mu_1 + 24\tan^4 \mu_1 + 2\varepsilon^2 \cos^2 \mu_1(3 + 4\tan^2 \mu_1) \\ \qquad - \varepsilon^4 \cos^4 \mu_1(3 - 4\tan^2 \mu_1) \end{array}\right)$$

$$T_2 = \frac{\tan \mu_1}{2}$$

$$T_4 = \frac{\tan \mu_1}{24}\left(5 + 3\tan^2 \mu_1 + \varepsilon^2 \cos^2 \mu_1\left(1 - 9\tan^2 \mu_1\right) - 4\varepsilon^4 \cos^4 \mu_1\right)$$

$$T_6 = \frac{\tan \mu_1}{720}\left(\begin{array}{l} 61 + 90\tan^2 \mu_1 + 45\tan^4 \mu_1 + \varepsilon^2 \cos^2 \mu_1\left(46 - 252\tan^2 \mu_1 - 90\tan^4 \mu_1\right) \\ \qquad - \varepsilon^4 \cos^4 \mu_1\left(3 + 66\tan^2 \mu_1 - 225\tan^4 \mu_1\right) \end{array}\right)$$

In (3.166), λ_0 is the longitude of the central meridian, μ_1 is the footprint latitude and

$$D = \frac{x}{R_{N1} k_0} \tag{3.167}$$

3.12   Conformal Latitude

At this juncture, it is appropriate to introduce yet another definition of latitude. *Conformal latitude* will be given the symbol χ and is related to geodetic latitude μ according to the following formula:

$$\tan\left(\frac{\pi}{4} + \frac{\chi}{2}\right) = \tan\left(\frac{\pi}{4} + \frac{\mu}{2}\right) \left(\frac{1 - e\sin\mu}{1 + e\sin\mu}\right)^{\frac{e}{2}} \tag{3.168}$$

With reference to (3.144), the relationship between conformal latitude (ζ) and isometric latitude (Ψ) is given by:

$$\psi = \ln\left[\tan\left(\frac{\pi}{4} + \frac{\chi}{2}\right)\right] \tag{3.169}$$

$$\chi = 2\tan^{-1}\left[e^{\psi}\right] - \frac{\pi}{2} \tag{3.170}$$

where 'ln' denotes a natural logarithm.

The following expressions permit numerical conversions between conformal latitude and geodetic latitude:

$$\begin{aligned}
\chi \approx \mu &- \frac{e^2}{2}\left(1 + \frac{5}{12}e^2 + \frac{3}{16}e^4 + \frac{281}{2880}e^6\right)\sin 2\mu \\
&+ \frac{e^4}{16}\left(\frac{1}{3} + \frac{7}{5}e^2 + \frac{697}{720}e^4\right)\sin 4\mu \\
&- \frac{e^6}{480}\left(13 + \frac{461}{28}e^2\right)\sin 6\mu + \frac{1237}{161280}e^8\sin 8\mu
\end{aligned} \tag{3.171}$$

$$\begin{aligned}
\mu \approx \chi &+ \frac{e^2}{2}\left(1 + \frac{5}{12}e^2 + \frac{1}{6}e^4 + \frac{13}{180}e^6\right)\sin 2\chi \\
&+ \frac{e^4}{48}\left(7 + \frac{29}{5}e^2 + \frac{811}{240}e^4\right)\sin 4\chi \\
&+ \frac{e^6}{40}\left(\frac{7}{3} + \frac{81}{28}e^2\right)\sin 6\chi + \frac{4279}{161280}e^8\sin 8\chi
\end{aligned} \tag{3.172}$$

The conformal co-latitude (χ') is defined as 90° minus the conformal latitude:

$$\chi' = \frac{\pi}{2} - \chi \tag{3.173}$$

Its tangent is determined (via half-angles) as follows:

$$\tan\frac{\chi'}{2} = \tan\left(\frac{\pi}{4} - \frac{\chi}{2}\right) = \tan\left(\frac{\pi}{2} - \left(\frac{\pi}{4} + \frac{\mu}{2}\right)\right) = \cot\left(\frac{\pi}{4} + \frac{\chi}{2}\right) = = \cot\left(\frac{\pi}{4} + \frac{\mu}{2}\right)\left(\frac{1 + e\sin\mu}{1 - e\sin\mu}\right)^{\frac{e}{2}}$$

$$\tan\frac{\chi'}{2} = \tan\left(\frac{\pi}{4} - \frac{\chi}{2}\right) = \tan\left(\frac{\pi}{4} - \frac{\mu}{2}\right)\left(\frac{1 + e\sin\mu}{1 - e\sin\mu}\right)^{\frac{e}{2}}$$

$$\tag{3.174}$$

3.13   Polar Stereographic Projection

3.13.1   *Basic Formulation*

The geometry of the polar stereographic projection is shown in Figure 3.35, for the purpose of mapping the north polar region. Here the plane of projection is tangential to the North Pole (point N) and the focal point is at the South Pole (point S). Any point of the surface is mapped by drawing a ray from point S through the point and extending it to the projection plane. Thus, points on the surface of the Earth that are labelled as A, B, C and D will appear as points on the map that are labelled A', B', C' and D', respectively. The method is simple but, of course, it is limited to the extent that it is not possible to project the entire Earth on to a finite map. An equivalent diagram could be drawn for the south polar region, with a plane of projection that is tangential to S and a focal point at N.

The equations of the polar stereographic projection for the *sphere* are:

$$x = +\rho \sin(\lambda - \lambda_0)$$
$$y = -\rho \cos(\lambda - \lambda_0) \; s_0 \tag{3.175}$$

where

$$\rho = 2 \; R \; k_0 \tan\left(\frac{\pi}{4} - s_0 \frac{\phi}{2}\right)$$

in which the central longitude is λ_0, the nominal scale factor is k_0 and the sign change is given by s_0 (such that $s_0 = 1$ for the north polar projection and $s_0 = -1$ for the south polar projection). The variation of scale (k) with geocentric latitude (ϕ) is:

$$k = \frac{2k_0}{1 + s_0 \sin \phi} \tag{3.176}$$

The inverse equations for the *sphere* are, as follows:

$$\phi = \sin^{-1}[s_0 \; \cos c]$$
$$\lambda = \lambda_0 + \tan^{-1}\left[\frac{x}{(-s_0 y)}\right] \tag{3.177}$$

where s_0 is the sign change and the angular distance from the central point is:

$$c = 2 \tan^{-1}\left[\frac{\rho}{2Rk_0}\right] \tag{3.178}$$

which is based on the radial distance from the centre of the map:

$$\rho = \sqrt{x^2 + y^2} \tag{3.179}$$

The projection of parallels of latitude will be circle of radius:

$$\rho = 2Rk_0 \; \frac{\cos \phi}{s_0 + \sin \phi} \tag{3.180}$$

The equations of the polar stereographic projection for the *ellipsoid* are:

$$
\begin{aligned}
x &= +\rho \sin(\lambda - \lambda_0) \\
y &= -\rho \cos(\lambda - \lambda_0) \, s_0
\end{aligned}
\tag{3.181}
$$

where

$$
\rho = \frac{2 a k_0}{\sqrt{(1+e)^{(1+e)}(1-e)^{(1-e)}}} \tan\left(\frac{\pi}{4} - s_0 \frac{\chi}{2}\right)
\tag{3.182}
$$

for a nominal scale factor k_0 and a sign change s_0 [$s_0 = 1$ for north and $s_0 = -1$ for south], where χ is conformal latitude. Recall (3.175):

$$
\tan\left(\frac{\pi}{4} - \frac{\chi}{2}\right) = \tan\left(\frac{\pi}{4} - \frac{\mu}{2}\right)\left(\frac{1 + e \sin \mu}{1 - e \sin \mu}\right)^{\frac{e}{2}}
$$

The variation of scale factor (k) with geodetic latitude is given by:

$$
k = \frac{\rho}{am}
\tag{3.183}
$$

where the radial distance is defined by (3.183) and where

$$
m = \frac{\cos \mu}{\sqrt{1 - e^2 \sin^2 \mu}}
\tag{3.184}
$$

The inverse equations for the *ellipsoid* are, as follows:

$$
\begin{aligned}
\chi &= \frac{\pi}{2} - \tan^{-1}\left[\frac{\rho}{2 a k_0} \sqrt{(1+e)^{(1+e)}(1-e)^{(1-e)}}\right] \\
\lambda &= \lambda_0 + \tan^{-1}\left[\frac{x}{(-s_0 y)}\right]
\end{aligned}
\tag{3.185}
$$

which is based on the radial distance from the centre of the map:

$$
\rho = \sqrt{x^2 + y^2}
\tag{3.186}
$$

Geodetic latitude can then be determined by applying (3.172).

3.13.2   *Universal Polar Stereographic (UPS) Projection*

Regions A, B, Y and Z of the UTM grid (cf. Figure 3.33) are determined by using the UPS projection. This conforms to the method describes in the preceding development, with a scale factor of 0.994 defined at each pole. The North Zone covers latitudes of 84 °N and above. Its origin as at the North Pole and it incorporates a false origin such that the true origin lies at coordinates 2000 kmN and 2000 kmE. The South Zone covers latitudes of 80 °S and below. Its origin as at the South Pole and it incorporates a false origin such that, in this case, the true origin also lies at coordinates 2000 kmN and 2000 kmE. In both cases, Grid North is defined such that it is directed towards the North Pole along the prime meridian. For illustration, the UPS grid system is depicted in Figure 3.36.

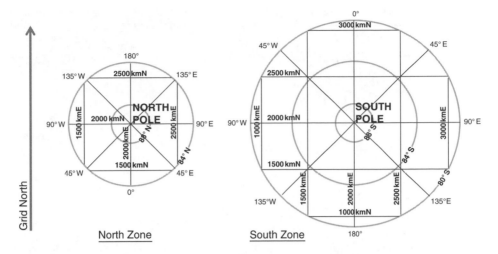

Figure 3.36 UPS grid system.

3.14   Three-Dimensional Mapping

Having considered the basics of two-dimensional mapping, it is useful to consider some facets of three-dimensional maps. These provide a topographical reference for air operations, such that a visual scene could be generated and navigation could be performed relative to ground features. The intention here is to give a very brief overview of the type of terrain representations that might be produced in pictorial and/or plan form and to raise a number of pertinent issues for implementation. It is certainly not intended to delve into the detail of data formats or real-time visualisation software.

The examples presented in this section have been created using the MATLAB Mapping Toolbox. Terrain data are readily accessible from the US National Imagery and Mapping Agency and the US Geological Survey. The source chosen is the Global Topography database called GTOPO30, giving 30 arc-second resolution. This can be applied to generate images like Figure 3.37, which would be consistent with a view of the Earth's surface from space.

The British Isles are covered by the W020N90 dataset from GTOPO30. Figure 3.38 shows an top-down view of the United Kingdom, Eire and part of Northern France. A colour scheme shows height variation from coastal areas up to mountain tops. Figure 3.39 zooms in on Scotland, shown in slant view. This is still a perspective from very high altitude and the Highlands are only visible because the height scale has been exaggerated by a factor of 10. Figure 3.40 zooms in further to show the Grampian Mountains. At this range, the nature of terrain rendering is becoming apparent; the relatively low resolution of the dataset (i.e. of the order of 1000m between data points) means that the continuous landscape is created by interpolation. The result is a smooth approximation to what should appear as real mountains, with jagged edges. Nonetheless this is a useful representation for the purpose of introduction and, indeed, it would provide the right sort of features for a flight environment.

As an alternative representation, Figure 3.41 presents a contour map of this same region, together with profile along the parallel at 57.08°N. From this particular dataset (at this

Figure 3.37 Long-range digital terrain map of Scandinavia.

particular resolution), this profile has been chosen because it passes through the highest elevation at about 1170 m. This occurs in the locality of the Cairngorm Mountains. Unfortunately, there are six or seven peaks in that close vicinity that exceed 1200 m, with Ben Macdui at 1306 m. Thus, a terrain map like this could be used for purposes of flight simulation but it would be indicative rather than necessarily accurate.

One important application of digital terrain models is to generate outside-world scenes, i.e. what can be observed from a given point defined by lat-lon and altitude. In flight simulation, this would be required for a given field of view subtended through an aperture in the aircraft (typically the visual scene from the pilot eye position). For illustrative purposes, the result of a full 360-degree line-of-sight analysis is presented in Figure 3.42, for lat-lon of 57.08 °N 4 °W and altitude of 1000 m. The position over ground is indicated by lines of longitude and latitude and the field of view is indicated by dark shading over a light terrain.

Finally, an all-round analysis of radar coverage is presented in Figure 3.43, centred on lat-lon of 57.7 °N 6 °W (which is roughly midway along a path between the northern tip of Skye and Loch Gairloch). Two images are given, corresponding to altitudes of 500 m and 2000 m. The lower altitude constrains the radar view between the mainland and the Western Isles whereas the higher altitude opens up the view into the North Atlantic and even into a small part of the North Sea (on the other side of the mainland). The essential difference between visual observation and radar observation is the effective radius of the Earth in each case, which can be stated as R and 4R/3 respectively. In other words, refraction near the surface is greater for radar than for light.

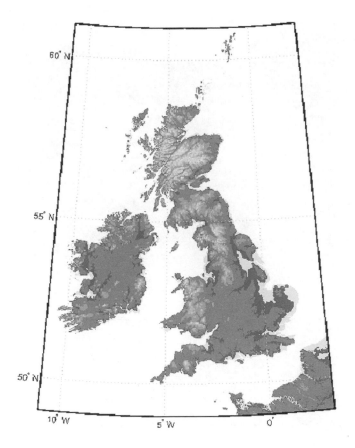

Figure 3.38 Map of the British Isles.

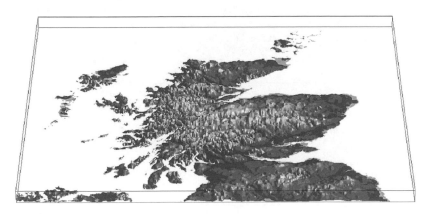

Figure 3.39 3-D Scotland.

Figure 3.40 3-D Grampian Mountains.

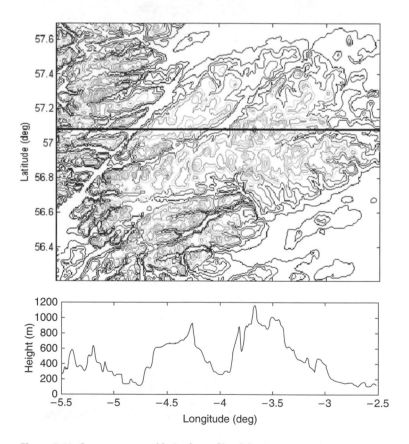

Figure 3.41 Contour map and latitude profile of the Grampian Mountains.

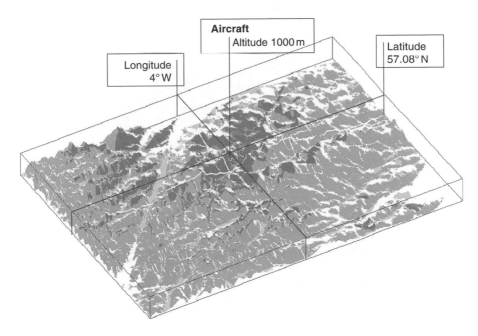

Figure 3.42 Example of visible terrain.

3.15   Actual Latitudes, Longitudes and Altitudes

In order to add an aviation context to the discussion on geospatial geometry, it is useful to introduce some real numbers that define the latitude, longitude and altitude of real places (e.g. airports and navigational aids). Various databases exist, including those that can be freely downloaded from Internet sites, such as The Global Airport Database[11] by Arash Partow. For this purpose, it is common to reference airports using ICAO[12] 4-letter codes, with the first two letters indicating the region and the last two designating the actual airport. In addition, there are IATA[13] 3-letter codes that are more familiar to air travellers as these are what appear on tickets and baggage tags.

For information and illustration, Table 3.7 gives the ICAO code designations for European countries and Table 3.8 gives the locations of various European airports. The Global Airport Database uses a simple string format for its data records, with a new-line character between records. Each record is punctuated with colons in order to delimit 14 fields of information. Invariably raw data needs to be parsed, i.e. separated into its syntactic components, and then reorganised for a specific application.

Algorithm 3.6 gives one method of parsing character data of this type. The principle is to tag the various methods of separation in turn, to derive the indices that mark the start

---

[11] International Civil Aviation Organisation.
[12] International Air Transport Association.
[13] Readers are urged to pay close attention to documentation on the MATLAB function strmatch.

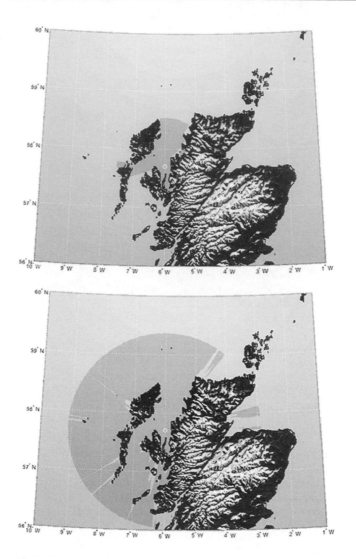

Figure 3.43 Example of radar coverage (at 500 m and 2000 m altitude).

and finish of each field (i.e. the first and last characters) and then to extract that field and perform data conversions if appropriate (e.g. converting character string data into numerical data). In this case, the raw data is held in a text file GlobalAirportDatabase.txt and the algorithm is designed to find airports associated with all or part of an ICAO code designation (i.e. one, two, three or all four letters of the code).[14] As an additional filter, only

---

[14] www.partow.net/miscellaneous/airportdatabase/index.html

Table 3.7 ICAO Designations for European Countries.

Belgium	**EB**	Albania	**LA**	Monaco	**LN**
Germany	**ED, ET**	Bulgaria	**LB**	Austria	**LO**
Estonia	**EE**	Cyprus	**LC**	Portugal	**LP**
Finland	**EF**	Spain	**LE**	Bosnia	**LQ**
United Kingdom	**EG**	France	**LF**	Romania	**LR**
Netherlands	**EH**	Greece	**LG**	Switzerland	**LS**
Ireland	**EI**	Hungary	**LH**	Turkey	**LT**
Denmark	**EK**	Italy	**LI**	Moldova	**LU**
Luxembourg	**EL**	Slovenia	**LJ**	Macedonia	**LW**
Norway	**EN**	Czech Republic	**LK**	Gibraltar	**LX**
Poland	**EP**	Israel	**LL**	Yugoslavia	**LY**
Sweden	**ES**	Malta	**LM**	Slovakia	**LZ**
Latvia	**EV**				
Lithuania	**EY**				

Table 3.8 Locations of Selected European Airports.

ICAO Code	IATA Code	Airport	LAT	LON	ALT	LAT	LON	ALT
EBBR	BRU	BRUSSELS	N50:54:08	E004:29:55	184ft	N50.9022°	E004.4986°	56m
EDDF	FRA	FRANKFURT	N50:01:35	E008:32:35	364ft	N50.0264°	E008.5431°	111m
EDDH	HAM	HAMBURG	N53:37:49	E009:59:17	53ft	N53.6303°	E009.9881°	16m
EDDM	MUC	MUNICH	N48:21:13	E011:47:09	1487ft	N48.3536°	E011.7858°	453m
EFHK	HEL	HELSINKI	N60:19:12	E024:57:22	179ft	N60.3200°	E024.9561°	55m
EGBB	BHX	BIRMINGHAM	N52:27:13	W001:44:52	325ft	N52.4536°	W001.7478°	99m
EGCC	MAN	MANCHESTER	N53:21:13	W002:16:29	257ft	N53.3536°	W002.2747°	78m
EGKK	LGW	GATWICK	N51:08:53	W000:11:25	196ft	N51.1481°	W000.1903°	60m
EGLL	LHR	HEATHROW	N51:28:39	W000:27:41	80ft	N51.4775°	W000.4614°	24m
EGPF	GLA	GLASGOW	N55:52:19	W004:25:59	26ft	N55.8719°	W004.4331°	8m
EGPH	EDI	EDINBURGH	N55:57:00	W003:22:21	135ft	N55.9500°	W003.3725°	41m
EHAM	AMS	SCHIPHOL	N52:18:31	E004:45:50	−11ft	N52.3086°	E004.7639°	−3m
EIDW	DUB	DUBLIN	N53:25:16	W006:16:12	242ft	N53.4211°	W006.2700°	74m
EKCH	CPH	COPENHAGEN	N55:37:25	E012:39:21	17ft	N55.6236°	E012.6558°	5m
ENGM	OSL	OSLO GARDERMOEN	N60:11:38	E011:06:01	681ft	N60.1939°	E011.1003°	208m
LEBL	BCN	BARCELONA	N41:17:49	E002:04:42	12ft	N41.2969°	E002.0783°	4m
LEMD	MAD	MADRID	N40:28:20	W003:33:39	2000ft	N40.4722°	W003.5608°	610m
LEVC	VLC	VALENCIA	N39:29:21	W000:28:53	225ft	N39.4892°	W000.4814°	69m
LFBO	TLS	TOULOUSE	N43:37:44	E001:21:49	499ft	N43.6289°	E001.3636°	152m
LFPG	CDG	CHARLES DE GAULLE	N49:00:46	E002:33:00	392ft	N49.0128°	E002.5500°	119m
LIRF	FCO	ROMA FIUMICINO	N41:48:46	E012:15:11	13ft	N41.8128°	E012.2531°	4m
LOWW	VIE	VIENNA	N48:06:37	E016:34:11	600ft	N48.1103°	E016.5697°	183m
LSGG	GVA	GENEVA	N46:14:17	E006:06:32	1411ft	N46.2381°	E006.1089°	430m
LSZH	ZRH	ZURICH	N47:27:53	E008:32:57	1416ft	N47.4647°	E008.5492°	432m

airports with a known name and with an IATA code are accepted in the search. For convenience, the end result is written to a spreadsheet using the MATLAB function xlswrite.

So, the database of interest here provides the following information fields:

1. **ICAO Code:** 4 character ICAO code
2. **IATA Code:** 3 character IATA code
3. **Airport Name:** string of varying length
4. **City, Town or Suburb:** string of varying length
5. **Country:** string of varying length
6. **Latitude Degrees:** 2 ASCII characters representing one numeric value
7. **Latitude Minutes:** 2 ASCII characters representing one numeric value
8. **Latitude Seconds:** 2 ASCII characters representing one numeric value
9. **Latitude Direction:** 1 ASCII character (N or S) representing compass direction
10. **Longitude Degrees:** 2 ASCII characters representing one numeric value
11. **Longitude Minutes:** 2 ASCII characters representing one numeric value
12. **Longitude Seconds:** 2 ASCII characters representing one numeric value
13. **Longitude Direction:** 1 ASCII character (E or W) representing compass direction
14. **Altitude:** varying sequence of ASCII characters representing a numeric value corresponding with the airport's altitude from mean sea level

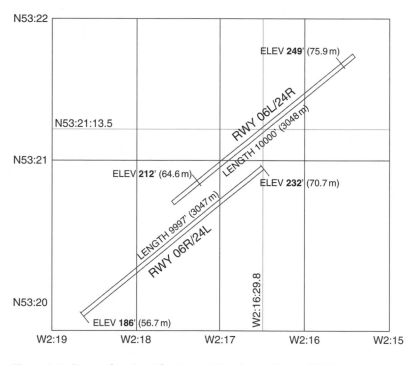

Figure 3.44 Runway locations/elevations at Manchester Airport (EGCC).

Figure 3.45 Approximate profile of EGCC Runway 06L/24R

A sample of the data content is as follows:

```
EGDL:LYE:LYNEHAM:LYNEHAM:U.K.:51:30:18:N:01:59:36:U:0513
EGDM:N/A:BOSCOMBE DOWN:BOSCOMBE DOWN:ENGLAND:51:09:07:N:01:44:50:U:0407
```

Note that these two UK airports are designated as 'UK' and 'England' respectively. This issue often arises when compiling large datasets and, to a certain extent, it is an unavoidable side-effect of having multiple designations, all of which are correct. It certainly does not detract from the usefulness of the data; it just underlines the need to apply filters in a comprehensive way (as with any database search engine) in order to ensure that a search requirement is satisfied fully.

Summary information on airport locations enables point-to-point navigation but says nothing about the actual layout of any individual airport or the management of its local airspace. As an example, Figure 3.44 gives a sketch of runway locations at Manchester Airport (EGCC), plus elevation data. Runway headings are 55°/235° relative to true North, which correspond to designations 06/24 respectively. Note that the magnetic variation in this locale is 4.1 °W, which defines the average direction of magnetic North (as will be discussed further in Chapter 4). As an added refinement, Figure 3.45 presents the approximate height profile of runway 06L/24R (Perkins and Parry (1996)).

```
function findAirports(ICAO_Code)

fid = fopen('GlobalAirportDatabase.txt') ;
GlobalAirportData = fread(fid) ;
fclose(fid) ;

GlobalAirportData = GlobalAirportData' ;
marker = 10 ;
tag = find(GlobalAirportData==marker) ;
tag = (0,tag) ;
nTags = length(tag) ;
for iTag = 1:nTags-1
 AirportData{iTag} = char(GlobalAirportData(tag(iTag)+1:tag(iTag+1)-1));
end
```

Continued

```
marker = ':';
nRecord = 0;
nAirports = length(AirportData);
for iAirport = 1:nAirports
 thisAirportData = AirportData{iAirport};
 tag = strfind(thisAirportData,marker);
 tag = (0,tag,length(thisAirportData)+1);
 nTags = length(tag);
 for iTag = 1:nTags-1
 AirportRecord{iTag} = thisAirportData(tag(iTag)+1:tag(iTag+1)-1);
 end

 isKnown = ~isempty(AirportRecord{3});
 isIATA = ~strcmp(AirportRecord{2},'N/A');
 if isKnown && isIATA
 nRecord = nRecord + 1;
 AirportDataList(nRecord,1:14) = AirportRecord(1:14);
 end
end

ICAO_List = AirportDataList(:,1);
if nargin==0 || isempty(ICAO_Code)
 Airports = AirportDataList(:,[1:3,5:14]);
else
 ICAO_Index = strmatch(ICAO_Code,ICAO_List);
 ICAO_Index = unique(ICAO_Index);
 Airports = AirportDataList(ICAO_Index,(1:3,6:14));
end

xlswrite(ICAO_Code,Airports);
```

Algorithm 3.6 'findAirports'.

Chapter 4

Positional Astronomy

4.1 Earth and Sun

The two dominant components of Earth motion are its rotation about the polar axis and its orbit about the Sun, as illustrated in Figure 4.1. Viewed from the 'north' side of the orbit (looking 'down'), both the rotation and the orbital motion are counter-clockwise.

For most general applications, the rotation rate is taken to be 7.292115×10^{-5} rad/s (as defined in WGS84 or GRS80), which is 15.04107°/hr or 360.98561°/day. This is related to the system of time measurement[1] and care is needed when interpreting what the relationship actually is. It would be convenient to choose a reference frame that is fixed across the celestial background and then use that as an absolute reference against which to measure the rotation of the earth. However this would not be so convenient to implement as the time base would slowly drift relative to the annual solar cycle that is commonly thought of as a 'year'.

As seen in Figure 4.2, the motion of Sun relative to the Earth can be drawn as an orbital path, called the *ecliptic*, that proceeds eastwards relative to the background. Including the effect of Earth rotation, the overall motion of the Sun proceeds westwards (rising in the East and setting in the West). Because the Earth's rotational axis is tilted approximately 23.44° from the orbital axis, the ecliptic is inclined by the same amount with respect to the celestial equator. This quantity is given the elaborate title of 'the obliquity of the ecliptic'.

The vernal equinox is the *time* at which the sun crosses the equator from south to north. The *position* at which this occurs is the First Point of Aries, marked by the zodiac symbol ⊥, although this is often referred to as the vernal equinox.[2] The equinoxes and solstices as marked in Figure 4.2 are appropriate to seasons in the northern hemisphere.

A *sidereal* year is defined as the period taken for the sun to complete a circuit of 360° against the celestial background, which is quoted as 365.25636 days.[3] The *calendar* year (also known

---

[1] Note the GRS67 rotation rate is $7.2921151467 \times 10^{-5}$ rad/s, which is the value employed in the definition of Universal Time (cf. Aoki *et al.*, 1982).

[2] When observations were formalised about 2000 years ago the vernal equinox did correspond with the first point of the constellation Aries but since that it has precessed almost 28° westwards.

[3] *Norton's Star Atlas*, 20th Edition, p. 21.

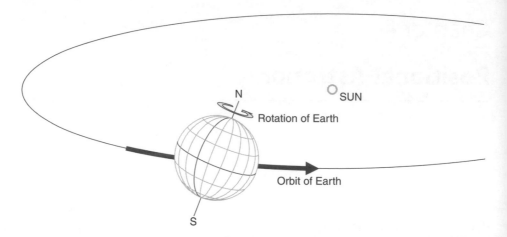

Figure 4.1 Motion of the Earth. Central intelligence Agency "The World Factbook" https://www.cia.gov/library/publications/the-world-factbook/docs/refmaps.html

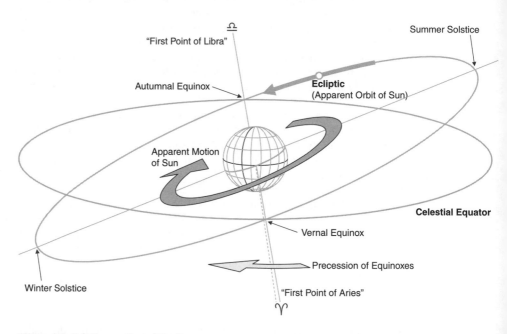

Figure 4.2 Relative motion of the Sun.

as the *solar* or *tropical* year) is the period to complete a circuit from one vernal equinox to the next. Orbital precession, mainly due to the gravitational pull of the Sun and Moon, results in a retrograde motion of 50.29″ per year, such that the actual circuit sweeps out an angle of $360°$ $- 50.29″ = 359.986031°$. Thus, the duration of 'one year' is 365.24219 days and the orbital precession has a cycle of about 25770 years.

Note 4.1

Traversing a distance of 359.986031° in 365.24219 days means that the orbital velocity of the Earth is equivalent to 0.98561°/day, which is consistent with the WGS84 Earth rotation rate, i.e. $360° + 0.98561°$.

Note 4.2

A solar day is the time between successive transits of the Sun at the local meridian and this corresponds with an Earth rotation of 360.98561°. A sidereal day is the time between transits of the vernal equinox and this corresponds with a rotation of 360°. Taking the ratio of these angles, the length of a sidereal day is found to be 23.93447207 hr = 23 h 56 m 4.099452 s = 86164.099452 s.

Note 4.3

The mean radius of the Earth's orbital path around the Sun is defined as one Astronomical Unit (AU). This has the numerical value given later in (4.60):

$$1\text{AU} = 149597\_870691 \text{ m}$$

where *underscore* characters have been inserted in order to make it easier to read large numbers. From this, the following equivalences can be found:

$$1\text{AU} = 149\_597\_871 \text{ km} = 81\_658\_227 \text{ nm} = 92\_948\_094 \text{ sm}$$

expressed in kilometres, nautical miles and statute miles, respectively. Note that

$$1 \text{ nm} = 1.852 \text{ km} = 6076 \text{ ft}$$
$$1 \text{ sm} = 5280 \text{ ft} = 1.609344 \text{ km}$$

With these dimensions and the various angular rates defined in this section, it is simple matter to calculate the average speed of the Earth relative to the Sun (V_1) and the average speed of a point on the Earth's equator relative to the polar axis (V_2). The results are as follows:

$$V_1 = 29\_785 \text{ ms}^{-1} = 107\_226 \text{ kph} = 66\_627 \text{ mph}$$
$$V_2 = 465 \text{ ms}^{-1} = 1\_674 \text{ kph} = 1\_040 \text{ mph}$$

where

$$\text{kph} = \text{kilometres per hour } and \text{ mph} = \text{(statute) miles per hour.}$$

In other words, objects that might ordinarily be thought of as static or slow (relative to the perception of a human observer) are actually moving very fast indeed. It is all a question of relativity. In fact, as more distant reference points are chosen, the numbers become very much larger!

4.2   Observational Reference Frames

Conceptually, stars are positioned on the celestial sphere, which is a background set at near-infinite range; for simplicity, the horizon is assumed to be a projection of the tangent plane (for the particular location) on to the celestial sphere. It is necessary to consider reference frames that

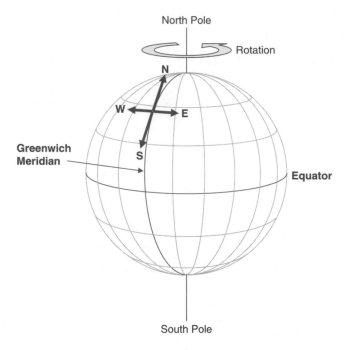

Figure 4.3 Local compass headings (for Manchester UK).

define the position (i.e. the perceived direction) of objects in the sky for navigational purposes. These are centred on an observation point, a horizon and a compass, as shown in Figure 4.3 (nominally for Manchester UK).

Thus, an observer is viewing a hemisphere with a directional reference that defines North. Points of interest are the Zenith (Z) (vertically above the observer), the North Celestial Pole (P) (vertically above the North Terrestrial Pole). The vertical circle passing through these points intersects the horizon plane at cardinal points that define North (N) and South (S). The angle between N and P corresponds exactly with the geodetic latitude of the observer; the angle between P and Z is the co-latitude. These reference points appear in the definition of three reference frames, as discussed below.

4.2.1 *Horizontal Frame*

The Horizontal Frame is defined in Figure 4.4. In this case, the vertical circle through Z and P is called the *principal vertical*. In order to define the position of a celestial object, draw a vertical circle from Z passing through the object. Azimuth (AZ) is the distance around the horizon from the north cardinal point to this vertical circle, measured eastwards. Altitude (α), in this context, is the distance along the vertical circle from the horizon to the object. The distance between the object and Z is called the zenith angle.

4.2.2 *First Equatorial Frame*

The First Equatorial Frame is defined in Figure 4.5. The basis for measurement is the celestial equator and the observer's meridian (i.e. the vertical circle through Z and P) which is called the

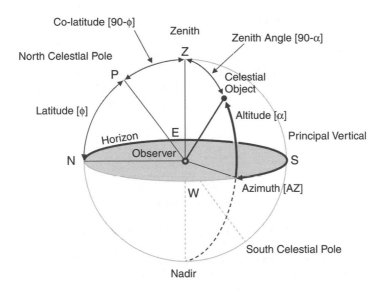

Figure 4.4 Horizontal frame.

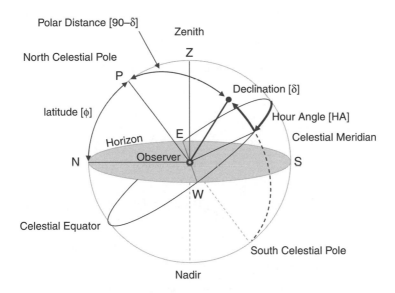

Figure 4.5 First equatorial frame.

celestial meridian. The origin of coordinates lies at the intersection between the southern half of the celestial equator and the celestial meridian. The position of an object is defined by drawing a meridian passing through it. Hour Angle (HA) is the distance from the celestial meridian to the object meridian, measured westwards in hours (from 0 to 24). Declination (δ) is the distance along the object meridian from the equator, measured in degrees (positive towards P). The distance from the object to P is called the polar distance.

4.2.3   Second Equatorial Frame

The Second Equatorial Frame is defined in Figure 4.6. The origin of coordinates is taken as the First Point of Aries (⊥) and, thus, this reference frame is independent of the observation place and time. The position of an object is defined by drawing a meridian passing through it. Right Ascension (RA) is the distance from the First Point of Aries to the object meridian, measured eastwards in hours (from 0 to 24). Declination (δ) is the same as in the First Equatorial Frame.

4.2.4   Frame Transformations

Without going into detail, it is important to recognise the method whereby horizontal coordinates can be converted to equatorial coordinates and vice versa. This is established using spherical trigonometry, as shown in Figure 4.7. Based on the North Celestial Pole, the Zenith and the Celestial Object, a triangle can be drawn with sides defined by the polar distance, the zenith distance and the co-latitude. The corresponding interior angles are complementary azimuth (360-AZ), hour angle (HA) and something called the parallactic angle (PA). Knowing some of these parameters, others can be derived by application of the sine and cosine rules.

Applying the cosine rule to this situation:

$$\cos(90 - \alpha) = \cos(90 - \phi)\cos(90 - \delta) + \sin(90 - \phi)\sin(90 - \delta)\cos HA$$

The standard formula for altitude is obtained, as follows:

$$\sin\alpha = \sin\phi\sin\delta + \cos\phi\cos\delta\cos HA \tag{4.1}$$

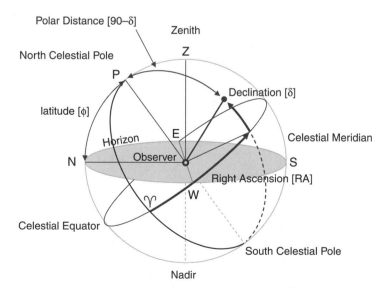

Figure 4.6 Second equatorial frame.

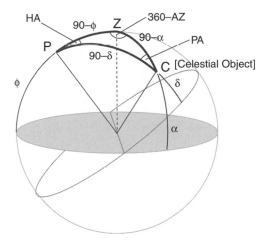

Figure 4.7 Relationship between horizontal and equatorial frames.

Where is North? For approximate navigation in the Northern hemisphere, it is referenced by the position of Polaris[4] (α UMi) in the night sky. A few thousand years ago, this closely marked the North Celestial Pole but, as a result of precession and nutation, this is no longer true although the deviation is small (in layman's terms). The position of Polaris at noon on 1 January 2000 (known as the J2000 epoch) was RA 2 h 13.8 m Dec 89°16′.

Where is South? During daylight, provided someone has a watch and the weather is not overcast, it is possible to estimate 'South' based on the knowledge that the Sun tracks from East to West at 15° per hour and it will transit the local meridian at noon (local time). Local time (say 13:30) implies the time increment after or before noon (+1.5 hours, in this example) and, from that, the angle east or west of the local meridian (22.5°); this is a horizontal coordinate which is crudely equivalent to an hour angle.

4.3   Measurement of Time

4.3.1   *Mean Time*

The perception of time is tied to the motion of the Sun across the sky, and the resulting transitions between 'day' and 'night'. This is the basis of the diurnal cycle, measured with respect to a twenty-four-hour clock that resets to 00:00 in the middle of each night and causes the date counter to increment by one day.

What is experienced is an *apparent solar day*, which is the interval between successive transits of the local meridian by the Sun (strictly, the centre of the Sun). The interval is not constant because the Earth's orbit around the Sun is not circular (i.e. its velocity is not constant) and because the Earth's axis of rotation is inclined (i.e. the Sun's apparent path is along the ecliptic and not the celestial equator). Even more important is the fact that 'local time' depends on the location of the observer; it was the arrival of the railways that necessitated a uniform system of time measurement.

---

[4] Polaris lies in the constellation of Ursa Minor.

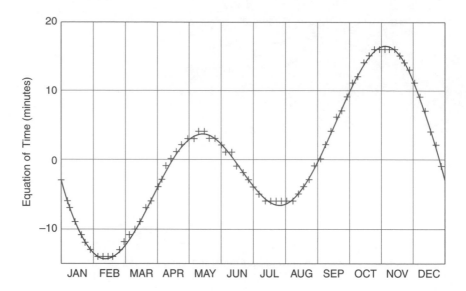

Figure 4.8 Equation of time.

In order to define a uniform time-scale, it is necessary to define a *mean Sun*. This is an imaginary Sun that moves around the equator with constant speed. The difference between the apparent Sun and the mean Sun is called the *Equation of Time*, where a positive difference indicates that the actual Sun is running ahead of the mean Sun. This is depicted in Figure 4.8, using data derived from solar system dynamics developed later in this chapter (in Section 4.8.3), overlaid with data from Norton's Star Atlas (Ridpath, 2004). A highly detailed analysis of solar position is provided by Reda and Andreas (2003). In contrast, an approximate curve-fitting solution is given in Example 4.1. This shows that the actual Sun is runs approximately two minutes behind the mean Sun at the Summer Solstice (on or around day 173).

Example 4.1

```
>> day = 173 ;
>> cyclic = 2*pi*(day/365) ;
>> equTime = − 7.4*sin(cyclic − 0.095) − 9.8*sin(2*cyclic + 0.33)
equTime =
−1.9206
```

4.3.2   *Diurnal Cycle*

Greenwich Mean Time (GMT) is defined by the transit of the mean Sun across the Greenwich meridian, such that the transit occurs at 12:00 (noon). In general, *apparent solar time* will differ from GMT because of (i) the equation of time and (ii) the longitude of an observer. For many practical purposes it is important to be able to reproduce the diurnal cycle and, especially, to be able to predict the times of sunrise and sunset, together with the transitional period called twilight.

An object rises above the horizon when its altitude becomes greater than zero and sets again when its altitude reduces to zero. The standard altitude formula (4.1) is:

$$\sin\alpha = \sin\phi\sin\delta + \cos\phi\cos\delta\cos HA$$

where α = altitude, ϕ = latitude, δ = declination and HA = hour angle. This is illustrated in Figure 4.9. When altitude is zero, this formula can be simplified:

$$\cos HA = - \tan\phi\tan\delta$$

In this simplification, the hour angle (HA) now defines the semi-diurnal arc, which equates to the time interval between the object crossing the local meridian and crossing the horizon.

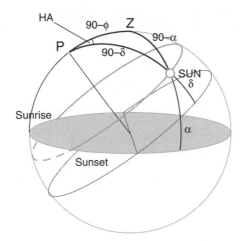

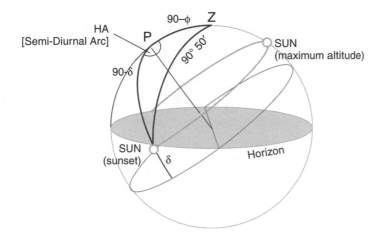

Figure 4.9 Diurnal cycle of the Sun.

Unfortunately, observations are dependent on light passing through the atmosphere; light is subject to increasing refraction as objects move closer to the horizon. Approximate expressions for refraction (R) (in degrees) are:

$$R = \left(\frac{P}{T}\right)\frac{0.00452}{\tan\alpha} \quad \text{for} \quad \alpha \geq 19.225°$$

$$R = \left(\frac{P}{T}\right)\frac{0.1594 + 0.0196\,\alpha + 0.00002\,\alpha^2}{1 + 0.505\,\alpha + 0.0845\,\alpha^2} \quad \text{for} \quad \alpha < 19.225°$$

where α = altitude (degrees), P = pressure (mbar) and T = temperature (K). For objects on the horizon, the refraction is calculated to be $35'$ although a figure of $34'$ is commonly used in calculations. The diameter of the Sun, viewed from Earth, is $32'$. Thus, sunrise and sunset occur when the Sun is at an altitude of $-0°50'$ (i.e. $-0°34'$ for atmospheric refraction plus $-0°16'$ for the Sun's radius).

The periods immediately before sunrise and immediately after sunset are called *twilight* and mark the transition in ambient illumination[5] between night and day. This is defined in three phases, as follows:

- *Civil twilight* is defined for altitudes from $-0°50'$ to $-6°$, at which point ambient illumination is sufficient to allow terrestrial objects to be distinguished.
- *Nautical twilight* is defined for altitudes from $-6°$ to $-12°$, at which point ambient illumination is sufficient to allow the outline of ground objects to be distinguished but the horizon is indistinct.
- *Astronomical twilight* is defined for altitudes from $-12°$ to $-18°$, at which point ambient illumination is effectively zero.

Example 4.2

Assume that the Sun has a declination of $23.44°$ at the Summer Solstice. It transits Manchester (approximately 53° N 2° W) approximately eight minutes after the Greenwich meridian. The Equation of Time (cf. Figure 8.3) says that the actual Sun is running two minutes behind the mean Sun at this time of year. Thus, the transit occurs at 12:10 GMT.

Sunset will occur when the centre of Sun has an altitude of $-0°50'$, i.e. the centre of the Sun is located geometrically $0°50' = 0.83°$ below the horizon. Using the altitude formula (4.1), the semidiurnal arc is derived as follows:

$$\cos\text{HA} = \frac{\sin\alpha - \sin\phi\sin\delta}{\cos\phi\cos\delta} = \frac{\sin(-0.83°) - \sin(53°)\,\sin(23.44°)}{\cos(53°)\,\cos(23.44°)}$$

Using MATLAB, the calculation can be undertaken as follows:

```
>> sine = @(degrees) sin(degrees*pi/180) ;
>> cosine = @(degrees) cos(degrees*pi/180) ;
>> degrees = @(radians) radians*180/pi ;
>> cosHA=(sine(-0.83)-sine(53)*sine(23.44))/(cosine(53)*cosine(23.44))
```

---

[5] Here, the term 'ambient illumination' implies the contribution of the Sun, in good weather, with no other sources of illumination (either natural or artificial).

```
cosHA =
−0.6016
>> angle = degrees(acos(cosHA))
angle
126.9845
>> hourAngle.H = fix(angle/15);
>> hourAngle.M = 60*(angle/15 − hourAngle.H)
hourAngle =
H: 8
M: 27.9380
```

This evaluates an hour angle (HA) of approximately 8h28, which means that the Sun will rise at 03:42 GMT and set at 20:38 GMT. Note that, since daylight saving time is in operation, times need to be expressed in BST, which is equal to GMT + 1. Thus, the Sun will rise at 04:42 BST and set at 21:38 BST.

Note that, by similar calculation, civil twilight will start at 03:50 BST in the morning and finish at 22:30 BST at night; nautical twilight will start at 02:21 BST in the morning and finish at 23:59 BST at night. The period of maximum darkness is spent in astronomical twilight because the Sun does not drop below an altitude of −13°30′.

4.3.3   Universal Time

In 1928, GMT was renamed as *Universal Time* (UT) for scientific purposes, on the recommendation of the IAU.[6] Subsequently the recognition of irregularities in Earth's rotation rate led to the introduction of several versions of UT, namely:

- UT0 is mean solar time determined from observation of star positions;
- UT1 is UT0 corrected for the variation between Earth's polar motion;
- UT2 is UT1 corrected for seasonal variations in Earth's rotation rate.

UT1 is used in navigation and astronomy and is usually referred to as 'UT'. UT2 has effectively been overtaken by International Atomic Time (TAI). TAI is maintained by the BIPM[7] and is a statistical time-scale based on clocks operated by standards laboratories around the world.

Coordinated Universal Time (UTC) is defined by CCIR[8] Recommendation 460-4 (1986) (available as ITU-R TF.460-4 'Standard-Frequency and Time-Signal Emissions'). This is transmitted by broadcast time signals and is now what is commonly thought of as 'GMT'. It is derived from atomic clocks that run at the same rate as TAI and is held within 0.9 second of UT1 by the introduction of a leap second[9] at the end of each quarter as required (with first preference given to June and December and second preference given to March and September). Variations in UTC are notified in Bulletin C of the IERS[10] (cf. Figure 4.10) about six months before their introduction.[11] A historical record of leap seconds is given in Table 4.1.

---

[6] International Astronomical Union (IAU).

[7] International Bureau on Weights and Measures or Bureau International des Poids et Mesures (BIPM).

[8] International Radio Consultative Committee (CCIR) is now incorporated within the International Telecommunications Union (ITU).

[9] For more information, go to tycho.usno.navy.mil/leapsec.html.

[10] International Earth Rotation Service (IERS) is now called the Internationa Earth Rotation and Reference System Service.

[11] The most recent bulletin can always be found at hpiers.obspm.fr/iers/bul/bulc/bulletinc.dat.

INTERNATIONAL EARTH ROTATION AND REFERENCE SYSTEMS SERVICE (IERS)

SERVICE INTERNATIONAL DE LA ROTATION TERRESTRE ET DES SYSTEMES DE REFERENCE

SERVICE DE LA ROTATION TERRESTRE
OBSERVATOIRE DE PARIS
61, Av. de l'Observatoire 75014 PARIS (France)
Tel.   : 33 (0) 1 40 51 22 26
FAX   : 33 (0) 1 40 51 22 91
Internet : services.iers@obspm.fr

Paris, 19 July 2006

Bulletin C 32
To authorities responsible
for the measurement and
distribution of time

INFORMATION ON UTC - TAI

NO positive leap second will be introduced at the end of December 2006.
The difference between Coordinated Universal Time UTC and the
International Atomic Time TAI is :

from 2006 January 1, 0h UTC, until further notice : UTC-TAI = -33 s

Leap seconds can be introduced in UTC at the end of the months of December or
June, depending on the evolution of UT1-TAI. Bulletin C is mailed every six
months, either to announce a time step in UTC, or to confirm that there will be
no time step at the next possible date.

Daniel GAMBIS
Director
Earth Orientation Center of IERS
Observatoire de Paris, France

Figure 4.10 IERS Bulletin C 31, dated 6 July 2006.

Table 4.1 Introduction of UTC leap seconds since 1972.

Date	TAI-UTC	Date	TAI-UTC	Date	TAI-UTC
1972 Jan 1	10	1980 Jan 1	19	1993 Jul 1	28
1972 Jul 1	11	1981 Jul 1	20	1994 Jul 1	29
1973 Jan 1	12	1982 Jul 1	21	1996 Jan 1	30
1974 Jan 1	13	1983 Jul 1	22	1997 Jul 1	31
1975 Jan 1	14	1985 Jul 1	23	1999 Jan 1	32
1976 Jan 1	15	1988 Jan 1	24	2006 Jan 1	33
1977 Jan 1	16	1990 Jan 1	25	2009 Jan 1	34
1978 Jan 1	17	1991 Jan 1	26		
1979 Jan 1	18	1992 Jul 1	27		

Note that an independent time-scale exists for the Global Positioning System (GPS). This is mechanised using atomic clocks and is synchronised with the US Naval Observatory atomic clock to within an accuracy of 1μs. The origin for GPS Time coincided with UTC on 6 January 1980 and so it differs from TAI by a constant offset of 19 seconds (i.e. GPS does not incorporate leap seconds). Thus, at the time of writing, the following offsets apply:

$$TAI - UTC = 34$$
$$TAI - GPS = 19$$
$$GPS - UTC = 15$$

GPS Time is of great importance because of the wide availability of commercial GPS receivers. Apart from position-fixes, GPS provides high-precision time signals, available worldwide.

4.3.4 *Time Zones*

Local time in countries and regions around the world is organised (broadly speaking) with respect to longitude. A formal classification is shown in Figure 4.11, with 25 time zones, each assigned a letter of the alphabet. All zones except one are 15° wide and have a central meridian; for instance, GMT is nominally allocated to zone 'Z' and often times will be written with suffix 'Z' (and quoted verbally with suffix 'Zulu').

Zones to the east are labelled A–L (with J omitted): zones to the west are labelled N–X. The exception is the zone around Meridian 180, which generally relates to the International Date Line. This is split in two, with the eastern segment designated as Zone M (GMT + 12) and the western segment designated as Zone Y (GMT − 12). Despite the adoption of the name 'Universal Time', time zones still tend to be expressed as GMT±n, where n is the

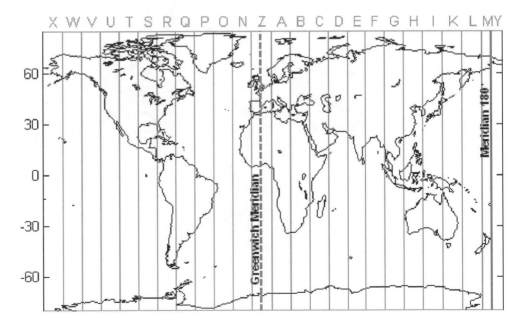

Figure 4.11 World time zones.

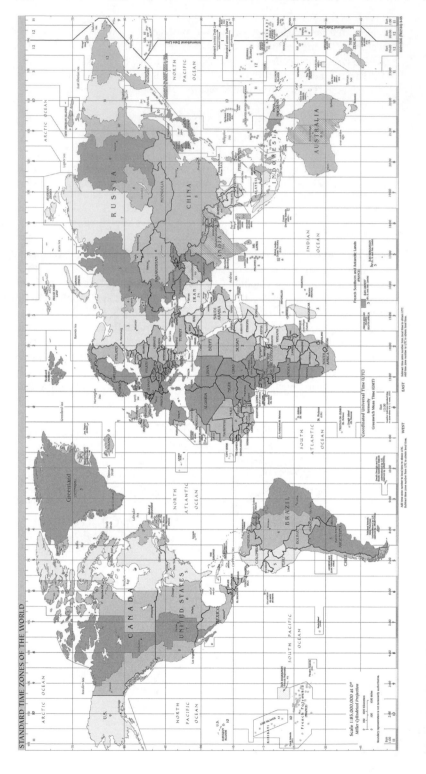

Figure 4.12 Standard time zones.

number of hours relative to GMT. For instance, Western Australia lies in Zone H (which is GMT + 8); in other words, 12:00 GMT corresponds with 20:00 local time, say, in Perth. It should also be noted that some countries adopt daylight-saving time during the summer months; in the UK this is called British Summer Time (BST) and it runs one hour ahead of GMT (which is GMT + 1).

The actual time standard that applies to a country or province is dictated by the appropriate government authority. Although there is a natural trend based on longitude, there are many variations[12] and, in some places, the regulation of time is subject to frequent change. The recent situation is illustrated in Figure 4.12,[13] showing Standard Time Zones as imposed in different regions. Generally, the increments relative to GMT are whole hours but, as might be recognised in the figure, a few countries opt for an intermediate time-scale (e.g. India uses GMT + 5.5). For more information, an authoritative source of time-zone data is the 'tz' database.[14]

Example 4.3

The MATLAB Mapping Toolbox provides a means of determining time zone for a given longitude, e.g. 120°E.

```
>> (timeDifference,timeZoneDesignator) = timezone(120,'degrees')
timeDifference =
−8
timeZoneDesignator =
H
```

4.3.5   *Sidereal Time*

Time can also be defined relative to astronomical observation. This is Local Sidereal Time (LST), which is defined as the time elapsed since the local meridian crossing of the Vernal Equinox. Similarly, Greenwich Sidereal Time (GST) is the time elapsed since the Vernal Equinox crossed the Greenwich Meridian.

The position of celestial objects as viewed from Earth can be defined by two angles, namely Right Ascension and Declination (cf. the 'Second Equatorial Frame' in Section 4.2.3). This is shown in Figure 4.13. It should be recognised that sidereal time is equal to the Right Ascension[15] of a star that is on the local meridian. The relationship between GST and LST is derived from the longitude of the observer. For every 15° east of Greenwich, add one hour to GST in order to obtain LST: for every 15° west, subtract one hour to GST in order to obtain LST.

---

[12] For an animation that highlights the relationship between natural time zones and regional time standards, go to www.radicalcartography.net/?timezoneslow.

[13] Image source: CIA World Factbook; for copyright information, refer to: www.cia.gov/cia/publications/factbook/docs/contributor_copyright.html.

[14] www.twinsun.com/tz/tz-link.htm.

[15] Right ascension can be measured either in degrees (optionally subdivided into minutes and seconds or arc) or in hours, minutes and seconds (while one hour is equivalent to 15°).

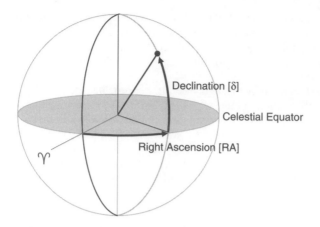

Figure 4.13 Right ascension and declination.

Sidereal time and right ascension are combined in order to derive hour angles (cf. the 'First Equatorial Frame' in Section 4.2.2), as follows:

$$\text{Hour Angle} = \text{Sidereal Time} - \text{Right Ascension}$$

Thus, the Local Hour Angle (LHA) and Greenwich Hour Angle (GHA) are, respectively:

$$\text{LHA} = \text{LST} - \text{RA}$$
$$\text{GHA} = \text{GST} - \text{RA}$$

The difference between sidereal time and solar time arises because they are based on different observations. A *sidereal day* is the time interval between successive transits of a chosen star across any given meridian: a *solar day* is the time interval between successive transits of the Sun. As a mental experiment, if the Earth were to rotate once on its axis for each orbit of the Sun, one side of the Earth would experience permanent daylight and the other would experience permanent darkness. Over one year, an observer on the dark side would see no transits of the Sun and only one transit of each star. Restoring Earth's rotation, the observer would now see 365 transits of the Sun and 366 transits of each star. Thus, 365 solar days correspond with 366 sidereal days, which leads to the conclusion that a sidereal day is about four minutes shorter than a solar day ... or stars rise four minutes earlier each evening.

4.3.6   *Terrestrial Time*

Terrestrial Time (TT) (formerly known as Terrestrial Dynamical Time) is the time standard on the surface of the Earth (strictly speaking, for an observer located on the geoid). It is a dynamic time-scale, defined in the context of the theory of relativity, in which the concept of time is dependent on the location and motion of the particular clock that is measuring 'time'.

TT is related to Geocentric Coordinate Time (TCG) and Barycentric Coordinate Time (TCB), which are related to nonrotating reference frames with their respective origins at the centre of the Earth and the barycentre of the solar system. These serve primarily as astronomical time standards and are of no real interest here.

The definition of Terrestrial Time can be traced through various IAU meetings but the key statements are Recommendation IV of the XXIst General Assembly and Resolution B1.9 of the XXIVth General Assembly. It can also be traced back to Ephemeris Time (ET), which was based on Newcomb's theory of Earth's motion (Newcomb, 1898) and was adopted in order to handle irregularities relative to the mean Sun. This produced a refined expression for the mean longitude of the Sun with respect to the Earth-Moon barycentre but this was much more difficult to measure than mean solar time. Implementation relied on comparisons of solar, lunar and planetary observations with predicted ephemerides and, because of its quicker rate of orbital motion, greatest emphasis was placed on the Moon.

The difference between ET and UT is called ΔT (literally 'Delta-T'), which varies over time. TAI was set equal to UT2 on 1 January 1958, when $\Delta T \approx 32.18$ seconds. The introduction of TT (and TCG) was designed to provide a smooth transition from ET on 1 January 1977 such that:

$$TT = ET = TAI + 32.184$$

Expressing dates in the format 'Year Month Day' (including fractions of a day), this relationship can be written as:

$$1077 \text{ January } 1.0003725 \text{ TT} = 1977 \text{ January } 1.0000000 \text{ TAI}$$

(*Note*: The difference TT-TAI = 32.184 is a constant value.)

Although ET has been discontinued, the suffix 'ET' still appears in documents from time to time. For any date in 1977 and after, this implies TT.

4.4   Calendars and the J2000 Reference Epoch

The Gregorian[16] calendar has been in use since AD 15 October 1582. In this, the average length of a year is defined as 365.2425 days. The mean solar year lasts for 365.2522 days, which implies an error of just less than 26 seconds per year. The accumulation of error is handled by introducing an extra day for each year number divisible by four, except for those that are divisible by 100 *and* not divisible by 400. This constitutes a leap year (with a so-called *intercalary* day on 29 February). Using the divisibility rule it should be seen that 2000 was a leap year but 1900 was not.

This is satisfactory for civilian time-keeping but not for systematic record-keeping. It is most appropriate to align aerospace environments with astronomical conventions and, accordingly, time is recorded on a chronological scale based on the so-called Julian Date (JD), which is the number of days since 1 January 4713 BC. For historical reasons, reckoning commences from Greenwich noon and is given in decimal form (*not* hours, minutes and seconds). The long time span is chosen so that all recorded astronomical events have positive Julian Day numbers. Under this convention, the Gregorian calendar starts[17] on JD 2299160.5. Note that one-second accuracy requires five decimal places in the Julian date.

---

[16] This is named after Pope Gregory XIII, who instituted reform of the previous calendar (based on Julian years comprising 365.25 days) because of the accumulation of errors relative to the observed solar cycle.
[17] What preceded the Gregorian calendar is not important here.

In passing, note that modern aerospace events commence on JD 2416465.94097, which is a date that has obvious historical significance!

In order to overcome the inconvenience of large numbers, an alternative is adopted for dates close to the present day, namely the Modified Julian Date (MJD), where

$$MJD = JD - 2400000.5 \qquad (4.2)$$

This reduces the number scale and also takes the opportunity of aligning the start of any day with Greenwich midnight.

The reference epoch[18] for current astronomical data is called J2000 and is defined to be Greenwich noon on 1 January 2000 (which is written as 2000 January 1.5 TT). This instant is also as JD 2451545.0 or MJD 51544.5. Elapsed time is commonly measured in days or centuries since J2000.

Note 4.4

For information, the previous reference epoch was B1950 and was based on Besselian years. A Besselian year starts when the ecliptic longitude[19] of a hypothetic mean Sun[20] is 280°; in order to simplify the calculation of start positions, the IAU recommended the use of Julian years in 1984.

A range of algorithms exist for conversion between calendar dates and Julian dates and many can be found from Internet searches, e.g. from the Jet Propulsion Laboratory. Without going into detail, the following algorithms are typical and work reliably. Because of the nature of the application here, there is no absolutely need to regress beyond the start of the Gregorian calendar! So, in effect, these represent a mapping between the present-day civilian calendar and relevant Julian dates for spatial reference data around the J2000 epoch.

A thorough discussion and exposition of date algorithms can be found on Peter Baum's website[21] and is recommended reading for enthusiasts. A straightforward procedure converts from calendar dates (expressed as an integer sequence denoting (Year Month Day)) to Julian dates. A rather more intricate procedure can be found that performs the conversion from Julian dates to calendar dates. In essence, these procedures are implemented in Algorithms 4.1 and 4.2, respectively. Their usage is illustrated in Example 4.4, showing the start of the Gregorian calendar.

Algorithms like these appear with variations and refinements, but they all do the same job. In addition, they all face the same basic problem and a word of warning is needed on the use of floating-point number representations in order to conduct integer arithmetic (evidenced in the use of MATLAB functions like 'floor' and 'fix'). It is advisable to confirm that algorithms do what they are supposed to do! For instance, the conversion of Julian dates can result in wrong calendar dates in January or February of leap years.

---

[18] 'Epoch' simply means an instant in time.
[19] Longitude (or angular distance) measured around the ecliptic.
[20] In this context, 'Mean Sun' is defined as traversing the ecliptic at a constant rate.
[21] www.vsg.cape.com/~pbaum/date/date0.htm.

```
function julianDate = getJulianDate(calendarDate)
year = calendarDate(1) ;
month = calendarDate(2) ;
day = calendarDate(3) ;

if month<= 2 % calculate from 1st March
 year = year-1 ;
 month = month+12 ;
end

calendarOffset = 1721118.5 ;
leapDays = floor(year/400) - floor(year/100) + floor(year/4) ;
monthStart = floor(30.6*(month+1) - 122) ;
julianDate = 365*year + monthStart + day + leapDays + calendarOffset ;

z = floor(julianDate) ;
r = julianData - z ;
integerString = num2str(z) ;
decimalString = num2str(r) ;

if isempty(decimalString)
 decimalString = '0';
else
 decimalString = decimalString(3:length(decimalString)) ;
end

fprintf(('JD: ',integerString,'.',decimalString,'[bsol]n'))
```

Algorithm 4.1 'getJulianDate'.

```
function calendarDate = getCalendarDate(julianDate)

calendarOffset = 1721118.5 ;
z = floor(julianDate-calendarOffset) ;
r = julianDate - calendarOffset - z ;
a = floor((z-0.25)/36524.25) ;
b = a - floor(a/4) ;
year = floor((b+z-0.25)/365.25) ;
c = b+z - floor(365.25*year) ;
month = fix((5*c+456)/153) ;
day = c - fix((153*month-457)/5) + r ;

if month>12
year = year+1 ;
month = month-12 ;
end

calendarDate =(year,month,day) ;

fprintf(('Date: ',num2str(calendarDate),'[bsol]n'))
```

Algorithm 4.2   'getCalendarDate'.

> ### Example 4.4
>
> ```
> >> jd = getJulianDate((1582 10 15)); % Start of the Gregorian Calendar
> JD: 2299160.5
> >> getCalendarDate(jd);
> Date: 1582 10 15
> ```

4.5   Chronological Scale

The chronological scale for astronomical calculations can be quoted in Julian days since J2000:

$$D = JD - 2451545.0 \qquad (4.3)$$

This would apply to fast-moving objects and, by implication, these would be close to Earth. For distant objects (e.g. stars and planets) this is quoted in Julian centuries since J2000:

$$T = \frac{JD - 2451545.0}{36525.0} \qquad (4.4)$$

Because it is so commonly employed, it is implemented in Algorithm 4.3. Note that chronology is expressed in Terrestrial Time (TT) (as discussed in Section 4.3.6).

```
function julianCenturies = getJulianCenturies(julianDate)

JD2000 = 2451545;
jCentury = 36525;
julianCenturies = (julianDate - JD2000)/jCentury;
```

Algorithm 4.3 'getJulianCenturies'.

4.6   Astrometric Reference Frames

4.6.1   *Inertial Frame*

A practical reference frame can be defined with 'x' aligned with the vernal equinox, 'y' rotated by 90° eastwards around the celestial equator and 'z' aligned with the Celestial North Pole. This is shown in Figure 4.14.

Aerospace textbooks that deal with the wider context of aircraft flight over the surface of Earth adopt this standard, which is fine for flights of any practical duration. In order to be consistent with astronomic data, as well as to enable long-term multivehicle operations and interoperability with satellite positioning systems, it is appropriate to align the x-axis with the Earth's equinox and mean equator for J2000; this gives a reference frame that is known as EME2000. Unless specified otherwise, this will be considered as the Earth-Centred Inertial (ECI) frame and, for aerospace applications, this can be considered to be a fixed reference frame.

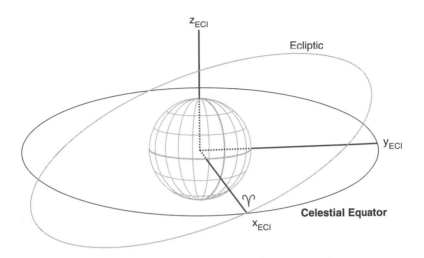

Figure 4.14 ECI reference frame.

4.6.2   *Rotating Frame*

Recalling the ECEF reference frame from Figure 2.3, its relationship with the ECI frame is shown in Figure 4.15. The Earth rotation rate is denoted by Ω and its direction of rotation is from West to East. Approximate formulae are provided by the IAU and IAG (Seidelmann *et al.*, 2002; Roncoli, 2005, section 3.2) for the movement of the *North Pole* and the *Prime Meridian*, while IERS standards are available for precise navigation (McCarthy & Petit, 2004).

Thus the approximate movement of the *Prime Meridian* (i.e. Greenwich) is:

$$RA_G = 280.147 + \Omega D \tag{4.5}$$

where RA is Right Ascension[22] and D is the number of Julian days after J2000. In this context, the GRS67 value of Earth rotation rate is quoted:

$$\Omega = 360.9856235 \text{ deg/day} \tag{4.6}$$

which differs from the GRS80 value (2.56) used in WGS84.

For completeness (and academic interest) the approximate movement of the *North Pole* is:

$$RA_N = 0 - 0.641T \tag{4.7}$$

$$\delta_N = 90 - 0.557T \tag{4.8}$$

where RA is Right Ascension, δ is Declination and T is the number of Julian centuries after J2000.

---

[22] The original formula measures right ascension from the Earth IAU-Node vector (which is the J2000 equinox plus 90 degrees)! Equation (9.1) uses the standard convention, measuring from the equinox.

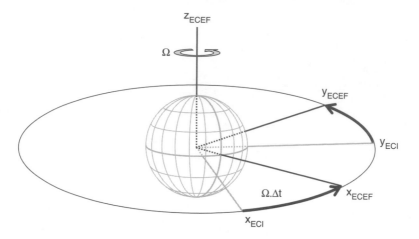

Figure 4.15 ECEF reference frame.

4.6.3   *Precession*

It is possible to expand the description of Earth's orbital precession that was provided in Section 4.1, based on the reorientation of the mean ecliptic and equator with respect to J2000 (as shown in Figure 4.16). This is more detailed than is required in many applications but it can arise in connection with inertial referencing (e.g. global positioning systems). A good summary is provided by Montenbruck and Gill (2000), based on the IAU 1976 precession model (Lieske *et al.*, 1977).

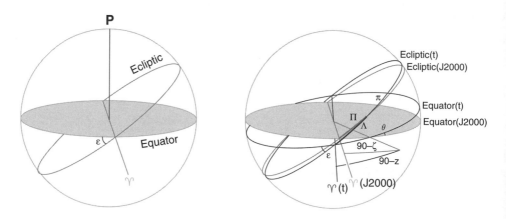

Figure 4.16 Precession of the Ecliptic, Equator and Vernal Equinox.

Precession of the equator refers to intersection of the mean equator of a given epoch and the mean ecliptic of J2000, which lags behind the vernal equinox of J2000 by an angle:

$$5038.8'T - 1.1'T^2 \tag{4.9}$$

The inclination of the mean equator relative to the J2000 ecliptic is almost constant:

$$23°26'21' + 0.05'T^2 \tag{4.10}$$

In both equations, T is the number of Julian centuries after J2000.

Precession of the ecliptic reveals a slight motion of the ecliptic plane and a slight change in its obliquity. Relative to the ecliptic of J2000, the ecliptic at another epoch is inclined at an angle:

$$\pi = 47.0029'T - 0.03302'T^2 + 0.000600'T^3 \tag{4.11}$$

and the line of intersection is given by the angle:

$$\Pi = 174.876383889° - 869.8089'T + 0.03536'T^2 \tag{4.12}$$

The obliquity is slightly decreasing:

$$\varepsilon = 23.4392911° - 46.8150'T - 0.00059'T^2 + 0.001813'T^3 \tag{4.13}$$

The combined precession of the equator and ecliptic (i.e. *general precession*) shows the following change in longitude:

$$p = \Pi - \Lambda = -5029.0966'T - 1.11113'T^2 + 0.000006'T^3 \tag{4.14}$$

The change in orientation of the mean equator and equinox is defined by three angles:

$$\varsigma = 2306.2181'T + 0.30188'T^2 + 0.017998'T^3 \tag{4.15}$$

$$\theta = 2004.3109'T - 0.42665'T^2 - 0.041833'T^3 \tag{4.16}$$

$$z = \varsigma + 0.79280'T^2 + 0.000205'T^3 \tag{4.17}$$

Note 4.5

Resolution 1 of the IAU General Assembly XXVI in Prague (2006) recommends (amongst other things) the terms 'precession of the equator' and 'precession of the ecliptic' as replacements for 'lunisolar precession' and 'planetary precession', respectively. This recognises that the gravitational attraction of planets has a significant effect on the motion of Earth's equator, as well as the ecliptic. Thus, the traditional distinction was held to be misleading.

4.6.4   *Coordinate Transformations*

The transformation of coordinates from the mean equator and equinox of J2000 ($\mathbf{r}_{ICRF}$) to the mean equator and equinox of another epoch ($\mathbf{r}_t$) is achieved using a sequence of elementary rotations:

$$\mathbf{r}_t = R_z(-90 - z)R_x(\theta)Rz(90 - \varsigma)\mathbf{r}_{ICRF} \qquad ((4.18))$$

where the subscript 'ICRF' refers to the International Celestial Reference Frame. This entire process is implemented in Algorithm 4.4, where the calculation of Julian centuries (as an input parameter) is given by Algorithm 4.3.

The ECI frame has already been introduced as an Earth-centred EME2000 frame; its orientation is the same as that of the ICRF. However, reference frames are never really as straightforward as they might first appear.

```
function precession = getPrecession(julianCenturies)

radians = @(degrees) degrees*pi/180;
Rx = @(x) ( 1 0 0; 0 cos(x) sin(x); 0 -sin(x) cos(x));
Ry = @(y) ( cos(y) 0 -sin(y); 0 1 0; sin(y) 0 cos(y));
Rz = @(z) ( cos(z) sin(z) 0; -sin(z) cos(z) 0; 0 0 1);

ecliptic = radians(23.4392911 - (46.8150 + 0.00059*T - 0.001813*T^2)*T/
  3600);
zeta = radians((2306.2181 + 0.30188*T + 0.017998*T^2)*T/3600);
theta = radians((2004.3109 - 0.42665*T - 0.041833*T^2)*T/3600);
z = zeta + radians((0.79280 + 0.000205*T)*T^2/3600);

precession = Rz(-pi+z)*Rx(theta)*Rz(pi-zeta)*Rx(-ecliptic);
```

Algorithm 4.4 'getPrecession'.

The ICRF is jointly maintained by the IERS and the IAU Working Group on Reference Frames (Arias *et al.*, 1995) and is an implementation of the International Celestial Reference System[23] (ICRS) that was adopted in 1998 (Feissel & Mignard, 1998). This was developed to overcome issues associated with dynamic variation of the ecliptic and equinox; being parallel with EME2000, it is cross-referenced to the FK5 catalogue (Fricke *et al.*, 1988), which gives precise positions and proper motions[24] of about 1500 stars for the J2000 epoch.

In contradistinction, the ICRS is orientated with respect to distant extragalactic radio objects, which supposedly have no proper motion, and so it will not rotate; its fundamental plane is aligned with the EME2000 equator but the origin of Right Ascension is referred to quasar 3C273. For completeness, note that the origin of the ICRS (and therefore the ICRF) is the solar system barycentre, whereas the ECI is, by definition, Earth-centred.

---

[23] A 'reference system' is a set of concepts and models that define the orientation of reference axes at any instant while a 'reference frame' is a particular implementation of that system.

[24] Proper motion is due to the transverse component of stellar motion and appears as an angular displacement over time relative to the celestial background.

Is this really necessary for an aerospace applications? From the narrow perspective of aircraft flight, the answer is 'No', and there is absolutely no need to observe quasars in order to define a practical reference frame. However, in the interests of harmonisation, established standards are used to support aircraft development programmes. Given the investment required to create unified models of this type, a more considered answer would obviously be 'Yes' ... but there is still no need to observe quasars!

Note 4.6

Precession models are subject to periodic review and update. IAU1976 is still widely used although there exists a more recent version (IAU2000). Consequent upon Resolution 1 at IAU General Assembly XXVI in Prague (2006), this will be superseded by the P03 precession model (Capitaine *et al.*, 2003, 2005) on 1 January 2009.

4.7   Orbital Mechanics

4.7.1   *Kepler's Laws*

The motion of an object in closed orbit around a central mass is described by Kepler's laws:

K1: the shape of the orbit is an ellipse
K2: the radius vector sweeps the area inside the ellipse at a constant rate
K3: the orbital period squared is proportional to the semi-major axis cubed

By implication, the orbiting mass is much smaller than the central mass and the orbit is unperturbed by any other objects of significant mass.

4.7.1.1 Elliptical Geometry

Before proceeding, it is appropriate to summarise the basic properties of an ellipse (cf. Figure 4.17). In cartesian coordinates (x,y), the equation of an ellipse is:

$$\left(\frac{x}{a}\right)^2 + \left(\frac{y}{b}\right)^2 = 1 \qquad (4.19)$$

where a = semi-major axis and b = semi-minor axis.

Recalling (2.61) and (2.62), and adjusting the notation for plane geometry, the eccentricity of an ellipse can be defined in two ways:

$$\text{First Eccentricity}: \ e = \sqrt{\frac{a^2 - b^2}{a^2}} \qquad (4.20)$$

$$\text{Second Eccentricity}: \ \varepsilon = \sqrt{\frac{a^2 - b^2}{b^2}} \qquad (4.21)$$

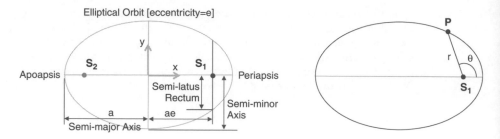

Figure 4.17 Elliptical geometry.

In most applications, the term 'eccentricity' refers to the first eccentricity and its value must lie in the range $0 \leq e < 1$.

Specifying the semi-major axis (a) and the eccentricity (e), the geometric construction of an ellipse is achieved by defining two focal points (or foci) $\mathbf{S_1}$ and $\mathbf{S_2}$, situated at points (ae,0) and (-ae,0) respectively, and then drawing the locus of a point $\mathbf{P}$ such that the sum of distances to each focus is equal to 2a, i.e. $\mathbf{S_1 P} + \mathbf{S_2 P} = 2a$.

It is possible to calculate the semi-minor axis (i.e. the value of y when x=0):

$$b = a\sqrt{1 - e^2} \tag{4.22}$$

and the semi-latus rectum (i.e. the value of y when x=ae):

$$p = a(1 - e^2) \tag{4.23}$$

Also, it is possible to modify the cartesian equation of an ellipse as follows:

$$y^2 = (a^2 - x^2)\ (1 - e^2) \tag{4.24}$$

In polar coordinates (r,θ), radiating from a single focus $\mathbf{S_1}$, the equation of an ellipse is derived by setting:

$$\begin{aligned} x &= ae + r\cos\theta \\ y &= r\sin\theta \end{aligned} \tag{4.25}$$

and then substituting these values into the modified cartesian equation:

$$\begin{aligned} &r^2[\sin^2\theta + \cos^2\theta\ (1 - e^2)] + r[2ae\cos\theta\ (1 - e^2)] - a^2(1 - e^2)^2 = 0 \\ &r^2[1 - e^2\cos^2\theta] + r[2ep\cos\theta] - p^2 = 0 \end{aligned}$$

Solving this quadratic equation yields the relationship between radius r and angle θ:

$$r = \frac{p}{1 + e\cos\theta} \tag{4.26}$$

Each extreme of radial distance is called an *apsis* and the line joining them (which contains the semi-major axis) is called the *line of apsides*. The minimum distance is called the *periapsis* (corresponding with $\theta = 0$) and the maximum distance is called the *apoapsis* (corresponding with $O = \pi$):

$$r_p = \frac{p}{1+e} \tag{4.27}$$

$$r_a = \frac{p}{1-e} \tag{4.28}$$

In the context of orbits, the maximum and minimum approaches are specifically named *perigee* and *apogee* for objects orbiting the Earth and *perihelion* and *apohelion* for objects orbiting the Sun.

4.7.1.2 Kepler's First Law

Anticipating the discussion in Chapter 4, it is noted that gravity is a central force[25] and that gravitational acceleration is defined by:

$$\mathbf{g} = -\frac{GM}{r^2}\hat{\mathbf{r}}$$

where the radius vector is defined for a closed orbit (as shown in Figure 4.18) in the context of polar coordinates (as shown in Figure 4.19).

Exploiting the properties of polar basis vectors:

$$\frac{d\hat{\boldsymbol{\theta}}}{d\theta} = -\hat{\mathbf{r}}$$

$$\Rightarrow \hat{\mathbf{r}} = -\frac{d\hat{\boldsymbol{\theta}}}{d\theta} = -1\dot{\theta}\,\frac{d\hat{\boldsymbol{\theta}}}{dt}$$

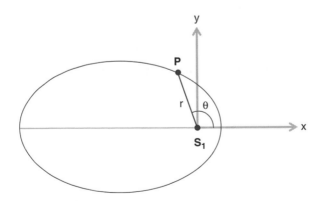

Figure 4.18 Orbital position.

---

[25] A 'central force' is a force towards a central point.

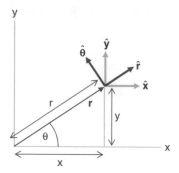

Figure 4.19 Polar coordinates.

Thus, the rate of change of velocity for the orbiting object is described by:

$$\mathbf{g} = \frac{d\mathbf{v}}{dt} = -\frac{GM}{r^2}\hat{\mathbf{r}} = \frac{GM}{r^2\dot\theta}\frac{d\hat\theta}{dt}$$

$$\mathbf{g} = \frac{GM}{h}\frac{d\hat\theta}{dt} \tag{4.29}$$

where $h = r^2\dot\theta$ is the angular momentum per unit mass of the satellite and, assuming that the object has fixed mass, this value must remain constant under the action of a central force (i.e. there is no applied torque). The derivation continues:

$$\frac{d\mathbf{v}}{dt} = \frac{GM}{h}\frac{d\hat\theta}{dt} \ \Rightarrow \mathbf{v} = \frac{GM}{h}\hat\theta + \mathbf{v}_0$$

When q = 0, the constant of integration v0 is aligned with the y-axis and so, without loss of generality, the expression becomes:

$$\mathbf{v} = \frac{GM}{h}\left(\hat\theta + k\hat{\mathbf{y}}\right) \tag{4.30}$$

Note that, when evaluated using polar coordinates, velocity is derived as follows:

$$\mathbf{r} = r\hat{\mathbf{r}}$$

$$\Rightarrow \mathbf{v} \equiv \dot{\mathbf{r}} = \dot{r}\hat{\mathbf{r}} + r\frac{d\hat{\mathbf{r}}}{dt} = \dot{r}\hat{\mathbf{r}} + r\dot\theta\frac{d\hat{\mathbf{r}}}{d\theta}$$

$$\Rightarrow \mathbf{v} = \dot{r}\hat{\mathbf{r}} + \frac{h}{r}\hat\theta \tag{4.31}$$

Therefore, the following equivalence exists:

$$\dot{r}\hat{\mathbf{r}} + \frac{h}{r}\hat\theta = \frac{GM}{h}\left(\hat\theta + k\hat{\mathbf{y}}\right)$$

$$\Rightarrow \left(\ddot{r}\hat{\mathbf{r}} + \frac{h}{r}\hat{\boldsymbol{\theta}} \right) \bullet \hat{\boldsymbol{\theta}} = \frac{GM}{h} \left(\hat{\boldsymbol{\theta}} + k\hat{\mathbf{y}} \right) \bullet \hat{\boldsymbol{\theta}}$$

$$\Rightarrow \frac{h}{r} = \frac{GM}{h} \left(1 + k\hat{\mathbf{y}} \bullet \hat{\boldsymbol{\theta}} \right) = \frac{GM}{h} \left(1 + k\cos\theta \right)$$

$$\Rightarrow r = \left(\frac{h^2}{GM} \right) \frac{1}{1 + k\cos\theta} \tag{4.32}$$

This is the equation for an ellipse and, by comparison with (4.26), it is seen that it has eccentricity $e = k$ and semi-latus rectum $p = h^2/GM$. This demonstrates Kepler's First Law.

Before moving on, the position and velocity of an orbiting object were found to be:

$$\begin{aligned} \mathbf{r} &= r\hat{\mathbf{r}} \\ \dot{\mathbf{r}} &= \dot{r}\hat{\mathbf{r}} + \frac{h}{r}\hat{\boldsymbol{\theta}} \end{aligned} \tag{4.33}$$

respectively. From these relationships, it should be obvious that:

$$h = \mid \mathbf{r} \times \dot{\mathbf{r}} \mid = \left| r\hat{\mathbf{r}} \times \left(\dot{r}\hat{\mathbf{r}} + \frac{h}{r}\hat{\boldsymbol{\theta}} \right) \right| \tag{4.34}$$

This shows that 'h' is an angular momentum (normalised with respect to mass).

4.7.1.3 Kepler's Second Law

In the last section, it was determined that angular momentum per unit mass is constant under the action of a central force. When the orbiting object is at position (r, θ), an infinitesimal change in angle θ (i.e. $d\theta$) will be associated with a swept area of $dS = r(r.d\theta)/2$ and thus the rate of sweep is given by:

$$\frac{dS}{dt} = \frac{1}{2}r^2\dot{\theta} \equiv \frac{h}{2} \tag{4.35}$$

This is constant, thus demonstrating Kepler's Second Law.

4.7.1.4 Kepler's Third Law

Using Kepler's Second Law, it is seen that:

$$dS = \frac{h}{2}dt \tag{4.36}$$

Integrating for one complete orbit (i.e. orbital period T and swept area $S = \pi ab$):

$$\left(\frac{h}{2} \right)T = \pi ab = \pi a^2\sqrt{1 - e^2} \Rightarrow \left(\frac{h}{2} \right)^2 T^2 = \pi^2 a^4 (1 - e^2) = \pi^2 a^3 p$$

where p is the semi-latus rectum, which was shown to be $p = h^2/GM$ in the proof of Kepler's First Law. Substituting this value, the derivation can be concluded:

$$\left(\frac{h}{2}\right)^2 T^2 = \pi^2 a^3 \frac{h^2}{GM} \Rightarrow T = 2\pi \sqrt{\frac{a^3}{GM}} \tag{4.37}$$

This demonstrates Kepler's Third Law.

4.7.2   *Orbital Energy and Velocity*

The total energy associated with an orbiting satellite is composed of kinetic energy and potential energy:

$$E = \frac{1}{2}mv^2 - \frac{GMm}{r} \tag{4.38}$$

At the periapsis, the velocity vector is perpendicular to the radius vector such that:

$$h = r_P v_P \tag{4.39}$$

where $r_P = a(1-e)$ and $h^2 = GMa(1-e^2)$. Thus, the velocity at the time of periapsis passage is:

$$v_P = \frac{\sqrt{GMa\ (1-e^2)}}{a\ (1-e)} = \sqrt{\frac{GM}{a}}\sqrt{\frac{1+e}{1-e}} \tag{4.40}$$

Incidentally, using similar reasoning, the velocity at apoapsis passage can be found to be:

$$v_A = \frac{\sqrt{GMa(1-e^2)}}{a(1+e)} = \sqrt{\frac{GM}{a}}\sqrt{\frac{1-e}{1+e}} \tag{4.41}$$

Since the orbital energy remains constant, the satellite states at periapsis are sufficient to calculate the total energy:

$$E = \frac{1}{2}m(v_P)^2 - \frac{GMm}{r_P} = \frac{1}{2}m\left(\frac{GM}{a}\right)\left(\frac{1+e}{1-e}\right) - \frac{GMm}{a(1-e)} E = -\frac{GMm}{2a} \tag{4.42}$$

Thus, the velocity at any position in the orbit can be established as follows:

$$-\frac{GM}{2a} = \frac{v^2}{2} - \frac{GM}{r} \Rightarrow v^2 = GM\left(\frac{2}{r} - \frac{1}{a}\right) \tag{4.43}$$

4.7.3   *Anomalies and Kepler's Equation*

The term 'anomaly' refers to the angular position of a satellite along its orbital path, measured from the periapsis. *True anomaly* is measured about the central body and *eccentric anomaly* is measured about the centre of the ellipse, as shown in Figure 4.20.

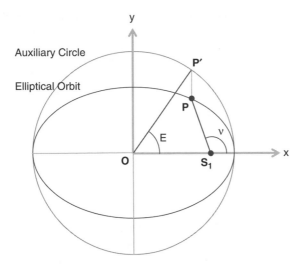

Figure 4.20 True anomaly (v) and eccentric anomaly (E).

The geometric construction requires the elliptical orbit, together with its auxiliary circle. Recalling that the focus S_1 is located at a point (ae,0), the orbit point **P** and its auxiliary point **P'** can be defined as:

$$\mathbf{P}:\ x = ae + r\cos v \qquad y = r\sin v$$

$$\mathbf{P'}:\ x' = a\cos E \qquad y' = a\sin E$$

where

$$x = x' \text{ and } y = \left(\frac{b}{a}\right)y'$$

The relationship between the radius and the eccentric anomaly is derived as follows:

$$r^2 = (x - ae)^2 + y^2$$
$$\Rightarrow r^2 = a^2(\cos E - e)^2 + a^2(1 - e^2)\sin^2 E$$
$$\Rightarrow r^2 = a^2(1 - 2e\cos E - e^2\cos^2 E)$$
$$\Rightarrow r = a(1 - e\cos E)$$

(4.44)

Compare and contrast this with equation (4.26), which would be obtained using true anomaly:

$$r = \frac{p}{1 + e\cos v}$$

(4.45)

where p is the semi-latus rectum, $p = a(1 - e^2) = h^2/GM$. Note that:

$$h = \sqrt{GMa\ (1 - e^2)}$$

(4.46)

Also, considering x-ordinates, the relationship between true anomaly and eccentric anomaly is found as follows:

$$x' = a \cos E = ae + r \cos v = x' \Rightarrow \cos E = e + (1 - e \cos E) \cos v$$

This gives two equivalent formulae:

$$\cos E = \frac{e + \cos v}{1 + e \cos v} \tag{4.47}$$

$$\cos v = \frac{\cos E - e}{1 - e \cos E} \tag{4.48}$$

Also, this can be recast in terms of half-angles in order to obtain an equivalent relationship:

$$\tan \frac{E}{2} = \sqrt{\frac{1 + e}{1 - e}} \tan \frac{v}{2} \tag{4.49}$$

Now, taking xy-coordinates relative to the focus S_1, it is possible to obtain a relationship between the angular momentum per unit mass (h) and the eccentric anomaly (E):

$$\mathbf{r} = r(\hat{\mathbf{x}} \cos v + \hat{\mathbf{y}} \sin v)$$

$$\Rightarrow \mathbf{r} = \hat{\mathbf{x}} a(\cos E - e) + \hat{\mathbf{y}} b \sin E$$

$$\Rightarrow \dot{\mathbf{r}} = -\hat{\mathbf{x}} a(\sin E)\dot{E} + \hat{\mathbf{y}} b(\cos E)\dot{E}$$

$$h = |\ \mathbf{r} \times \dot{\mathbf{r}}\ |$$

$$\Rightarrow h = a(\cos E - e)b(\cos E)\ \dot{E} - b(\sin E)a(\sin E)\dot{E}$$

$$\Rightarrow h = ab(1 - e \cos E)\dot{E}$$

$$\Rightarrow h = a^2 \sqrt{1 - e^2}(1 - e \cos E)\dot{E}$$

From the value of h developed previously, this can be simplified as follows:

$$\sqrt{GMa(1 - e^2)} = a^2 \sqrt{1 - e^2}(1 - e \cos E)\dot{E}$$

$$\Rightarrow (1 - e \cos E)\dot{E} = \sqrt{\frac{GM}{a^3}} \tag{4.50}$$

It is usual to express this equation in the form:

$$(1 - e \cos E)\dot{E} = n \tag{4.51}$$

where n is the *mean motion*.

Integration with respect to time yields Kepler's equation:

$$E(t) - e \sin E(t) = n(t - t_P) \tag{4.52}$$

where t_P denotes the instant at which the satellite passes through the periapsis (which is commonly known as the 'time of periapsis passage'). The quantity on the left-hand side of Kepler's equation is called the *mean anomaly* (M):

$$M = E - e \sin E \tag{4.53}$$

It traverses 360° in each revolution and, unlike the true and eccentric anomalies, it increases uniformly with time. A computational method for solving (4.53) is provided by Algorithm 4.5. This uses an iterative calculation of eccentric anomaly for given values of mean anomaly and orbital eccentricity.

By convention, the time of periapsis passage is substituted for a mean anomaly (M_0) at some reference epoch (t_0) such that, at any other epoch, the mean anomaly is found to be:

$$M = M_0 + n(t - t_0) \equiv n(t - t_P) \tag{4.54}$$

Note that the orbital period is $T = 2\pi/n$ where $n = \sqrt{(GM/a^3)}$, which reasserts Kepler's Third Law.

```
function E = kepler(M,e)
M = mod(M,2*pi);

E = pi;
tolerance = 0.0001;
converged = false;
while ~converged
 func = E - e*sin(E) - M;
 grad = 1 - e*cos(E);
 E = E - func/grad;
 converged = abs(func)<=tolerance;
end
```

Algorithm 4.5 'kepler'.

4.7.4   *Orbital Elements*

From the basic properties of planar orbits, it is straightforward step to reorientate the orbital plane is three-dimensional space and then reorientate the orbit within that plane. The full definition is contained in six *orbital elements*:

1. Semi-major axis (a)
2. Eccentricity (e)
3. Right Ascension of the Ascending Node (Ω)
4. Inclination of the Orbital Plane (i)
5. Argument of Periapsis (ω)
6. Time of Periapsis Passage (t_P) or Mean Anomaly at some reference epoch (M_0) or True Anomaly at some reference epoch (v_0)

For a satellite in Earth orbit, the orbital geometry is visualised in Figure 4.21. The actual parameters in this example were chosen as follows:

1. a = 5 Earth equatorial radii
2. e = 0.5
3. $\Omega = \tilde{4}°$
4. i = 25°
5. $\omega = 120°$
6. $v_0 = 140°$

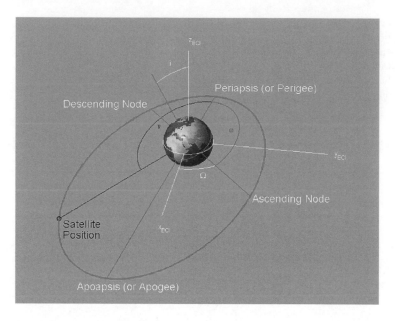

Figure 4.21 Satellite in Earth orbit.

The semi-major axis and the eccentricity define the orbital shape. Right ascension is defined in this book with respect to the J2000 equinox but different source material may opt for a different equinox (which would require appropriate compensation prior to any calculation). The right ascension of the ascending node marks the direction from the centre of the central body at which the satellite crosses the equator from south to north; the opposite direction points to the descending node and the line joining the two nodes is called simply the *Line of Nodes*. The orbital plane is rotated about this line, executing a positive rotation in the direction of the ascending node; the angle of rotation is called the *inclination*.

With the plane of the orbit now fixed, the orbit is rotated so that the line of apsides lies in the required direction. This direction is specified by an angle drawn from the ascending node to the periapsis and this angle is called the *Argument of Periapsis*. All that is needed to complete the orbit definition is to the satellite position at some reference epoch (usually expressed as a mean anomaly) or, equivalently, the epoch at which the satellite occupies a predefined position (usually the time of periapsis passage). With this information, the satellite can be tracked in its orbit and its position can be calculated in Earth-centred coordinates, noting that progress of the orbit must be expressed as a true anomaly. Note that the parameter $u = \omega + v$ is known as the *Argument of Latitude*.

Applying the xy-coordinate frame shown in Figure 4.22 (which is the same as Figure 4.17 except that the focus is Earth ($\mathbf{E}$) angular displacement is formally shown as true anomaly v), the satellite position in the orbital plane at any epoch is:

$$\mathbf{r}_{orbit} = \begin{pmatrix} r\cos v \\ r\sin v \\ 0 \end{pmatrix} \tag{4.55}$$

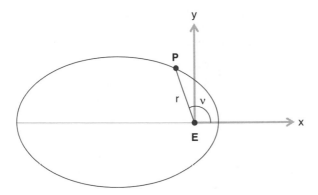

Figure 4.22 Orbital position.

Applying elementary rotations, the satellite position can be re-expressed in Earth equatorial coordinates, with respect to the ECI frame, as follows:

$$\mathbf{r}_{ECI} = R_z(-\Omega)R_x(-i)R_z(-\omega)\mathbf{r}_{orbit} \tag{4.56}$$

This was the method employed in order to generate the visualisation shown in Figure 4.21.

4.8   Solar System Orbit Models

As an example of how to process real orbit data, this section will determine the approximate positions of the major planets and the Moon. In the context of the navigational environment this provides the basis for calculating the positions of the Sun, Moon and Venus[26] relative to the Earth.

One of the few newsworthy developments at any IAU meeting occurred during the writing of this book. IAU General Assembly XXVI in Prague (2006) agreed a rigorous definition of the term 'planet'. This became necessary because of the huge number of objects that had been discovered since 1990 beyond the orbit of Neptune (in a region known as the Kuyper Belt). These spawned a variety of descriptions (e.g. Trans-Neptunian Object (TNO), Kuyper Belt Object (KBO)) and questioned whether Pluto was sufficiently distinct from other small objects to justify its original designation as the 'ninth planet'.

Resolution 5A at that meeting established the following definitions:

- A 'planet' is a celestial body that (a) is in orbit around the Sun, (b) has sufficient mass for its self-gravity to overcome rigid body forces so that it assumes a hydrostatic equilibrium (nearly round) shape, and (c) has cleared the neighbourhood around its orbit.
- A 'dwarf planet' is a celestial body that (a) is in orbit around the Sun, (b) has sufficient mass for its self-gravity to overcome rigid body forces so that it assumes a hydrostatic equilibrium

---

[26] Note that the planet Venus is often a conspicuous object in the morning or evening sky.

(nearly round) shape, (c) has not cleared the neighbourhood around its orbit, and (d) is not a satellite.

- All other objects except satellites orbiting the Sun shall be referred to collectively as 'Small Solar-System Bodies'.

The practical consequence of this resolution (which is why it became news-worthy) was that Pluto was reclassified as a 'dwarf planet', amid some controversy.

4.8.1   *Planetary Data*

Planetary data available from the JPL Solar System Dynamics group[27] is expressed using the following orbital elements (cf. Section 4.7.4):

1. Semi-major axis (a)
2. Eccentricity (e)
3. Inclination (i)
4. Mean Longitude (L)
5. Longitude of Perihelion (ϖ)
6. Longitude of the Ascending Node (Ω)

Here, *Mean Longitude* (L) is defined as an alternative to mean anomaly by setting the inclination of the orbital plane were zero:

$$L = M + \Omega + \omega \qquad (4.57)$$

where M is the mean anomaly, Ω is the longitude of the ascending node and ω is the argument of periapsis (or, in this case, perihelion). The equivalent longitude of periapsis is:

$$\varpi = \Omega + \omega \qquad (4.58)$$

which allows the mean longitude to be written more concisely, in the form:

$$L = M + \varpi \qquad (4.59)$$

Data can be readily extracted for the major planets, with the Earth and Moon system being treated as a combined system located at its barycentre. Table 4.2 reproduces the particular data set for the period AD 1900–2050, with values for each of the orbital elements at J2000 and their rates of change (otherwise called *secular change*). The reference frame is the mean ecliptic and equinox of J2000. Distances are measured in astronomical units (AU), where

$$1\text{AU} = 14\_95978\_70691 \text{ m} \qquad (4.60)$$

This is the mean distance between the Earth and the Sun.

The value of each orbital element (using the generic symbol σ) at a given epoch is:

$$\sigma = \sigma_{J2000} + \dot{\sigma}T \qquad (4.61)$$

---

[27] ssd.jpl.nasa.gov.

Table 4.2 Orbital elements for major planets: AD 1900–2050.

	a AU, AU/Cy	e [], []/Cy	i deg, deg/Cy	L deg, deg/Cy	ϖ deg,deg/Cy	Ω deg, deg/Cy
Venus	0.38709927	0.20563593	7.00497902	252.25032350	77.45779628	48.33076593
	0.00000037	0.00001906	−0.00594749	149472.67411175	0.16047689	−0.12534081
Venus	0.72333566	0.00677672	3.39467605	181.97909950	131.60246718	76.67984255
	0.00000390	−0.00004107	−0.00078890	58517.81538729	0.00268329	−0.27769418
EM	1.00000261	0.01671123	−0.00001531	100.46457166	102.93768193	0.0
	0.00000562	−0.00004392	−0.01294668	35999.37244981	0.32327364	0.0
Mars	1.52371034	0.09339410	1.84969142	−4.55343205	−23.94362959	49.55953891
	0.00001847	0.00007882	−0.00813131	19140.30268499	0.44441088	−0.29257343
Jupiter	5.20288700	0.04838624	1.30439695	34.39644051	14.72847983	100.47390909
	−0.00011607	−0.00013253	−0.00183714	3034.74612775	0.21252668	0.20469106
Saturn	9.53667594	0.05386179	2.48599187	49.95424423	92.59887831	113.66242448
	−0.00125060	−0.00050991	0.00193609	1222.49362201	−0.41897216	−0.28867794
Uranus	19.18916464	0.04725744	0.77263783	313.23810451	170.95427630	74.01692503
	−0.00196176	−0.00004397	−0.00242939	428.48202785	0.40805281	0.04240589
Neptune	30.06992276	0.00859048	1.77004347	−55.12002969	44.96476227	131.78422574
	0.00026291	0.00005105	0.00035372	218.45945325	−0.32241464	−0.00508664

where secular change is defined per century and T is the number of centuries since J2000. For epochs reasonably close to J2000, the argument of perihelion (ω) and mean anomaly (M) are found from (4.58) and (4.59):

$$\omega = \varpi - \Omega \tag{4.62}$$

$$M = L - \varpi \tag{4.63}$$

where $-\pi < M \le \pi$. Then, the eccentric anomaly is derived from (4.53):

$$M = E - e \sin E$$

The heliocentric coordinates for each planet, using an xy-coordinate orbital frame similar to that shown in Figure 4.22, are calculated:

$$\mathbf{r}_{\text{orbit}} = \begin{pmatrix} r \cos v \\ r \sin v \\ 0 \end{pmatrix} = \begin{pmatrix} a(\cos E - e) \\ a \sin E \sqrt{1 - e^2} \\ 0 \end{pmatrix} \tag{4.64}$$

These can then be transformed into coordinates relative to the ecliptic:

$$\mathbf{r}_{\text{ecliptic}} = R_z(-\Omega) R_x(-i) R_z(-\omega) \ \mathbf{r}_{\text{orbit}} \tag{4.65}$$

The coordinates can be transformed to an Earth-centred equatorial frame based on the radius vector of Earth and obliquity of the ecliptic at the given epoch:

$$\mathbf{r}_{\text{equator}} = R_x(-\varepsilon)(\mathbf{r}_{\text{ecliptic}} - \mathbf{r}_{\text{earth}}) \tag{4.66}$$

where

$$\varepsilon = 23.4392911° - 46.8150'T - 0.00059'T^2 + 0.001813'T^3 \tag{4.67}$$

and $\mathbf{r}_{\text{earth}}$ is the position of the Earth in ecliptic coordinates.

Table 4.3 Orbital elements for the Moon.

	a km	e	i deg	M deg, deg/Cy	ω deg, deg/Cy	Ω deg, deg/Cy
Moon	384400.0	0.0554	5.16	135.27 478983.35	318.15 6003.00	125.08 −1935.48
	a AU	e	i deg	L deg, deg/Cy	ω deg, deg/Cy	Ω deg, deg/Cy
Moon	0.00256956	0.0554	5.16	218.50 478983.87	443.23 4067.52	125.08 −1935.48

JPL also provides orbital elements for planetary satellites. Interestingly, these data do not have the same format as that used for planetary data. Applying the appropriate conversions, the orbital elements of the Moon are summarised in Table 4.3. Note that regression of the ascending node gives a cycle of the orbital pole every 18.6 years; also, precession of the argument of periapsis gives a cycle of just less than six years.

In order to complete the dataset, the gravitational parameters (GM) for the main objects of interest in the Solar System are given in Table 4.4. These are taken from DE403[28] (Standish *et al.*, 1995). For comparison, some alternative values are provided from WGS84, LP150Q (Roncoli, 2005; Konopliv *et al.*, 2001); these are presented in Table 4.5. Incidentally, the approximate gravitational constant is defined as $G \approx 6.67 \times 10^{11} \text{ m}^3\text{kg}^{-1}\text{s}^{-2}$.

Table 4.4 Gravitational parameters for the Sun and major planets.

Object	GM (m3s2)
Sun	$13271244001.7987 \times 10^{10}$
Mercury	$2203.2080486 \times 10^{10}$
Venus	$32485.8598826 \times 10^{10}$
Earth	$39860.0435608 \times 10^{10}$
Mars[*]	$4282.8314258 \times 10^{10}$
Jupiter[*]	$12671276.7857796 \times 10^{10}$
Saturn[*]	$3794062.6061137 \times 10^{10}$
Uranus[*]	$579454.9007072 \times 10^{10}$
Neptune[*]	$683653.4063879 \times 10^{10}$
Moon	$490.2799108 \times 10^{10}$

[*] includes the gravitational parameters for planetary satellites

Table 4.5 Alternative gravitational parameters for the Earth and Moon.

Object	GM (m^3s^2)
Earth	$39860.0435608 \times 10^{10}$
Earth (WGS84)	$39860.04418 \times 10^{10}$
Moon	$490.2799108 \times 10^{10}$
Moon (LP150Q)	$490.2801076 \times 10^{10}$
Moon (Konopliv et al.)	$490.28000 \pm 0.00003 \times 10^{10}$

---

[28] DE403/LE403 (Development Ephemeris 403/Lunar Ephemeris 403); usually referred to as 'DE403'.

4.8.2   *Planetary Trajectories*

Following the sequence of calculation developed in Section 4.8.1, from (4.61) to (4.67), it is possible to build a computational model of solar system dynamics based on the Keplerian orbital elements given in Table 4.2. This is a two-part process; the first part establishes the heliocentric coordinates of designated objects relative to the ecliptic (4.65) and the second part establishes geocentric coordinates relative to EME2000.

The first calculation is implemented in Algorithm 4.6, utilising date Algorithms 4.1 and 4.3. In this code, orbital elements are held in 'solarSystemObject', which is declared as *persistent* so that it is initialised once (by Algorithm 4.7) and thereafter is available for all subsequent orbital calculations. This generates the planetary positions shown in Figure 4.23. Necessarily, because of the distances involved, separate figures are needed for the inner and outer planets. Two views are given in each case, perpendicular (giving a top-down projection on to the ecliptic) and oblique (revealing slight differences in orientation for each orbital plane).

The second calculation is implemented in Algorithm 4.8. In essence this performs the calculation specified by (4.66) and (4.67). For completeness, it implements precessional components for equator and ecliptic, as specified by (4.18). Also, it reformats the positional data for subsequent processing and removes the combined 'EarthMoon' system. This algorithm uses the heliocentric coordinates produced by Algorithm 4.6 and produces equivalent coordinates relative to the ECI reference frame (cf. Section 4.6.1).

The application of these algorithms is illustrated in Example 4.5. This calculates planetary positions for the start of 23 July 2006, with respect to the ECI frame, and derives the radius

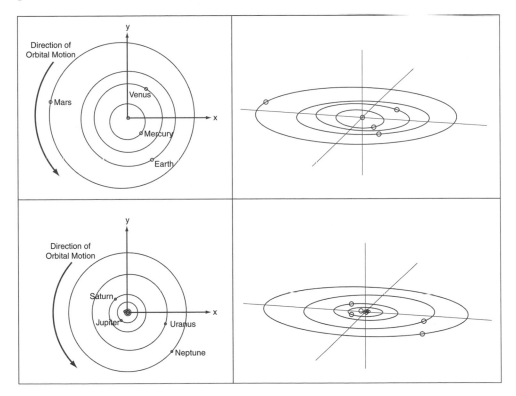

Figure 4.23 Orbital trajectories and positions of planets for 23 July 2006.

vector for each object, with respect to the Earth's centre. Each vector is expressed as a radial distance (r), declination (δ) and right ascension (RA), according to the following equations:

$$r = \sqrt{x^2 + y^2 + z^2}$$

$$\delta = \tan^{-1}\left(\frac{z}{w}\right) \tag{4.68}$$

$$RA = \tan^{-1}\left(\frac{y}{x}\right)$$

where

$$w = \sqrt{x^2 + y^2} \tag{4.69}$$

```
function xyz = getSolarSystemObjectPositions(calendarDate)
persistent solarSystemObject

2if isempty(solarSystemObject)
 solarSystemObject = defineSolarSystemObjects;
end

T = getJulianCenturies(getJulianDate(calendarDate));

elementName = {'a','e','i','meanLongitude','periapsis','node'};

nElements = length(elementName);
nObjects = length(solarSystemObject.name);
for iObject = 1:nObjects
 thisObject = solarSystemObject.name{iObject};
 elementArray = solarSystemObject.element;
 j2000 = 2*iObject - 1;
 jRate = j2000 + 1;

 for iElement = 1:nElements
 thisElement = elementName{iElement};
 thisValue = ...
 elementArray(j2000,iElement) + elementArray(jRate,iElement)*T;
 switch thisElement
 case {'a','e'}
 orbit.(thisElement) = thisValue;
 otherwise
 orbit.(thisElement) = radians(thisValue);
 end
 end

 orbit.omega = orbit.periapsis - orbit.node;
 orbit.M = orbit.meanLongitude - orbit.periapsis;
 orbit.E = kepler(orbit.M,orbit.e);

 orbitPosition = ...
 orbit.a*( cos(orbit.E)-orbit.e; sin(orbit.E)*sqrt(1-orbit.e^2); 0 );
 xyz.(thisObject) = ...
 Rz(-orbit.node)*Rx(-orbit.i)*Rz(-orbit.omega)*orbitPosition;
end
```

```
gm.Sun = 1 ;
gm.Earth = 3.003489664917443e-006 ;
gm.Moon = 3.694303261499454e-008 ;
barycentre = gm.Earth/(gm.Earth + gm.Moon) ;

xyz.Earth = xyz.EarthMoon - xyz.Moon*(1-barycentre) ;
xyz.Moon = xyz.EarthMoon + xyz.Moon*barycentre;
xyz.Sun = ( 0; 0; 0 ) ;
```

Algorithm 4.6   'getSolarSystemObjectPositions'.

```
function solarSystemObject = defineSolarSystemObjects

object.element = ( ...
 0.38709927, 0.20563593, 7.00497902, 252.25032350, 77.45779628,
  48.33076593;
 0.00000037, 0.00001906,-0.00594749,149472.67411175, 0.16047689,
  -0.12534081 ;
 0.72333566, 0.00677672, 3.39467605, 181.97909950, 131.60246718,
  76.67984255 ;
 0.00000390,-0.00004107,-0.00078890, 58517.81538729, 0.00268329,
  -0.27769418 ;
 1.00000261, 0.01671123,-0.00001531, 100.46457166, 102.93768193, 0.0 ;
 0.00000562,-0.00004392,-0.01294668, 35999.37244981, 0.32327364, 0.0 ;
 1.52371034, 0.09339410, 1.84969142, -4.55343205, -23.94362959,
  49.55953891 ;
 0.00001847, 0.00007882,-0.00813131, 19140.30268499, 0.44441088,
  -0.29257343 ;
 5.20288700, 0.04838624, 1.30439695, 34.39644051, 14.72847983,
  100.47390909 ;
-0.00011607,-0.00013253,-0.00183714, 3034.74612775, 0.21252668,
  0.20469106 ;
 9.53667594, 0.05386179, 2.48599187, 49.95424423, 92.59887831,
  113.66242448 ;
-0.00125060,-0.00050991, 0.00193609, 1222.49362201, -0.41897216,
  -0.28867794 ;
 19.18916464, 0.04725744, 0.77263783, 313.23810451, 170.95427630,
  74.01692503 ;
-0.00196176,-0.00004397,-0.00242939, 428.48202785, 0.40805281,
  0.04240589 ;
 30.06992276, 0.00859048, 1.77004347, -55.12002969, 44.96476227,
  131.78422574 ;
 0.00026291, 0.00005105, 0.00035372, 218.45945325, -0.32241464,
  -0.00508664 ;
 0.00256956, 0.05540000, 5.16000000, 218.50000000, 443.23000000,
  125.0800000 ;
 0.0 , 0.0 , 0.0 ,478983.87000000,4067.52000000,-1935.48000000 ) ;
object.name = { ...
'Mercury','Venus','EarthMoon','Mars','Jupiter','Saturn','Uranus','
  Neptune','Moon'} ;
```

Algorithm 4.7   'defineSolarSystemObjects'.

```
function eci = getEarthCentredObjectPositions(xyz)

T = getJulianCenturies(getJulianDate(calendarDate)) ;
ecliptic2EME = getPrecession(T) ;

objectList = fieldnames(xyz) ;
nObjects = length(objectList) ;

for iObject = 1:nObjects
 thisObject = objectList{iObject} ;
 switch thisObject
 case 'EarthMoon'
 otherwise
 dr = xyz.(thisObject) - xyz.Earth ;
 eci.position(1:3,iObject) = ecliptic2EME*dr ;
 eci.object{iObject} = thisObject ;
 end
end
```

Algorithm 4.8 'getEarthCentredObjectPositions'.

Example 4.5

```
>> degrees = @(radians) radians*180/pi ;
>>
>> xyz = getSolarSystemObjectPositions((2006 7 23)) ;
JD 2453939.5
>> eci = getEarthCentredObjectPositions(xyz) ;
>> x = eci.position(1,:) ;
>> y = eci.position(2,:) ;
>> z = eci.position(3,:) ;
>>
>> radius = sqrt(x.^2+y.^2+z.^2)
radius =
1.0160 0.6107 1.4945 2.4610 5.1440 10.1287 19.3441 29.0900 0.0027
>>
>> equatorialRadius = sqrt(x.^2+y.^2) ;
>> latitude = degrees(atan2(z,equatorialRadius))
latitude =
20.1924 16.8402 22.8574 12.5002 -13.6332 17.7888 -6.9624 -15.3931
-25.7819
>> longitude = degrees(atan2(y,x))
longitude =
-57.8729 -66.4189 -85.0032 30.5293 -27.4627 37.3065
-44.6505 165.7704 141.3451
```

The end result is given in Table 4.6 and in Figure 4.24. For reference, the position of the Greenwich Meridian is shown, as given by (4.5):

$$RA_G = 280.147 + \Omega D$$

Table 4.6 Earth-centred position data for 23 July 2006.

	Sun	Mercury	Venus	Moon	Mars	Jupiter	Saturn	Uranus	Neptune
r	1.0159	0.6106	1.4945	0.0025	2.4610	5.1440	10.1286	19.3442	29.0900
δ	20.1924	16.8402	22.8574	−25.7819	12.5002	−13.6332	17.7888	−6.9624	−15.3931
RA	122.1253	113.5779	94.9952	297.4425	152.5370	217.3068	135.3494	345.7704	321.3451

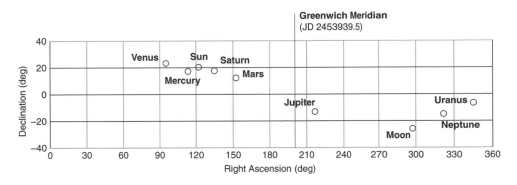

Figure 4.24 Celestial positions of planets for 23 July 2006.

with respect to the J2000 equinox, where D is the number of days after J2000. Note that two Earth rotation rates were quoted earlier (the GRS67 value in (4.6) and the GRS80/WGS84 value in (2.56)):

$$\Omega = 360.9856235 \text{ deg/day (GRS67)}$$

$$\Omega = 360.98561 \text{ deg/day (GRS80/WGS84)}$$

Here, the chronological time-scale is defined by:

$$D = JD - 2451545.0$$

Thus, the position of the Greenwich Meridian on 23 July 2006 is calculated as:

$$RAG = 20.075471° \text{ (using the GRS67 rotation rate)}$$

$$RAG = 20.043145° \text{ (using the GRS80/WGS84 rotation rate)}$$

This serves to illustrate the nuisance of having slightly different standards with which to work but, equally, it shows that the numerical error is so slight as to be of no real concern in most aerospace applications (i.e. in this case, approximately 0.03° in 6.5 years).

Repeating the entire calculation for 18 August 2006 produces a new set of position data, showing the proper motion of these objects; the result is given in Table 4.7 and in Figure 4.25. Not surprisingly, objects closer to Earth show greater motion over time. In order to put these numbers into context, a daily update has been calculated for the Sun, the Moon and Venus (as the most prominent planet viewed from Earth). This covers virtually a full orbit of the Moon

Table 4.7 Earth-centred position data for 19 August 2006.

	Sun	Mercury	Venus	Moon	Mars	Jupiter	Saturn	Uranus	Neptune
r	1.0121	1.1888	1.6050	0.0025	2.5540	5.5556	10.1417	19.1176	29.0489
δ	12.9367	18.0552	19.0474	−27.0093	5.9794	−14.4527	16.8161	−7.3047	−15.6226
RA	148.0664	135.4371	130.0234	290.3571	168.4052	219.5163	138.8176	344.9703	320.6325

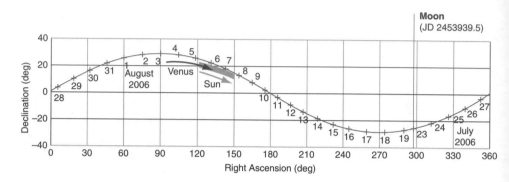

Figure 4.25 Proper motion of Sun, Moon and Venus.

around the Earth and shows the migration across the celestial sphere that would have to be replicated in aerospace applications, for navigational or observational purposes.

4.8.3   *Mean Sun and the Equation of Time*

Recalling Section 4.3.1, the discussion of mean solar time was based on the definition of a mean Sun that moves around the equator with constant speed. The difference between the apparent Sun and the mean Sun is determined by the so-called *Equation of Time*. Having developed the theory of orbital mechanics and having applied it to the planets in the Solar System, it is now possible to construct an algorithm that will evaluate the equation of time, as was shown in Figure 4.8.

In essence, the position of the Earth in its orbit around the Sun provides the basis for calculating the right ascension of the Sun. The orbital elements given in Table 4.2 enable this parameter to be tracked as a function of mean longitude, which defines the progression around a hypothetical circular orbit in the equatorial plane of the Earth. The difference between mean longitude and right ascension leads directly to the equation of time, noting that an *angular* difference is equivalent to a *time* difference in solar observation.

The actual computation is organised in two parts, the first being the calculation of the Sun's parameters (Algorithm 4.9) and the second being the numerical and graphical tasks associated with plotting the 'equation of time' (Algorithm 4.10). This process makes use of Algorithm 4.5. The end result is shown in Figure 4.26.

Note that the orbital elements provided in Table 4.2 actually refer to the combined Earth-Moon system, rather than the Earth, and so the calculation ignores the slight oscillations caused by the Moon in its orbit around the Earth.

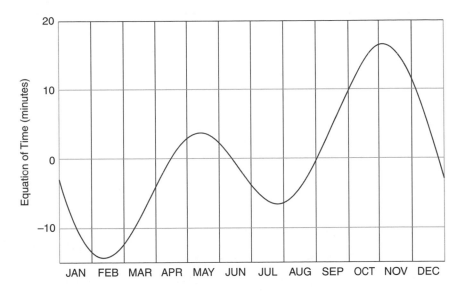

Figure 4.26 Equation of time.

```
function sun = getSunParameters(calendarDate)

radians = @(degrees) degrees*pi/180 ;
Rx = @(x) ( 1 0 0; 0 cos(x) sin(x); 0 -sin(x) cos(x)) ;
Ry = @(y) ( cos(y) 0 -sin(y); 0 1 0; sin(y) 0 cos(y)) ;
Rz = @(z) ( cos(z) sin(z) 0; -sin(z) cos(z) 0; 0 0 1) ;

AU = 149597870691 ;
GM = 13271244001.7987e10 ;

element = ( ...
 1.00000261, 0.01671123, -0.00001531, 100.46457166, 102.93768193, 0.0 ;
 0.00000562, -0.00004392, -0.01294668, 35999.37244981, 0.32327364, 0.0 ) ;
elementName = { ...
 'a','e','i','meanLongitude','periapsis','node'}
nElements = length(elementName) ;

JD = getJulianDate(calendarDate) ;
JD2000 = 2451545 ;
julianCentury = 36525 ;

sun.year = 1:365 ;

T = (JD - JD2000 + sun.year - 1)/julianCentury ;
ecliptic = ...
 radians(23.4392911 - (46.8150*T + 0.00059*T.^2 - 0.001813*T.^3)/3600) ;
```

Continued

```
zeta = radians((2306.2181*T + 0.30188*T.^2 + 0.017998*T.^3)/3600) ;
theta = radians((2004.3109*T - 0.42665*T.^2 -0.041833*T.^3)/3600) ;
z = zeta + radians((0.79280*T.^2 + 0.000205*T.^3)/3600) ;

for iDay = sun.year
 ecliptic2EME = ...
 Rz(-pi + z(iDay))*Rx(theta(iDay))*Rz(pi-zeta(iDay))*Rx(-
 ecliptic(iDay));

 for iElement = 1:nElements
 thisElement = elementName{iElement} ;
 thisValue = ...
 element(1,iElement) + element(2,iElement)*T(iDay) ;
 switch thisElement
 case 'e'
 orbit.(thisElement) = thisValue ;
 case 'a'
 orbit.(thisElement) = thisValue*AU ;
 otherwise
 orbit.(thisElement) = radians(thisValue) ;
 end
 end

 orbit.omega = orbit.periapsis - orbit.node ;
 orbit.M = orbit.meanLongitude - orbit.periapsis ;
 orbit.E = kepler(orbit.M,orbit.e) ;
 orbit.b = orbit.a*sqrt(1-orbit.e^2) ;

 orbitPosition = ...
 ( orbit.a*(cos(orbit.E)-orbit.e); orbit.b*sin(orbit.E); 0 ) ;
 earthPosition = ...
 Rz(-orbit.node)*Rx(-orbit.i)*Rz(-orbit.omega)*orbitPosition ;
 sunPosition = -ecliptic2EME*earthPosition ;

 x = sunPosition(1,:) ;
 y = sunPosition(2,:) ;
 z = sunPosition(3,:) ;
 w = sqrt(x.^2 + y.^2) ;

 sun.distance(iDay) = sqrt(w.^2 + z.^2) ;
 sun.declination(iDay) = atan2(z,w) ;
 sun.rightAscension(iDay) = atan2(y,x) ;
 sun.meanLongitude(iDay) = pi-mod(-m,2*pi) ;
end
```

Algorithm 4.9 'getSunParameters'.

```
function plotEqnOfTime(sun)
degrees = @(radians) radians*180/pi ;
radians = @(degrees) degrees*pi/180 ;
```

```
meanLongitude = degrees(unwrap(sun.meanLongitude));
rightAscension = degrees(unwrap(sun.rightAscension));

eqt = 4*(meanLongitude - rightAscension);
plot(sun.year,eqt,'k');
ylabel('Equation of Time (minutes)');

set(gca, ...
 'GridLineStyle','-', ...
 'XTick',(0 31 59 90 120 151 181 212 243 273 304 334 ), ...
 'XTickLabel',( ...
 'JAN' ; 'FEB' ; 'MAR' ;
 'APR' ; 'MAY' ; 'JUN' ;
 'JUL' ; 'AUG' ; 'SEP' ;
 'OCT' ; 'NOV' ; 'DEC' ), ...
 'Xcolor',(1 1 1)*0.6, ...
 'Xgrid','on', ...
 'Ytick',(-20 -10 0 10 20), ...
 'Ycolor',(1 0.5 0.5), ...
 'Ygrid','on', ...
 'XLim',(0 365))
```

Algorithm 4.10 'plotEqnOfTime'.

4.9   GPS Orbit Models

As a further example of real orbit data, this section will consider satellite vehicles in the Global Positioning System (GPS). These are available in two representations, namely Almanac and Ephemeris models, which support position fixes of low accuracy and high accuracy respectively. Further information can be found in the Navstar GPS Interface Control Document (ICD) (ICD-GPS-200, 1997). Incidentally, the following constants are specified in the GPS ICD:

$$\omega = 7.2921151467 \times 10^{-5} \text{ rad.s}^{-1} \tag{4.70}$$

$$GM = 39860.05 \times 10^{10} \text{ m}^3\text{s}^{-2} \tag{4.71}$$

Note that these are WGS84 constants that predate the latest release (NIMA, 2004) as stated in (2.58) and (5.25) (also stated in Table 4.5), respectively. These values have been retained because it was totally impracticable to recall the hundreds of thousands of GPS receivers around the world, just to implement a relatively tiny change in the value of a constant (no matter how fundamental).

Time is measured according to GPS system time (cf. Section 4.3.3), with an origin defined as the standard epoch 1980 January 6.0 GPS (i.e. JD(GPS) 2444244.5). From this origin, the GPS week number (WN) can be calculated using Algorithm 4.11 and the start date for a given week can be calculated using Algorithm 4.12. A GPS week starts Sunday 0:00 GPS time (with Week 0 commencing at the origin of GPS time) and, within a GPS week, times are specified in elapsed seconds (up to maximum of 604800).

```
function WN = getGpsWeekNumber (jd_GPS)
gpsTimeOrigin = 2444244.5 ;
WN = floor ((jd_GPS [ndash] gpsTimeOrigin) /7) ;
```

Algorithm 4.11 'getGpsWeekNumber'.

```
function jd_GPS = getGpsWeekStartDate (WN)
gpsTimeOrigin = 2444244.5 ;
jd_GPS = gpsTimeOrigin + 7*WN ;
```

Algorithm 4.12 'getGpsWeekStartDate'.

4.9.1   *GPS Almanac Model*

The GPS almanac message from each satellite comprises eleven parameters, including seven orbital parameters (cf. Section 4.7.4), as follows:

1. Almanac Reference Epoch (part 1): GPS Week Number (*modulo 256*)
2. Almanac Reference Epoch (part 2): Fraction of current GPS Week (*in seconds*)
3. Square root of semi-major axis ($\sqrt{a}$)
4. Eccentricity (e)
5. Offset inclination (Δi) from reference value $i_{ref} = 0.3\pi$
6. Longitude of the Ascending Node relative to the weekly epoch (Ω_0)
7. Rate of change of the Ascending Node ($d\Omega/dt$)
8. Argument of Perigee (ω)
9. Mean Anomaly at reference epoch (M_0)
10. Satellite clock offset from GPS time (a_0) (*in seconds*)
11. Clock frequency offset (a_1) (*in seconds per second*)

The unit of angular measurement is a semi-circle (i.e. π radians). The reference value for orbital inclination is set at 0.3π radians (i.e. $54°$), which is slightly less than the nominal inclination for satellites in the GPS constellation. In a significant departure from standard orbital elements, the ascending node is defined with respect to the Greenwich meridian at the start of the GPS week; so the corresponding right ascension is:

$$\Omega(t_a) = \Omega_0 + \Theta(t_0) \qquad (4.72)$$

where Ω is Right Ascension, Θ is Greenwich Sidereal Time and the two epochs are start of the GPS week (t_0) and the almanac reference epoch (t_a). Note that the almanac reference epoch is specified in two parts, i.e. the GPS Week Number plus the fraction of the current GPS week (in seconds).

The secular movement in the ascending node is specified, enabling the calculation of its longitude at any instant:

$$\lambda(t) = \Omega(t) - \Theta(t)$$

$$\lambda(t) \approx \Omega(t_a) + (d\Omega/dt)(t - t_a) - \Theta(t_0) - \omega(t - t_0) \qquad (4.73)$$

$$\lambda(t) \approx \Omega_0 + (d\Omega/dt)(t - t_a) - \omega(t - t_0)$$

The eccentric anomaly at any instant is derived from Kepler's Equation (cf. 4.52):

$$E(t) - e \sin E(t) = M_0 + n(t - t_a) \qquad (4.74)$$

where n is the mean motion:

$$n = \sqrt{\frac{GM}{a^3}} \qquad (4.75)$$

The geocentric coordinates for the particular satellite, using an xy-coordinate orbital frame similar to that shown in Figure 9.4, are calculated:

$$\mathbf{r}_{orbit} = \begin{pmatrix} r \cos v \\ r \sin v \\ 0 \end{pmatrix} = \begin{pmatrix} a(\cos E - e) \\ a \sin E \sqrt{1 - e^2} \\ 0 \end{pmatrix} \qquad (4.76)$$

These can then be transformed into the ECEF frame, as follows:

$$\mathbf{r}_{ECEF} = R_z(-\lambda)R_x(-i_{ref} - \Delta i)R_z(-\omega)\mathbf{r}_{orbit} \qquad (4.77)$$

4.9.2   *GPS Ephemeris Model*

The GPS ephemeris model is essentially the same as the GPS almanac model, except that it provides correction terms that enable a much more accurate description of satellite trajectory around its orbital path.

The GPS ephemeris message from each satellite comprises 18 parameters, including 13 orbital parameters (cf. Section 4.7.4), as follows:

1. Ephemeris Reference Epoch (part 1): GPS Week Number (modulo 1024)
2. Ephemeris Reference Epoch (part 2): Fraction of current GPS Week (in seconds)
3. Square root of semi-major axis ($\sqrt{a}$)
4. Eccentricity (e)
5. Correction to mean motion (Δn)
6. Inclination at reference epoch (i_0)
7. Rate of change of inclination (di/dt)
8. Longitude of the Ascending Node relative to the weekly epoch (Ω_0)
9. Rate of change of the Ascending Node ($d\Omega/dt$)
10. Argument of Perigee (ω)
11. Mean Anomaly at reference epoch (M_0)
12. Sine correction to orbital radius (in metres) (C_{rs})
13. Cosine correction to orbital radius (in metres) (C_{rc})
14. Sine correction to argument of latitude (in radians) (C_{us})
15. Sine correction to argument of latitude (in radians) (C_{us})
16. Sine correction to inclination (in radians) (C_{is})

17. Sine correction to inclination (in radians) (C_{is})
18. Satellite clock offset from GPS time (a_{f0}) (*in seconds*)
19. Clock frequency offset (a_{f1}) (*in seconds per second*)
20. Clock frequency drift (a_{f2}) (*in seconds per second per second*)

The unit of angular measurement is a semi-circle (i.e. π radians) unless stated otherwise. Two epochs are defined, namely the start of the GPS week (t_0) and the ephemeris reference epoch (t_e). Note that the ephemeris reference epoch is specified in two parts, i.e. the GPS Week Number plus the fraction of the current GPS week (in seconds).

The eccentric anomaly at any instant is derived from Kepler's Equation (cf. 4.52):

$$E(t) - e \sin E(t) = M_0 + n(t - t_e) \tag{4.78}$$

where n is the mean motion:

$$n = \sqrt{\frac{GM}{a^3}} + \Delta n \tag{4.79}$$

Without any correction terms, the geocentric coordinates for the particular satellite, using an xy-coordinate orbital frame similar to that shown in Figure 9.4, are calculated:

$$\mathbf{r}'_{orbit} = \begin{pmatrix} r \cos v \\ r \sin v \\ 0 \end{pmatrix} = \begin{pmatrix} a(\cos E - e) \\ a \sin E \sqrt{1 - e^2} \\ 0 \end{pmatrix} \tag{4.80}$$

from which an uncorrected Argument of Latitude is obtained:

$$u' = \omega + v \tag{4.81}$$

using the Argument of Perigee supplied in the ephemeris message.

The orbital corrections are calculated as follows:

$$\begin{aligned}
\Delta r &= C_{rs} \sin(2u') + C_{rc} \cos(2u') \\
\Delta u &= C_{us} \sin(2u') + C_{uc} \cos(2u') \\
\Delta i &= C_{is} \sin(2u') + C_{ic} \cos(2u')
\end{aligned} \tag{4.82}$$

and then applied as follows:

$$\begin{aligned}
r &= a(1 - e \cos E) + \Delta r \\
u &= u' + \Delta u \\
i &= i_0 + (di/dt)\ (t - t_e) + \Delta i
\end{aligned} \tag{4.83}$$

The longitude of the ascending node is calculated as follows:

$$\lambda(t) \approx \Omega_0 + (d\Omega/dt)(t - t_a) - \omega(t - t_0) \tag{4.84}$$

The satellite position can be transformed into the ECEF frame (cf. Section 3.1):

$$\mathbf{r}_{ECEF} = R_z(-\lambda)\ R_x(-i) \begin{pmatrix} r \cos u \\ r \sin u \\ 0 \end{pmatrix} \tag{4.85}$$

Here the argument of latitude has been applied directly, based on the equivalence:

$$\begin{pmatrix} r\cos u \\ r\sin u \\ 0 \end{pmatrix} = R_z(-\omega) \begin{pmatrix} r\cos v \\ r\sin v \\ 0 \end{pmatrix} \qquad (4.86)$$

4.9.3   *YUMA Almanac*

As an example of published almanac data for GPS satellites, it is useful to review the content and format of a Yuma almanac. This is compiled for each GPS week number (modulo 1024) and is available for free download from the Internet. Typical information is shown in Figure 4.27. This example is the Week 358 almanac for a given satellite in the GPS constellation (referenced as PRN-02). By way of basic configuration data, it contains an identification number (ID), a health status (000 = usable), a week number (358) and a time of applicability or time-stamp (233472 seconds into the week). Given that the GPS week number is counted modulo 1024, Week 0 on this cycle starts at 1999 August 22.0 GPS (i.e. JD(GPS) 2451412.5). The seven orbit-related parameters are those specified within the GPS Almanac Model; likewise the two clock offset values. For illustration, the short-term perturbations in the orbit of PRN-002 from mid June to early July 2006 are shown in Table 4.8.

4.9.4   *Two-Line Elements*

Another popular source of satellite orbit data is available[29] in so-called Two-Line Element (TLE) format. Examples are given in Figure 4.28 (for GPS PRN-02) and

Mean Anom(rad):	-0.2932968270E+001
Af0(s):	0.1049041748E-004
Af1(s/s):	0.3637978807E-011
week:	358

Figure 4.27 Typical GPS almanac data in Yuma format.

Table 4.8 Perturbations in orbital parameters for GPS PRN-002.

Parameter [unit]		WN 356 2006 June 18.5 GPS	WN 357 2006 July 25.5 GPS	WN 358 2006 July 2.5 GPS
e	–	0.8984088898E-002	0.8947849274E-002	0.8935928345E-002
i	rad	0.9502615502	0.9500398420	0.9500158735
$d\Omega/dt$	rad/s	−0.7634603726E-008	−0.8091765626E-008	−0.8023191341E-008
$\sqrt{a}$	$m^{1/2}$	5153.651367	5153.639648	5153.6474610
Ω_0	rad	−0.7547515043E-001	−0.2007421338E+000	−0.3232152947E+000
ω	rad	2.092907061	2.097881638	2.099749305
M_0	rad	−0.1714383070E+001	−0.1469815750E+001	−0.2932968270E+001

---

[29] celestrak.com/NORAD/elements/master.asp.

GPS BIIR-13 (PRN 02)

1 28474U 04045A   06203.21992904 .00000025 00000-0 10000-3 0 6948

2 28474 54.4087 260.5992 0087501 121.3495 239.5887 2.00560373 12612

Figure 4.28 Two-line element data for GPS PRN-02.

ISS (ZARYA)

1 25544U 98067A   06204.17674769 .00008060 00000-0 53683-4 0 3547

2 25544 51.6350 2.0624 0010946 105.9460 239.4825 15.76444825438836

Figure 4.29 Two-line element data for the International Space Station

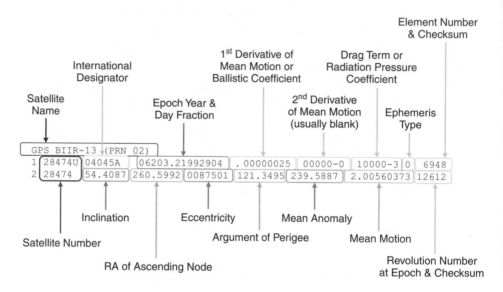

Figure 4.30 Two-line element format.

Figure 4.29 (for the International Space Station); the reference epoch for these data is 23 July 2006.

An explanation of the format is presented in Figure 4.30. The main information is essentially the same as would be provided by any standard almanac and this is of most interest here. Inclination, RA of Ascending Node, Eccentricity and Argument of Perigee, collectively, define the shape and orientation of the satellite orbit. Mean Anomaly is defined with respect to the reference Epoch; the Epoch is declared by the last two digits of the respective Year, the day number and the fraction of that day (i.e. '06203.21992904' translates to 5:16:41.87AM on 23 July 2003). The mean anomaly at another epoch is determined from the mean anomaly at the reference epoch, the time difference between the two epochs and the Mean Motion (declared as the number of complete orbits in one solar day or 86400s). The semi-major axis is not stated explicitly but is implied by the mean motion (n):

$$T = \frac{2\pi}{n} \qquad (4.87)$$

where (from (4.75)):

$$n = \sqrt{\frac{GM}{a^3}}$$

The semi-major axis can be derived using the formula:

$$a = \sqrt[3]{GM \left(\frac{T}{2\pi}\right)^2} \qquad (4.88)$$

This can be rearranged in order to reproduce (4.37):

$$T = 2\pi \sqrt{\frac{a^3}{GM}}$$

4.9.5   *GPS Constellation*

The GPS constellation is based on a nominal distribution of 24 satellites in six orbit planes (separated by 60°), with several spares. The inclination is each orbital is approximately 55°. Using the YUMA almanac for Week 358, the actual distribution can be determined as shown in Figure 4.31, where the allocation of satellites to orbit planes is given by GPS Operational Advisory 204 on GPS Status[30] (23 July 2006). From these data, the following mean values are established:

$$a = 26560028 \text{ m}$$
$$e = 0.008254$$
$$i = 54.88114°$$

In addition, the mean Right Ascension of each orbit plane is as follows:

$$A : 159.476$$
$$B : 220.331$$
$$C : 280.218$$
$$D : 341.900$$
$$E : 43.139$$
$$F : 102.551$$

As can be seen, the actual satellite positions conform closely to this distribution.

Incidentally, GPS orbits are almost circular (i.e. with very small eccentricity), which means that mean anomaly is almost exactly the same as true anomaly. Also, the orbital period is chosen to equal half a sidereal day (43082s) (cf. Note 4.2).

---

[30] www.navcen.uscg.gov/Ftp/gps/status.txt.

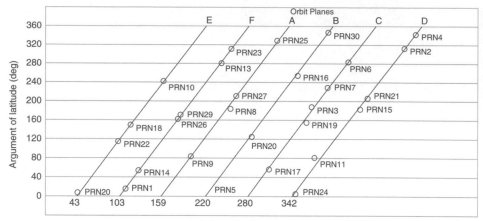

Figure 4.31 GPS constellation positions.

Applying (4.88), the semi-major axis is found to be:

$$a = \sqrt[3]{GM \left(\frac{T}{2\pi} \right)^2} = \left[\sqrt[3]{0.3986004418 \text{ x } 47.01451671} \right] \text{ x } 10^7 = 26561744 \text{ m}$$

where WGS84 defines GM = $39860.04418 \times 10^{10}$ (cf. Table 9.3).

Thus, the orbital radius is approximately 26562 km. Given that the Earth's mean radius is roughly 6371 km, the mean altitude of each satellite is 20191 km, which is usually rounded up to 20200 km for purposes of general discussion.

4.10   Night Sky

In order to provide reference points for celestial navigation, the brightest stars are listed in Table 4.9 for epoch J2000.0 (using data from Norton's Star Atlas (20th edition), Table 48). As a reminder, the Greek alphabet is listed in Table 4.10. A complete listing of the 9110 stars with magnitude greater than 6.5 is given in the Bright Star Catalogue[31] (from the Astronomical Data Center[32] at NASA Goddard Space Flight Center). In theory, these stars are visible to the naked eye although light pollution is known to be a serious problem in urban areas.

Of the many star catalogues that exist, the widely quoted standard for aerospace applications is the 5th Catalogue of Fundamental Stars (FK5) (Fricke *et al.*, 1988). Provided that the source data can be acquired,[33] this catalogue can be accessed using the MATLAB Mapping Toolbox, via a function called 'readfk5'. In its published form, this

---

[31] A convenient source can be found at www.alcyone.de/SIT/bsc/index.html.

[32] Note that the Astronomical Data Center has discontinued maintenance and updates to its website.

[33] A convenient source for FK5 is cdsweb.u-strasbg.fr/viz-bin/ftp-index?/ftp/cats/I/149A.

Table 4.9 Brightest stars.

STAR	NAME	CONSTELLATION	Right Ascension h m	Declination o /	Apparent Magnitude
α CMa	**Sirius**	Canis Major	06 45.2	−16 43	−1.44
α Car	**Canopus**	Carina	06 24.0	−52 42	−0.62
α Cen	**Rigil Centaurus**	Centaurus	14 39.7	−60 50	−0.28
α Boo	**Arcturus**	Boötes	14 15.7	19 11	−0.05
α Lyr	**Vega**	Lyra	18 36.9	38 47	0.03
α Aur	**Capella**	Auriga	05 16.7	46 00	0.08
β Ori	**Rigel**	Orion	05 14.5	−08 12	0.18
α CMi	**Procyon**	Canis Minor	07 39.3	05 14	0.40
α Eri	**Achernar**	Eridanus	01 37.7	−57 14	0.45
α Ori	**Betelgeuse**	Orion	05 55.2	07 24	0.45
β Cen	**Hadar**	Centaurus	14 03.8	−60 22	0.61
α Aql	**Altair**	Aquila	19 50.8	08 52	0.76
α Cru	**Acrux**	Crux	12 26.6	−63 06	0.77
α Tau	**Aldebaran**	Taurus	04 35.9	16 31	0.87
α Vir	**Spica**	Virgo	13 25.2	−11 10	0.98
α Sco	**Antares**	Scorpius	16 29.4	−26 26	1.05
β Gem	**Pollux**	Gemini	07 45.3	28 02	1.16
α PsA	**Fomalhaut**	Piscis Austrinus	22 57.6	−29 37	1.16
β Cru	**Mimosa**	Crux	12 47.7	−59 41	1.25
α Cyg	**Deneb**	Cygnus	20 41.4	45 17	1.25
α Leo	**Regulus**	Leo	10 08.4	11 58	1.36
ε CMa	**Adhara**	Canis Major	06 58.6	−28 58	1.50
α Gem	**Castor**	Gemini	07 34.6	31 53	1.58
γ Cru	**Gacrux**	Crux	12 31.2	−57 07	1.59
λ Sco	**Shaula**	Scorpius	17 33.6	−37 06	1.62

Table 4.10 Greek alphabet.

A	α	Alpha	H	η	Eta	N	ν	Nu	T	τ	Tau			
B	β	Beta	Θ	θ	Theta	Ξ	ξ	Xı	Y	υ	Upsilon			
Γ	γ	Gamma	I	ι	Iota	O	o	Omicron	Φ	φ	Phi			
Δ	δ	Delta	K	κ	Kappa	Π	π	Pi	X	χ	Chi			
E	ε	Epsilon	Λ	λ	Lambda	P	ρ	Rho	Ψ	ψ	Psi			
Z	ζ	Zeta	M	μ	Mu	Σ	σ	Sigma	Ω	ω	Omega			

is rather too long to reproduce here but, as an alternative, a more compact implementation is offered in Algorithm 4.13.

This gives a subset of the FK5 catalogue data for 1535 stars, an illustration of which is shown in Example 4.6. This is drawn from a data file that has been conveniently re-named 'fk5.dat'. Other parameters are provided for each star and an FK5 extension is available that contains data for a further 3117 stars.

A visual plot of star positions in the northern hemisphere can be constructed as a scatter map, with (exaggerated) sizes indicating brightness, as implemented in Algorithm 4.14. Rather than attempt a line-by-line interpretation of the MATLAB code in that algorithm, it is recommended that the calculation be developed step-by-step in order to observe the purpose of each line of code.

Example 4.6

```
>> fk5 = getFk5('fk5.dat');
>> fk5(1)
ans =
FK5: 1
RAh: 0
RAm: 8
RAs: 23.2650
pmRA: 1.0390
DEd: 29
DEm: 5
DEs: 25.5800
pmDE: -16.3300
EpRA_1900: 43.3100
EpDE_1900: 33
Vmag: 2.0600
n_Vmag: ''
SpType: 'A0p'
plx: 0.0240
```

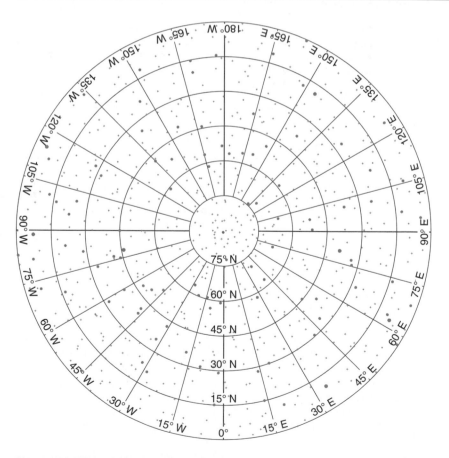

Figure 4.32 FK5 stars (Northern Hemisphere).

The end result is shown in Figure 4.32, after applying a 180° rotation. (*Note*: Imagine that you are standing under the centre of the hemisphere looking towards the origin of right ascension; looking to east or west implies a shift of view to left or right, respectively, which is consistent with what the map is showing.) Features of interest include the Plough (i.e. the principal stars of Ursa Major (UMa)), located between 165 °W and 150 °E at a declination around 60 °N. Note the pointers to the pole star, Polaris, (i.e. α UMa and β UMa) located almost exactly on 165°W. Also the 'W' of Cassiopeia is located between 0° and 30°W, also at around 60°N.

Star positions in the southern hemisphere can be rendered by a similar function to that implemented in Algorithm 4.12, with the following graph properties applied:

```
setm(gca, ...'
Origin',(-90,180,0), ...
'PLabelLocation',-15:-15:-75, ...
'PLabelMeridian',180) ;
```

The end result is shown in Figure 4.33, again after applying a 180° rotation. Features of interest include the Southern Cross (i.e. the principal stars of Crux), located between 165

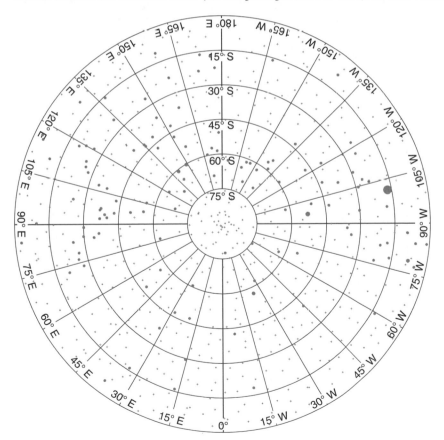

Figure 4.33 FK5 stars (Southern Hemisphere).

°E and 180° at a declination around 60 °S, and Sirius (α CMa) in the constellation of Canis Major, located at approximately 100 °E and 20 °S, which is the brightest star in the night sky.

```
function fk5 = getFk5Data(filename,d)

fk5Format = { ...
01,01,04,'%d','FK5', 'FK5 number' ; ...
02,06,07,'%d','RAh', 'Right ascension hours (J2000)' ; ...
03,09,10,'%d','RAm', 'Right ascension minutes (J2000)' ; ...
04,12,17,'%f','RAs', 'Right ascension seconds (J2000)' ; ...
05,19,25,'%f','pmRA', 'Proper motion: right ascension (J2000)' ; ...
06,27,29,'%d','DEd', 'Declination degrees (J2000)' ; ...
07,31,32,'%d','DEm', 'Declination arc-minutes (J2000)' ; ...
08,34,38,'%f','DEs', 'Declination arc-seconds (J2000)' ; ...
09,40,46,'%f','pmDE', 'Proper motion in declination (J2000)' ; ...
11,90,94,'%f','EpRA_1900', 'Mean Epoch of observed RA' ; ...
12,107,111,'%f','EpDE_1900','Mean Epoch of observed DE' ; ...
13,124,128,'%f','Vmag', 'Apparent visual magnitude' ; ...
14,129,129,'%s','n_Vmag', '(VvD) Magnitude flag' ; ...
15,131,137,'%s','SpType', 'Spectral type(s)' ; ...
16,139,144,'%f','plx', 'Parallax' } ;

(nFields,nElements) = size(fk5Format) ;
(lhs,rhs,form,label) = deal(2,3,4,5) ;

fid = fopen(filename,'r') ;

switch fid
 case -1
 error('FK5 file could not be opened') ;
 otherwise
 fseek(fid,0,1); eof = ftell(fid);
 fseek(fid,0,-1); pos = ftell(fid);

 iStar = 0 ;
 while pos ~ = eof
 fk5Star = fgetl(fid) ;

 iStar = iStar + 1 ;
 for iField = 1:nFields
 thisField = fk5Format(iField,:) ;
 thisLabel = thisField{label} ;
 thisString = fk5Star(thisField{lhs}:thisField{rhs}) ;
 thisFormat = thisField{form} ;

 switch thisFormat
 case '%s'
 fk5(iStar).(thisLabel) = ...
  deblank(leadblnk(thisString)) ;
 otherwise
 thisValue = str2num(thisString) ;
 fk5(iStar).(thisLabel) = thisValue ;
```

```
 end
 end
 pos = ftell(fid);
 end
end

fclose(fid);
```

Algorithm 4.13 'getFk5Data'.

```
function plotNorthernHemisphere(database)
hr2lon = @(hr) 360*(hr/24);

lat = dms2deg((database.DEd),(database.DEm),(database.DEs))';
lon = hr2lon(...
hms2hr((database.RAh),(database.RAm),(database.RAs)))';
relativeIntensity = ((database.Vmag)'+2).^(-100^0.2);

axesm eqdazim;
framem;
gridm;
mlabel;
plabel;

setm(gca,...
 'Origin',(90,180,0),...
 'FLatLimit',(-Inf 90),...
 'MLineLocation',15,...
 'MLabelLocation',-180:15:165,...
 'MLabelParallel',0,...
 'MLineLimit',(-75 75),...
 'PLabelLocation',15:15:75,...
 'PLabelMeridian',0,...
 'LabelRotation','on',...
 'GLineWidth',0.01,...
 'GLineStyle','-');
set(handlem('alltext'),...
 'HorizontalAlignment','center',...
 'VerticalAlignment','bottom');
set(gcf,...
 'Color','white')
set(gca,...
 'XColor','white',...
 'YColor','white');
scatterm(lat,lon,75*sqrt(relativeIntensity),'filled');
```

Algorithm 4.14 'plotNorthernHemisphere'.

Chapter 5

Geopotential Fields

5.1   Potential Fields

Flying vehicles, like all terrestrial objects, are influenced by Earth's potential fields. *Gravity* provides the fundamental force which all airborne vehicles must overcome and *geomagnetism* provides the environment for magnetic instruments. Because of their importance, it is appropriate to consider how to create geopotential models. This chapter develops the common method that underpins the Earth Gravitational Model (1996) (EGM96) and the World Magnetic Model (2005) (WMM2005).

5.1.1   *Gauss's Theorem*

Consider the flow of material or energy (called a *flux*) at a point in space (x,y,z) surrounded by an infinitesimal volume with dimensions (dx,dy,dz). If $\mathbf{F}(x,y,z)$ is the instantaneous flow vector at that position, then the change in flow across this volume is defined by:

$$d\Phi = \left(\frac{\partial F_x}{\partial x} + \frac{\partial F_y}{\partial y} + \frac{\partial F_z}{\partial z} \right) dx.dy.dz = (\nabla \cdot \mathbf{F}) dV \tag{5.1}$$

where $dV = dxdydz$ (i.e. the size of the volume). Integrating the flow across a finite volume, V, gives a nett outflow:

$$\Phi = \iiint (\nabla \cdot \mathbf{F}) dV \tag{5.2}$$

Integrating the flow across the external surface of V must yield the same result:

$$\Phi = \oiint \mathbf{F} \cdot \hat{\mathbf{n}} dS \tag{5.3}$$

where the surface integral is performed across infinitesimal elements with an area dS and a normal vector $\hat{\mathbf{n}}$ (pointing outwards). The dot product calculates flow perpendicular to each surface element. The combined result is Gauss's Theorem:

$$\iiint (\nabla \cdot \mathbf{F}) dV = \oiint \mathbf{F} \cdot \hat{\mathbf{n}} dS \tag{5.4}$$

Computational Modelling and Simulation of Aircraft and the Environment   D.J. Diston
© 2009 John Wiley & Sons, Ltd

5.1.2 *Applications of Gauss's Theorem*

Gauss's Theorem **c**an be applied to the study of gravity by considering the cumulative effect of all mass elements contained by the Earth's surface. Initially, just consider the gravitational flux due to a point mass Δm as shown in Figure 5.1. The generic concept of 'flux' might not always be intuitively obvious; it is a vector quantity that varies with position. In this case, the flux is the force exerted by the Earth on a unit mass. It is probably better to think of this as lines of force spreading out with distance, thereby reducing the gravitational attraction.

Figure 5.1 shows the effect of a small mass inside the Earth on a small surface element, and the implication is that the gravitational attraction at that surface element is proportional to the number of lines of force passing through it. So the appropriate surface integral is:

$$\Phi = \oiint \mathbf{g} \cdot \hat{\mathbf{n}} \, dS \tag{5.5}$$

where

$$\mathbf{g} \cdot \hat{\mathbf{n}} = -G\Delta m \frac{\cos \theta}{r^2} \tag{5.6}$$

Now the surface element dS subtends a solid angle at the point mass, which is evaluated as:

$$dA = dS \frac{\cos \theta}{r^2} \tag{5.7}$$

Introducing these two pieces of information:

$$\Phi = -G\Delta m \oiint dA = -4\pi G\Delta m \tag{5.8}$$

Considering the entire Earth, the total gravitational field of all point masses is simply the summation of their individual gravitational fields. Thus, the surface integral becomes:

$$\Phi = -4\pi G \sum_{i} \Delta m_i \tag{5.9}$$

For a continuous mass distribution, this equation has to be rewritten as:

$$\Phi = -4\pi G \oiiint \rho \, dV \tag{5.10}$$

where ρ = density and dV = infinitesimal volume.

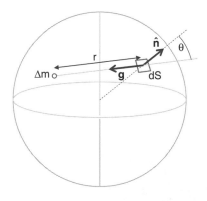

Figure 5.1 Gravitational flux due to a point mass.

At this juncture, Gauss's Theorem enables the following rationalisation to occur:

$$\oiiint(\nabla \cdot \mathbf{g} + 4\pi G\rho)dV = 0 \tag{5.11}$$

and, since this equality has to be true for any volume, it must follow that:

$$\nabla \cdot \mathbf{g} = -4\pi G\rho \tag{5.12}$$

Note that this is the result that enables gravitational field intensity (**g**) to be calculated for any point inside the Earth. The same procedure is valid for a point that is outside the Earth but the result would be a total cancellation of surface effects:

$$\nabla \cdot \mathbf{g} = 0 \tag{5.13}$$

This is because the front and rear faces subtend that same solid angle but, viewed from the external point, present external and internal surfaces respectively (meaning that the normal vectors are effectively pointing in opposite directions).

5.1.3   *Poisson's Equation and Laplace's Equation*

Now, it is conventional to describe gravitational field intensity (**g**) in terms of a scalar potential (U), where

$$\mathbf{g} = -\nabla U \tag{5.14}$$

The boundary condition is set so that gravity tends to zero as distance (r) increases, i.e.

$$\mathbf{g} \to \mathbf{0} \text{ as } r \to \infty$$

Using spherical coordinates, as defined in Figure 5.2, the relationship between field intensity and field potential can be readily established:

$$\mathbf{g} = -\nabla U = -\frac{\partial U}{\partial r}\hat{\mathbf{r}} \equiv -\frac{GM}{r^2}\hat{\mathbf{r}} \Rightarrow U(r) = -\frac{GM}{r} \tag{5.15}$$

which conforms with the convention interpretation of the potential function U(r) being negative and becoming increasingly negative on approach to the field source, which is often regarded as a *potential well* for a force of attraction.

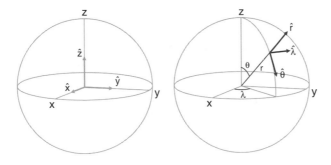

Figure 5.2 Cartesian and spherical coordinates.

From the earlier application of Gauss's Theorem, field equations were derived for points internal and external to the Earth. Applying the concept of gravitational potential, these can be rewritten as follows:

$$\nabla \cdot \mathbf{g} = -4\pi G\rho \quad \Rightarrow \quad \nabla^2 U = -4\pi G\rho \qquad \text{Poisson's Equation} \qquad (5.16)$$

$$\nabla \cdot \mathbf{g} = 0 \qquad \Rightarrow \quad \nabla^2 U = 0 \qquad \text{Laplace's Equation} \qquad (5.17)$$

5.1.4   *Generic Solution of Laplace's Equation*

Because Laplace's Equation effectively defines an unforced oscillator, its solutions are called *harmonics*. Using spherical coordinates (as in Figure 5.2), Laplace's Equation becomes:

$$\nabla^2 U = \frac{1}{r^2}\frac{\partial}{\partial r}\left(r^2\frac{\partial U}{\partial r}\right) + \frac{1}{r^2 \sin\theta}\frac{\partial}{\partial\theta}\left(\sin\theta\frac{\partial U}{\partial\theta}\right) + \frac{1}{r^2 \sin^2\theta}\frac{\partial^2 U}{\partial\lambda^2} = 0 \qquad (5.18)$$

The solutions of this equation are called *spherical harmonics*.

It is well-known that Laplace's Equation can be solved by a separation of variables. So assume a solution of the form:

$$U(r,\theta,\lambda) = R(r)T(\theta)L(\lambda) \quad \Rightarrow \quad \frac{TL}{r^2}\frac{\partial}{\partial r}\left(r^2\frac{\partial R}{\partial r}\right) + \frac{LR}{r^2 \sin\theta}\frac{\partial}{\partial\theta}\left(\sin\theta\frac{\partial T}{\partial\theta}\right) + \frac{RT}{r^2 \sin^2\theta}\frac{\partial^2 L}{\partial\lambda^2} = 0$$

$$\Rightarrow \quad \frac{\sin^2\theta}{R}\frac{\partial}{\partial r}\left(r^2\frac{\partial R}{\partial r}\right) + \frac{\sin\theta}{T}\frac{\partial}{\partial\theta}\left(\sin\theta\frac{\partial T}{\partial\theta}\right) + \frac{1}{L}\frac{\partial^2 L}{\partial\lambda^2} = 0$$

Only the last term on the left depends on λ, which indicates that it must be constant. It is convenient to express this as follows:

$$\frac{1}{L}\frac{\partial^2 L}{\partial\lambda^2} = -m^2$$

where m is a constant in this context (... *not* mass!).

This is an ordinary differential equation whose solution is:

$$L = A \cos m\lambda + B \sin m\lambda$$

where A and B are constants and m is an integer.

Substituting this part of the solution, Laplace's equation becomes:

$$\frac{\sin^2\theta}{R}\frac{\partial}{\partial r}\left(r^2\frac{\partial R}{\partial r}\right) + \frac{\sin\theta}{T}\frac{\partial}{\partial\theta}\left(\sin\theta\frac{\partial T}{\partial\theta}\right) - m^2 = 0$$

$$\Rightarrow \quad \frac{1}{R}\frac{\partial}{\partial r}\left(r^2\frac{\partial R}{\partial r}\right) + \frac{1}{T\sin\theta}\frac{\partial}{\partial\theta}\left(\sin\theta\frac{\partial T}{\partial\theta}\right) = \frac{m^2}{\sin^2\theta}$$

Now, the two terms on the left-hand side are independent. The first term is the only term which depends on r and so it must be constant. It transpires that the appropriate solution to this part of the problem is:

$$\frac{1}{R}\frac{\partial}{\partial r}\left(r^2\frac{\partial R}{\partial r}\right)=n^2+n$$

$$\Rightarrow \frac{1}{R}\left[2r\frac{\partial R}{\partial r}+r^2\frac{\partial^2 R}{\partial r^2}\right]=n^2+n$$

$$\Rightarrow R = Cr^n + Dr^{n-1}$$

where C and D are constants.

Substituting this part of the solution, Laplace's equation becomes:

$$\frac{1}{R}\frac{\partial}{\partial r}\left(r^2\frac{\partial R}{\partial r}\right)=\frac{m^2}{\sin^2\theta}-n^2-n$$

which is Legendre's Equation. The solutions to this particular equation have the form:

$$R = K_n^m P_n^m(\cos\theta)$$

where the K's are constants and the P's are so-called *Legendre* polynomials.

The culmination of this work is the generic solution of Laplace's Equation:

$$U = -\frac{1}{a}\sum_{n=0}^{\infty}\sum_{m=0}^{n}\left[\left(C_n^m\left(\frac{a}{r}\right)^{n+1}+\overline{C}_n^m\left(\frac{r}{a}\right)^n\right)\cos m\lambda\right.$$
$$\left.+\left(S_n^m\left(\frac{a}{r}\right)^{n+1}+\overline{S}_n^m\left(\frac{r}{a}\right)^n\right)\sin m\lambda\right]P_n^m(\cos\theta)$$

(5.19)

Note that this is a summation of terms of *degree* n and *order* m.

The Legendre polynomial of *degree* n can be derived from Rodrigues' formula:

$$P_n(x)=\frac{1}{2^n n!}\frac{d^n}{dx^n}(x^2-1)$$

(5.20)

The associated Legendre polynomial of *degree* n and *order* m is derived as follows:

$$P_n^m(x)=\left(1-x^2\right)^{m/2}\frac{d^m}{dx^m}P_n(x)$$

(5.21)

For reference, the first few Legendre polynomials are given below:

$$P_0(x)=1 \quad P_1(x)=x \quad P_2(x)=\frac{1}{2}\left(3x^2-1\right) \quad P_3(x)=\frac{1}{2}\left(5x^3-3x\right)$$

Note 5.1

Although it is common to talk generically about Legendre polynomials of any degree and any order, the formal use of the term 'Legendre polynomial' implies that $m=0$; the nth degree Legendre polynomial is written as $P_n(x)$. Otherwise $m\neq 0$ and the term 'associated Legendre polynomial' should be used. Just to compound the confusion, alternative notations are sometimes applied in the literature, namely $P_{nm}(x)$ or $P_{n,m}(x)$.

5.2   Gravitation

5.2.1   *Gravitation Attraction*

For approximate calculations or for calculations that are insensitive to gravitational fluctuations, a single value is often quoted for gravitational acceleration. The classical textbook value is $g = 9.8 \, \text{m.s}^{-1}$ (often rounded up to $g = 10 \, \text{m.s}^{-1}$ for purposes of estimation). The ICAO[1] International Standard Atmosphere (ISA) defines a mean gravity of $g = 9.80665 \, \text{m.s}^{-2}$ (ICAO, 1993); this value is also recommended by NIST[2] (Mohr & Taylor, 1998).

The gravitational attraction between two objects, with respective masses of m_1 and m_2, varies according to the inverse square law:

$$F = \frac{G m_1 m_2}{r^2} \tag{5.22}$$

where G is the gravitational constant and r is the distance between the centres of gravity of the two objects.

For a small object located on or near to the Earth, it is usual to write this law as follows:

$$F = \frac{GMm}{r^2} \tag{5.23}$$

where M is the mass of the Earth and m is the mass of the object. Knowing that the gravitational force acting on the small object is $F = mg$, it should be clear that the gravitational acceleration (g) is given by:

$$g = \frac{GM}{r^2} \tag{5.24}$$

Satellite measurements have provided accurate estimates of the product GM, which are considerably more accurate than estimates of G or M individually. WGS84 assigns the following value:

$$GM = 3.986004418 \times 10^{14} \, \text{m}^3 \text{s}^{-2} \tag{5.25}$$

Note 5.2

For information, the individual components are estimated crudely as follows:
$$G \approx 6.67 \times 10^{11} \, \text{m}^3 \text{kg}^{-1} \text{s}^{-2} \text{ and } M \approx 5.98 \times 10^{24} \, \text{kg}$$

5.2.2   *Apparent Gravity: Spherical Earth*

Taking a point on the surface of a spherical Earth, gravitational acceleration is given by (5.24). Actual gravity acts towards the centre of the Earth but, in addition, there is a

---

[1] International Civil Aviation Organisation.
[2] National Institute of Standards and Technology.

centrifugal force acting because of the Earth's rotation. This is shown in Figure 5.3 for a point P located at a (geocentric) latitude ϕ, with xz-coordinates ($r cos\phi$, $r sin\phi$). The nett force is established by a vector addition and represents the *apparent* force of gravity for a spherical Earth.

Note 5.3

Based on the dimensions of the WGS84 ellipsoid given in Section 6.2 (i.e. a = 6378137 m and c = 6356752.3 m), the volume of the Earth is:

$$V = \frac{4}{3}\pi a^2 b = \left(\frac{4}{3}\pi\right) 2.585966984 \times 10^{20} = 1.083207317 \times 10^{21}$$

Defining a sphere of equal volume, the *mean* radius is determined as follows:

$$V = \frac{4}{3}\pi r^3 \Rightarrow r = \sqrt[3]{2.585966984 \times 10^{20}} = 6371000.785$$

This is often approximated to the more convenient value of r = 6371000 m.

Example 5.1

For a point P at a (geocentric) latitude of 53 °N, the xz-position is (3834163.5, 5088106.8) and the radius is 6371000. Thus, gravitational acceleration is g = 9.820251 m.s^{-2} and centrifugal acceleration is $x\omega^2 = 0.02038814$ m.s^{-2}. Performing the necessary trigonometry, it is found that this creates a nett acceleration given by:

$$\mathbf{g'} = 9.807994627 \angle 53.09511934°$$

... which reveals a misalignment of about 0.1° in comparison with the geometric vertical.

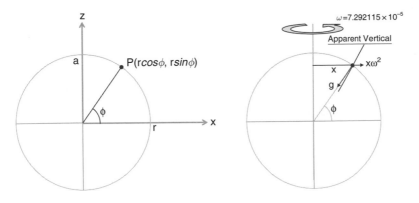

Figure 5.3 Components of apparent gravity (spheroid).

5.2.3   *Apparent Gravity: WGS84 Ellipsoid*

Repeating the procedure outlined in Section 5.1.2 for an ellipsoid, it is necessary to adopt the geometrical definitions appropriate to elliptical coordinates (as developed in Section 3.6). As before, actual gravity acts towards the centre of the Earth but, in addition, there is a centrifugal force acting because of the Earth's rotation. This is shown in Figure 5.4 for a point P located at a (geodetic) latitude μ, with xz-coordinates $(Ncos\mu, N(1-e^2)sin\mu)$, where the parameters N and e are given by (6.38) and (6.35), respectively:

$$N = \frac{a}{\sqrt{1 - e^2 \sin^2 \mu}}$$

$$e = \sqrt{\frac{a^2 - c^2}{a^2}}$$

Recalling (3.60) and (3.61), the elliptical dimensions are $a = 6378137$ and $c = 6356752.3$ and, thus, the eccentricity of the WGS84 ellipsoid is $e = 0.081819218$. The nett force, established by vector addition, represents the *apparent* force of gravity for an ellipsoidal Earth. Note that the local radius at P (i.e. the radial distance from the Earth's centre) is derived as:

$$r = N\sqrt{1 - (2 - e^2)e^2 \sin^2 \mu} \tag{5.26}$$

Example 5.2

For a point P at a (geodetic) latitude of 53°N, the xz-position is (3846679.8, 5070543.5) and the local radius is 6364539. Thus, gravitational acceleration is $g = 9.84401988$ m.s^{-2} and centrifugal acceleration is $x\omega^2 = 0.02045470$ m.s^{-2}. Performing the necessary trigonometry, it is found that this creates a nett acceleration given by:

$$\mathbf{g}' = 9.827902418 \angle 53.09523651°$$

... which reveals almost the same misalignment as far the equivalent spherical Earth calculation (Example 5.1), together with a marginal increase in gravitational acceleration.

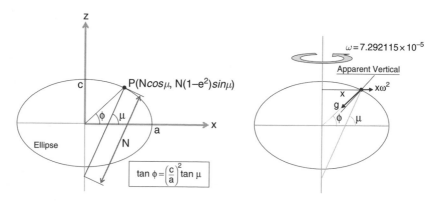

Figure 5.4 Components of apparent gravity (ellipsoid).

5.2.4   ***Gravitational Moments***

Using spherical coordinates, as defined in Figure 5.2, gravitational acceleration, or field intensity, (g) is expressed as a vector quantity:

$$\mathbf{g} = -\frac{GM}{r^2}\hat{\mathbf{r}} \qquad (5.27)$$

The relationship between field intensity and field potential can be readily established (5.15):

$$\mathbf{g} = -\nabla U = -\frac{\partial U}{\partial r}\hat{\mathbf{r}} \equiv -\frac{GM}{r^2}\hat{\mathbf{r}} \Rightarrow U(r) = -\frac{GM}{r}$$

This conforms to the conventional interpretation of the potential function U(r) being negative and becoming increasingly negative on approach to the field source, which is often regarded as a *potential well* for a force of attraction.

From the overview of potential field theory presented in Section 5.1, it is noted that gravitational fields are governed by Laplace's Equation (5.17):

$$\nabla^2 U = 0$$

Its general solution is based on spherical harmonics (5.19):

$$U = -\frac{1}{a}\sum_{n-0}^{\infty}\sum_{m-0}^{n}\left[\left(C_n^m\left(\frac{a}{r}\right)^{n+1} + \overline{C}_n^m\left(\frac{r}{a}\right)^n\right)\cos m\lambda\right.$$
$$\left. + \left(S_n^m\left(\frac{a}{r}\right)^{n+1} + \overline{S}_n^m\left(\frac{r}{a}\right)^n\right)\sin m\lambda\right]P_n^m(\cos\theta)$$

The boundary condition of zero potential at infinity requires that

$$\overline{C}_n^m = 0 \quad \text{and} \quad \overline{S}_n^m = 0$$

for all values of n and m, thereby simplifying the solution:

$$U = -\frac{1}{a}\sum_{n=0}^{\infty}\sum_{m=0}^{n}\left(\frac{a}{r}\right)^{n+1}\left[C_n^m\cos m\lambda + S_n^m\sin m\lambda\right]P_n^m(\cos\theta) \qquad (5.28)$$

Assuming axial symmetry, it is possible to remove the dependence on λ i.e. by setting $m = 0$, thereby further simplifying the solution:

$$U = -\frac{1}{a}\sum_{n=0}^{\infty}C_n\left(\frac{a}{r}\right)^{n+1}P_n(\cos\theta) \qquad (5.29)$$

Note that the superscripts 'm' have been removed.

A more convenient form of this equation is generally used, which is:

$$U = -\frac{GM}{r}\left(J_0 P_0 - \sum_{n=1}^{\infty}J_n\left(\frac{a}{r}\right)^n P_n(\cos\theta)\right) \qquad (5.30)$$

where the constants J_k are called gravitational moments.

As r increases, the gravitational potential approximates to the standard expression given in (5.15), namely $U = -GM/r$. This means that $J_0 P_0 = 1$ and, thus, $J_0 = 1$ (given that $P_0 = 1$).

Also, if the coordinate system has its origin at the centre of mass then $J_1 = 0$. This produces a more compact equation:

$$U = -\frac{GM}{r}\left(1 - \sum_{n=2}^{\infty} J_n \left(\frac{a}{r}\right)^n P_n(\cos\theta)\right) \tag{5.31}$$

where the first few gravitational moments (derived from satellite measurements) are:

$$J_2 = 1.082635 \times 10^{-3}$$
$$J_3 = -2.531 \times 10^{-6}$$
$$J_4 = -1.600 \times 10^{-6}$$
$$J_5 = -0.246 \times 10^{-6}$$

The dominant effect is J_2 which relates to Earth's rotation, leading to its flattened shape, as will be seen in Section 5.2.6. A secondary effect is seen in J_3, which indicates that the southern hemisphere contains slightly more mass than the northern hemisphere. Note that, if the Earth were symmetrical about the equator then $J_n = 0$ for all odd values of n.

For most flight applications, it is sufficient to apply the most approximate definition and, thus, the frequently quoted form of the gravitational potential is, as follows:

$$U \approx -\frac{GM}{r}\left(1 - J_2\left(\frac{a}{r}\right)^2 P_2(\cos\theta)\right) \tag{5.32}$$

Using the explicit form of $P_2(\cos\theta)$ derived from (5.20), this becomes:

$$U \approx -\frac{GM}{r}\left(1 - \frac{1}{2}\left(\frac{a}{r}\right)^2 J_2\left(3\cos^2\theta - 1\right)\right) \tag{5.33}$$

5.2.5   *Earth Gravitational Model (EGM96)*

WGS84 incorporates a much more accurate definition of the gravitational field, called EGM96, which is based on spherical harmonics up to order $n = 360$ and degree $m = 360$ (comprising 130317 coefficients). In this context, the generic solution of Laplace's Equation (5.19) can be recast as follows:

$$U = -\frac{1}{a} \sum_{n=0}^{\infty} \sum_{m=0}^{n} \left[\left(C_n^m\left(\frac{a}{r}\right)^{n+1} + \overline{C}_n^m\left(\frac{r}{a}\right)^n\right)\cos m\lambda\right.$$

$$\left. + \left(S_n^m\left(\frac{a}{r}\right)^{n+1} + \overline{S}_n^m\left(\frac{r}{a}\right)^n\right)\sin m\lambda\right] P_n^m(\cos\theta) \tag{5.34}$$

$$U = -\frac{GM}{r}\left[1 + \sum_{n=2}^{\infty} \sum_{m=0}^{n} \left(\frac{a}{r}\right)^n \left[C_n^m\cos m\lambda + S_n^m\sin m\lambda\right]\right] P_n^m(\cos\theta)$$

where (i) a factor GM has been introduced, (ii) terms have been deleted in order to impose zero potential at infinity and (iii) terms of degree 0 and 1 have been rationalised as in the previous section. Unlike the previous section, this retains the dependence on λ which means that the solution uses all Legendre polynomials.

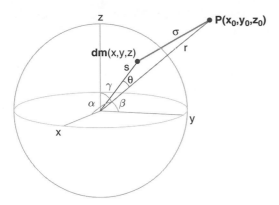

Figure 5.5 Reference geometry for gravitation.

Table 5.1 EGM96 Coefficients (truncated at n=m=6).

Degree	Order	Normalised Coefficients	
n	m	$\bar{c}_n^m$	$\bar{s}_n^m$
2	0	$-0.484165\_371736 \times 10^{-3}$	
2	1	$-0.186987\_635955 \times 10^{-9}$	$0.119528\_012031 \times 10^{-8}$
2	2	$0.243914\_352398 \times 10^{-5}$	$-0.140016\_683654 \times 10^{-5}$
3	0	$0.957254\_173792 \times 10^{-6}$	
3	1	$0.202998\_882184 \times 10^{-5}$	$0.248513\_158716 \times 10^{-6}$
3	2	$0.904627\_768605 \times 10^{-6}$	$-0.619025\_944215 \times 10^{-6}$
3	3	$0.721072\_657057 \times 10^{-6}$	$0.141435\_626958 \times 10^{-5}$
4	0	$0.539873\_863789 \times 10^{-6}$	
4	1	$-0.536321\_616971 \times 10^{-6}$	$-0.473440\_265853 \times 10^{-6}$
4	2	$0.350694\_105785 \times 10^{-6}$	$0.662671\_572540 \times 10^{-6}$
4	3	$0.990771\_803829 \times 10^{-6}$	$-0.200928\_369177 \times 10^{-6}$
4	4	$-0.188560\_802735 \times 10^{-6}$	$0.309953\_169333 \times 10^{-6}$
5	0	$0.685323\_475630 \times 10^{-7}$	
5	1	$-0.621012\_128528 \times 10^{-7}$	$-0.944226\_127525 \times 10^{-7}$
5	2	$0.652438\_297612 \times 10^{-6}$	$-0.323349\_612668 \times 10^{-6}$
5	3	$-0.451955\_406071 \times 10^{-6}$	$-0.214847\_190624 \times 10^{-6}$
5	4	$-0.295301\_647654 \times 10^{-6}$	$0.496658\_876769 \times 10^{-7}$
5	5	$0.174971\_983203 \times 10^{-6}$	$-0.669384\_278219 \times 10^{-6}$
6	0	$-0.149957\_994714 \times 10^{-6}$	
6	1	$-0.760879\_384947 \times 10^{-7}$	$0.262890\_545501 \times 10^{-7}$
6	2	$0.481732\_442832 \times 10^{-7}$	$-0.373728\_201347 \times 10^{-6}$
6	3	$0.571730\_990516 \times 10^{-7}$	$0.902694\_517163 \times 10^{-8}$
6	4	$-0.862142\_660109 \times 10^{-7}$	$-0.471408\_154267 \times 10^{-6}$
6	5	$-0.267133\_325490 \times 10^{-6}$	$-0.536488\_432483 \times 10^{-6}$
6	6	$0.967616\_121092 \times 10^{-8}$	$-0.237192\_006935 \times 10^{-6}$

The actual form of gravitational potential in EGM96 makes use of semi-normalised polynomials and can be expressed as:

$$U = -\frac{GM}{r}\left[1 + \sum_{n=2}^{\infty}\sum_{m=0}^{n}\left(\frac{a}{r}\right)^n\left[\bar{c}_n^m\cos m\lambda + \bar{s}_n^m\sin m\lambda\right]\bar{p}_n^m(\cos\theta)\right] \qquad (5.35)$$

where

$$\bar{c}_n^m = C_n^m A_n^m$$

$$\bar{s}_n^m = S_n^m A_n^m$$

$$\bar{p}_n^m(\cos\theta) = P_n^m(\cos\theta)/A_n^m$$

The normalisation factors are defined by:

$$A_n^m = \sqrt{\frac{(n+m)!}{(n-m)!\,k(2n+1)}} \qquad (5.36)$$

where

$$k = 1 \ \text{if} \ \ m = 0$$
$$k = 2 \ \text{if} \ \ m \neq 0$$

The EGM96 normalised coefficients up to $n = m = 6$ are given in Table 5.1. The WGS84 documentation (NIMA, 2004) gives these data up to $n = m = 18$. It also notes the theoretical validity of this model for $r \geq a$ although it can be used with 'probably negligible error near or on the Earth's surface' (ibid., p. 5–3). It definitely should not be used for points beneath the Earth's surface, which hopefully should never be necessary in aerospace applications!

This method is completely general for all solar system bodies. Equivalent coefficients are available for the Moon (Roncoli, 2005) based on the LP150Q gravitational field model and for Mars (Lemoine et al., 2001) based on GMM2B. Also, summary is readily available from the Jet Propulsion Laboratory for the gravitational field of the outer planets.

5.2.6   *MacCullagh's Formula*

The gravitational field due to a large object can be established by considering a general point P situated outside the object and integrating the effect of all mass elements (dm) into the object. This is constructed geometrically as shown in Figure 5.5, noting that point P is at a radial distance r from the origin and the sample mass element is at distance s. By implication, the object is the Earth, it is almost spherical and its origin of coordinates is at the centre (i.e. the centre of mass). Thus, the gravitational field is derived as follows:

$$U = -G\int\frac{dm}{\sigma} = -G\int\frac{dm}{\sqrt{r^2 + s^2 - 2rs\cos\theta}} \qquad (5.37)$$

where θ is the angle between the two radius vectors, pointing towards P and towards dm. This can be written in an equivalent form:

$$U = -\frac{G}{r}\int\left[1 - 2\frac{s}{r}\cos\theta + \frac{s^2}{r^2}\right]^{-1/2}dm$$

where $s < r$, which can then be approximated:

$$U \approx -\frac{G}{r} \int \left[1 + \frac{s}{r}(\cos\theta) + \frac{s^2}{2r^2}(3\cos^2\theta - 1) \right] dm$$

Separating the individual terms in this equation gives:

$$U \approx -\frac{G}{r} \int dm - \frac{G}{r^2} \int s(\cos\theta)dm - \frac{G}{2r^3} \int 2s^2 dm + \frac{3G}{2r^3} \int s^2(\sin^2\theta)dm$$

where each term relates to an aspect of the object's inertia.

The first term is simply the object mass:

$$\int dm = M$$

The second term gives the position of the centre of mass, which (by definition here) is located at the origin of coordinates:

$$\int s(\cos\theta)dm \equiv 0$$

The third term is the sum of the principal moments of inertia A, B and C (relative to the x, y and z axes, respectively):

$$\int 2s^2 dm = \int \left[(x^2 + y^2) + (y^2 + z^2) + (z^2 + x^2) \right] dm \equiv A + B + C$$

The last term is based on the perpendicular distance of mass elements from the line joining the origin to point P; the actual quantity obtained in the moment of inertia about that line:

$$\int s^2(\sin^2\theta)dm = \int \Delta^2 dm = I$$

Thus, the gravitational potential can be simplified to the following form:

$$U \approx -\frac{GM}{r} - \frac{G}{2r^3}(A + B + C - 3I) \tag{5.38}$$

A spherical Earth $(A = B = C = I)$ has the simplest form of gravitational model, namely $U \approx -GM/r$.

For an ellipsoid, it is conventional to associate the z-axis with the polar radius (which is smaller than the equatorial radius). Assuming axial symmetry, therefore, $A = B$. The moment of inertia about any line through the centre of mass can be derived from the principal moments and inertia, together with the direction cosines of the line:

$$I = A\cos^2\alpha + B\cos^2\beta + C\cos^2\gamma \tag{5.39}$$

This reduces to a simpler expression, as follows:

$$I = A(\cos^2\alpha + \cos^2\beta) + C\cos^2\gamma$$
$$I = A(1 - \cos^2\gamma) + C\cos^2\gamma$$
$$I = A\sin^2\gamma + C\cos^2\gamma$$

where the direction angle (γ) from the z-axis is equal to the (geocentric) co-latitude (θ) of the particular point. For consistency with the derivations in previous sections, it is appropriate to standardise notation as follows:

$$I = A\sin^2\theta + C\cos^2\theta \qquad (5.40)$$

Using all of this information, the combination of inertia is now expressed as:

$$A + B + C - 3I = 2A + C - 3\left(A\sin^2\theta + C\cos^2\theta\right) = (C - A)\left(1 - 3\cos^2\theta\right)$$

Substituting this into the equation for gravitational potential gives:

$$U \approx -\frac{GM}{r} - \frac{G}{2r^3}(C - A)\left(1 - 3\cos^2\theta\right)$$

This is rearranged to give *MacCullagh's Formula*:

$$U \approx -\frac{GM}{r}\left[1 - \frac{(C - A)}{2Mr^2}\left(3\cos^2\theta - 1\right)\right] \qquad (5.41)$$

Recall the earlier approximation to the Earth's gravitational potential (5.33):

$$U \approx -\frac{GM}{r}\left[1 - \frac{1}{2}\left(\frac{a}{r}\right)^2 J_2\left(3\cos^2\theta - 1\right)\right]$$

By comparison with MacCullagh's Formula, it is clear that

$$J_2 = \frac{(C - A)}{Ma^2} \qquad (5.42)$$

which demonstrates that J_2 is related to the flattening of the Earth. Incidentally, measurements show that the numerical value of J_2 is gradually decreasing, a trend that is interpreted as evidence of so-called *glacial rebound*, i.e. a continuing readjustment to the reduction of mass at the poles after the last ice age.

Note 5.4

WGS84 (NIMA 2004, Table 3.5) quotes Earth's principal moments of inertia as:

$$A = 8.0091029 \times 10^{37}\,\text{kg.m}^2$$
$$B = 8.0092559 \times 10^{37}\,\text{kg.m}^2$$
$$C = 8.0354872 \times 10^{37}\,\text{kg.m}^2$$

Take the mean value of A and B to be

$$E = \frac{A + B}{2} = 8.0091794 \times 10^{37}$$

A verification of the second gravitational moment can be performed as follows:

$$\frac{C - E}{Ma^2} = \frac{2.63078 \times 10^{35}}{\left(5.98 \times 10^{24}\right)\left(6.378137 \times 10^6\right)} = 1.0814231 \times 10^{-3} \approx 1.082635 \times 10^{-3} = J_2$$

5.2.7   *Earth Flattening or `Oblateness'*

The effective gravitational field of the Earth should include the effect of Earth rotation, as shown in Figure 5.6. The components of acceleration are due to gravity and centripetal force (shown as centifugal acceleration, $p\omega^2$, for a perpendicular distance, p). Denoting these field components by U and Ω, their mathematical definitions are:

$$\mathbf{g} = -\nabla U$$
$$\mathbf{p} = -\nabla\Omega = \omega^2(x\hat{\mathbf{x}} + y\hat{\mathbf{y}})$$

respectively. The first expression has been developed already; the second is obtained simply in the following form:

$$\Omega = -\frac{1}{2}\omega^2(x^2 + y^2) = -\frac{1}{2}\omega^2 p^2$$

Replacing the perpendicular distance with the appropriate trigonometry:

$$\Omega = -\frac{1}{2}r^2\omega^2\sin^2\theta \tag{5.44}$$

So, the total potential that acts at the Earth's surface is:

$$V(r,\theta) = U(r,\theta) + \Omega(r,\theta) = -\frac{GM}{r}\left[1 - \frac{1}{2}\left(\frac{a}{r}\right)^2 J_2(3\cos^2\theta - 1)\right] - \frac{1}{2}r^2\omega^2\sin^2\theta \tag{5.44}$$

This is an equipotential surface such that the potential at the equator is the same as that at the poles, i.e.

$$\Omega\left(a, \frac{\pi}{2}\right) = \Omega(c, 0) \tag{5.45}$$

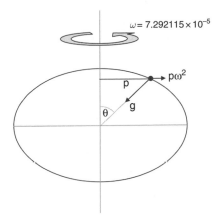

Figure 5.6 Components of acceleration on Earth's surface.

Making the relative substitutions, the following equality is found:

$$-\frac{GM}{a}\left[1 + \frac{1}{2}J_2\right] - \frac{1}{2}a^2\omega^2 = -\frac{GM}{c}\left[1 - \left(\frac{a}{c}\right)^2 J_2\right]$$

From this, the following derivation can be performed:

$$-c\left[1 + \frac{1}{2}J_2\right] - \frac{1}{2}c\frac{a^2\omega^2}{GM} = -a\left[1 - \left(\frac{a}{c}\right)^2 J_2\right]$$

$$\Rightarrow a - c = a\left(\frac{a}{c}\right)^2 J_2 + \frac{c}{c}J_2 + \frac{1}{2}c\frac{a^2\omega^2}{GM}$$

$$\Rightarrow \frac{a-c}{a} = \left[\left(\frac{a}{c}\right)^2 + \frac{1}{2}\left(\frac{c}{a}\right)\right]J_2 + \frac{1}{2}c\frac{a^2\omega^2}{GM}$$

With reference to (3.62), this leads to an expression for Earth 'flattening' or 'oblateness' (f):

$$f = \left[1 + \varepsilon^2 + \frac{1}{2}(1 - f)\right]J_2 + \frac{1}{2}(1 - f)q \tag{5.46}$$

where q is the nondimensional centripetal acceleration, defined by:

$$q = \frac{\omega^2 a^3}{GM} = \frac{\omega^2 a}{\left(\dfrac{GM}{a^2}\right)} \tag{5.47}$$

and ε is the second eccentricity, which can be approximated as:

$$\left(\frac{a}{c}\right)^2 = \frac{1}{(1-f)^2} = 1 + \varepsilon^2 \Rightarrow 1 - \varepsilon^2 \approx 1 - 2f + f^2$$

This gives the following result:

$$\varepsilon^2 \approx 2f \tag{5.48}$$

Thus, the Earth flattening can be approximated as:

$$f \approx \left[1 + 2f + \frac{1}{2}(1 - f)\right]J_2 + \frac{1}{2}(1 - f)q \Rightarrow \left[1 - \left(\frac{3}{2}J_2 - \frac{1}{2}q\right)\right]f \approx \frac{3}{2}J_2 + \frac{1}{2}q$$

$$f \approx \frac{3}{2}J_2 + \frac{1}{2}q \tag{5.49}$$

Note 5.5

Based on numerical values $J_2 = 1.082635 \times 10^{-3}$ and $q = 3.461392 \times 10^{-3}$, the approximation of Earth flattening (or ellipticity) can be verified by two simple calculations:

$$\#1 : \left[1 - \left(\frac{3}{2}J_2 - \frac{1}{2}q\right)\right]\left(\frac{a-c}{a}\right) = [1 - 1.067534 \times 10^{-4}](3.352813 \times 10^{-3}) = 0.003352455$$

$$\#2 : \frac{3}{2}J_2 + \frac{1}{2}q = 0.003354648$$

The gravitational moment J_2 is directly related to the Earth's flattened shape while the nondimensional acceleration q is directly related to the Earth's rotation. Thus, the flattening parameter (f) can be thought of as the additional potential caused by the distortion $(3J_2/2)$ plus the potential causing the distortion $(q/2)$. The ratio of these two components $(3J_2/q)$ is known as the second Love Number and has the value 0.938323. The equivalent value for tidal movements is less than 0.3. Thus the Earth is 'rigid' but behaves like a viscous fluid over long time-scales.

5.2.8 *Cartesian Components of Gravity*

The gravitational field is radial in direction, acting (to a very good approximation) as if the entire mass of the Earth were concentrated at the centre of mass. What is needed for application to aircraft flight is a representation that resolves gravity into cartesian components along the x, y and z axes.

By general consideration of the geometry presented in Figure 5.2, together with basic calculus, the following simple definitions can be stated:

$$\cos\theta = \frac{z}{r}$$

$$r^2 = x^2 + y^2 + z^2$$

$$\frac{\partial r}{\partial x} = \frac{x}{r}$$

$$\frac{\partial r}{\partial y} = \frac{y}{r}$$

$$\frac{\partial r}{\partial z} = \frac{z}{r}$$

The gravitational potential needs to be restated explicitly in terms of vertical displacement (z) and radial distance (r) in order for derivatives to be calculated:

$$U = -\frac{GM}{r}\left[1 - \frac{1}{2}\left(\frac{a}{r}\right)^2 J_2(3\cos^2\theta - 1)\right] = -GM\left[r^{-1} - \frac{a^2}{2}J_2(r^{-3} - 3z^2r^{-5})\right] \qquad (5.50)$$

Now, the x, y and z components of gravity can be established through partial differentiation:

$$\frac{\partial U}{\partial x} = GMr^{-2}\left[1 + \frac{3}{2}a^2 J_2 r^{-2}\left(1 - 5z^2 r^{-2}\right)\right]\left(\frac{x}{r}\right) \tag{5.51}$$

$$\frac{\partial U}{\partial y} = GMr^{-2}\left[1 + \frac{3}{2}a^2 J_2 r^{-2}\left(1 - 5z^2 r^{-2}\right)\right]\left(\frac{y}{r}\right) \tag{5.52}$$

$$\frac{\partial U}{\partial z} = GMr^{-2}\left[1 + \frac{3}{2}a^2 J_2 r^{-2}\left(1 - 5z^2 r^{-2}\right) + 3a^2 J_2 r^{-2}\right]\left(\frac{z}{r}\right) \tag{5.53}$$

$$= GMr^{-2}\left[1 + \frac{3}{2}a^2 J_2 r^{-2}\left(3 - 5z^2 r^{-2}\right)\right]\left(\frac{z}{r}\right)$$

Finally, gravitational acceleration can be written as a simple, consolidated expression using cartesian coordinates:

$$\mathbf{g} = -\nabla U = -\hat{\mathbf{x}}\frac{\partial U}{\partial x} - \hat{\mathbf{y}}\frac{\partial U}{\partial y} - \hat{\mathbf{z}}\frac{\partial U}{\partial z} \tag{5.54}$$

where

$$\frac{\partial U}{\partial x} = \frac{GM}{r^2}\left(1 - \frac{3}{2}\left(\frac{a}{r}\right)^2 J_2\left(5\cos^2\theta - 1\right) - \frac{5}{8}\left(\frac{a}{r}\right)^4 J_4\left(63\cos^4\theta - 42\cos^2\theta + 3\right)\right)\left(\frac{x}{r}\right)$$

$$\frac{\partial U}{\partial y} = \frac{GM}{r^2}\left(1 - \frac{3}{2}\left(\frac{a}{r}\right)^2 J_2\left(5\cos^2\theta - 1\right) - \frac{5}{8}\left(\frac{a}{r}\right)^4 J_4\left(63\cos^4\theta - 42\cos^2\theta + 3\right)\right)\left(\frac{y}{r}\right)$$

$$\frac{\partial U}{\partial z} = \frac{GM}{r^2}\left(1 - \frac{3}{2}\left(\frac{a}{r}\right)^2 J_2\left(5\cos^2\theta - 3\right) - \frac{5}{8}\left(\frac{a}{r}\right)^4 J_4\left(63\cos^4\theta - 70\cos^2\theta + 15\right)\right)\left(\frac{z}{r}\right)$$

where these gravity components correspond with (5.51), (5.52) and (5.53) and, for interest, they have been expanded to include the J_4 term in definition of gravitational potential (5.31).

5.2.9   *WGS84 Gravity Formula*

Gravity formulae give moderately accurate methods for calculating theoretical (or 'normal') gravity (γ) on the surface of an ellipsoid at a given latitude (μ). WGS84 adopts the following calculation:

$$\gamma = \gamma_e\left(\frac{1 + k\sin^2\mu}{\sqrt{1 - e^2\sin^2\mu}}\right) \tag{5.55}$$

where

$$k = \frac{c\gamma_p}{a\gamma_e} - 1$$

Relevant numerical values are as follows:

Equatorial radius	$a = 6378137.0\,\text{m}$
Polar radius	$c = 6356752.3142\,\text{m}$
Normal gravity at equator	$\gamma_e = 9.7803253359\,\text{m.s}^{-2}$
Normal gravity at pole	$\gamma_p = 9.8321849378\,\text{m.s}^{-2}$
First eccentricity	$e = 0.081819190842622$

For 'small' heights (h) above the reference ellipsoid (which effectively encompasses the service ceiling of operational aircraft), WGS84 represents normal gravity (γ_h) using a truncated Taylor series expansion:

$$\gamma_h = \gamma + \left(\frac{\partial Y}{\partial h}\right) h + \frac{1}{2}\left(\frac{\partial^2 Y}{\partial h^2}\right) h^2 \tag{5.56}$$

which is then quoted as:

$$\gamma_h = \gamma\left[1 - \frac{2}{a}\left(1 + f + q - 2f\sin^2\mu\right) h + \frac{3}{a^2}h^2\right] \tag{5.57}$$

where

$$q = \frac{\omega^2 a^2 c}{GM}$$

Relevant numerical values are as follows:

Equatorial radius	$a = 6378137.0\,\text{m}$
Polar radius	$c = 6356752.3142\,\text{m}$
Earth flattening parameter	$f = 1/298.257223563$
Earth rotation rate	$\omega = 7.292115 \times 10^{-5}\,\text{rad.s}^{-1}$
Earth gravitational constant	$GM = 3.986004418 \times 10^{14}\,\text{m}^3\text{s}^{-2}$

5.2.10   Geoid

The concept of a *geoid* has already been introduced in Section 3.5.2. It was discussed in that place because of its relevance to geospatial geometry, in respect of reference ellipsoids that define horizontal datum levels. In the context of this chapter, the important parameter is the gravitational potential on the surface of the Earth. Referring to GRS80, this is defined as $U_0 = 6.263686085 \times 10^7\,\text{m}^2\text{s}^{-2}$.

5.3   Geomagnetism

5.3.1   Earth's Magnetic Field

Geomagnetism is a combination of magnetic fields from different sources, which interact with each other through inductive processes. Ninety-five per cent of the total field is internal. This is the *main* field generated within the Earth's core, which remains reasonably constant over short time-scales (days) but changes over long time-scales (years). This slow change is known as secular variation. The *external* field is generated in the ionosphere and magnetosphere, driven by solar radiation and solar wind particles. It is much weaker and is usually variable over short time-scales. The remaining component is an *induced* field resulting from magnetisation in the crust driven by the main field and, to a lesser extent, the external field.

In general, the Earth's magnetism is highly dynamic but the dominant characteristics of the main field are predictable. To a first approximation, the Earth acts as a very large dipole, with a line of best-fit (defining the geomagnetic poles) that is only approximately aligned with the geographic poles.[3] The magnetic poles[4] (otherwise known as 'dip' poles) are the points at which the actual field lines are nominally vertical although this is subject to variability over short time-scales. Thus, a magnetic compass cannot give a truly accurate bearing . . . but has proved sufficiently accurate for practical navigation over many centuries. Knowledge of how the main field varies with location and time enables compass readings to be corrected and enables an effective navigational reference to be maintained over several years, before an update is required.

5.3.2   Magnetic Attraction

The force of attraction between two (opposite) magnetic poles, with strengths m_1 and m_2, varies according to the inverse square law:

$$F = \frac{\mu_0 m_1 m_2}{4\pi r^2} \tag{5.58}$$

where μ_0 is the permeability of free space and r is the distance between the two poles. Magnetic poles are either positive (attracted to North) and negative (attracted to South); similar (or 'like') poles repel and opposite poles attract.

As in gravitation, the motivation is to be able to calculate the magnetic force exerted by the Earth on a (weak) pole. Given the relative pole strengths, it is appropriate to write:

$$F = \frac{\mu_0 M m}{4\pi r^2} \tag{5.59}$$

where M is the strong pole (associated with the Earth) and m is the weak pole (associated with the object). By analogy with the concept of gravitational acceleration, the force on a unit pole is given by:

$$B = \frac{\mu_0 M}{4\pi r^2} \tag{5.60}$$

The quantity B is called the *magnetic induction*, which can be thought of as the field intensity. It is, in fact, a vector quantity B expressed in terms of magnetic potential U, where

$$\mathbf{B} = -\nabla U \tag{5.61}$$

---

[3] At the time of writing, the best-fit dipole gives the Geomagnetic North Pole at 79.74 °N 71.78 °W and the Geomagnetic South Pole at 79.74 °S 108.22 °E.

[4] At the time of writing, the North Magnetic Pole is estimated at 83.21 °N 118.32 °W and the South Magnetic Pole is estimated at 64.53 °S 137.86 °E.

Using spherical coordinates (as defined in Figure 5.2), the relationship between field intensity and field potential can be readily established:

$$\mathbf{B} = -\nabla U = -\frac{\partial U}{\partial r}\hat{\mathbf{r}} \equiv -\frac{\mu_0 M}{4\pi r^2}\hat{\mathbf{r}} \Rightarrow U(r) = -\frac{\mu_0 M}{4\pi r} \tag{5.62}$$

This defines the field around a magnetic monopole, which does not exist in nature. However, this can be applied twice in order to derive the field around a magnetic dipole, as shown in Figure 5.7. By convention, a unit pole is placed at a point P with polar coordinates (r, θ); this pole is positive, so that it is attracted to the top of the dipole (i.e. North).

The dipolar field is simply the sum of the individual fields due to the North and South poles:

$$U_1(r_1) = -\frac{\mu_0 M}{4\pi r_1} \quad U_2(r_2) = \frac{\mu_0 M}{4\pi r_2} \tag{5.63}$$

Applying the cosine rule, the distances from each pole to point P are:

$$r_1^2 = r^2 + \left(\frac{L}{2}\right)^2 - Lr\cos\theta \quad r_2^2 = r^2 + \left(\frac{L}{2}\right)^2 + Lr\cos\theta \tag{5.64}$$

Thus, the combined field is:

$$\begin{aligned} U(r, \theta) &= U_1(r, \theta) + U_2(r, \theta) \\ &= -\frac{\mu_0 M}{4\pi r}\left[\left(1 + \left(\frac{L}{2r}\right)^2 - \left(\frac{L}{r}\right)\cos\theta\right)^{-\frac{1}{2}} - \left(1 + \left(\frac{L}{2r}\right)^2 + \left(\frac{L}{r}\right)\cos\theta\right)^{-\frac{1}{2}}\right] \end{aligned} \tag{5.65}$$

This can be approximated as:

$$U(r, \theta) \approx -\frac{\mu_0 ML}{4\pi r}\cos\theta$$

Defining the dipole moment to be $m = ML$, this can be simplified as follows:

$$U(r, \theta) \approx -\frac{\mu_0 m}{4\pi r^2}\cos\theta \tag{5.66}$$

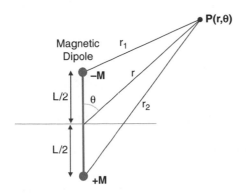

Figure 5.7 Dipolar magnetic field.

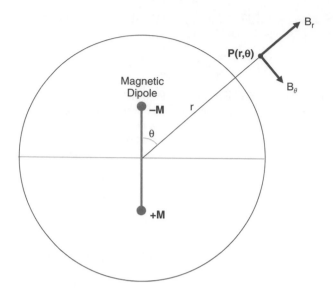

Figure 5.8 Radial and tangential field components.

Based on (5.61), the radial and tangential field components (shown in Figure 5.8) can be established as:

$$\mathbf{B} = B_r\hat{\mathbf{r}} + B_\theta\hat{\boldsymbol{\theta}} = -\hat{\mathbf{r}}\frac{\partial U}{\partial r} - \hat{\boldsymbol{\theta}}\frac{1}{r}\frac{\partial U}{\partial \theta} \tag{5.67}$$

This yields the following relationships:

$$B_r = 2B_0 \cos\theta \tag{5.68}$$

$$B_\theta = B_0 \sin\theta \tag{5.69}$$

where

$$B_0 = -\frac{\mu_0 m}{4\pi r^3} \tag{5.70}$$

Thus, $B_r = 0$ at the geomagnetic equator and $B_\theta = 0$ at a geomagnetic pole. Note that B_0 is the strength of the dipole at its equator ($\theta = \pi/2$). Also note that the field has axial symmetry.

The field intensity is given by:

$$B = \sqrt{B_r^2 + B_\theta^2} \tag{5.71}$$

$$B = \frac{\mu_0 m}{4\pi r^3}\sqrt{1 + 3\cos^2\theta} \tag{5.72}$$

This shows the field is strongest along the dipole axis and weakest around the equator.

5.3.3   *World Magnetic Model (WMM2005)*

WMM2005 (McLean, 2004) introduces an accurate definition of the main field, based on spherical harmonics (akin to EGM96) from the generic solution of Laplace's Equation (5.19):

$$U = -\frac{1}{a}\sum_{n=0}^{\infty}\sum_{m=0}^{n}\left[\left(C_n^m\left(\frac{a}{r}\right)^{n+1} + \overline{C}_n^m\left(\frac{r}{a}\right)^n\right)\cos m\lambda + \left(S_n^m\left(\frac{a}{r}\right)^{n+1} + \overline{S}_n^m\left(\frac{r}{a}\right)^n\right)\sin m\lambda\right]P_n^m(\cos\theta)$$

Unlike gravity, the coefficients are nonzero but they are neglected because they have small magnitude and tend to average out over short time-scales. Terms for which n=0 are deleted because magnetic monopoles do not exist in nature. From this, this solution can be simplified:

$$U = -\frac{1}{a} \sum_{n=1}^{\infty} \sum_{m=0}^{n} \left(\frac{a}{r}\right)^{n+1} \left[\ C_n^m \cos m\lambda + S_n^m \sin m\lambda\right] P_n^m(\cos\theta) \tag{5.73}$$

The actual form of magnetic potential in WMM2005 makes use of semi-normalised polynomials and so-called *Gauss* coefficients 'g' and 'h' (which are time-varying):

$$U(t) = a \sum_{n=1}^{\infty} \sum_{m=0}^{n} \left(\frac{a}{r}\right)^{n+1} \left[\ g_n^m(t)\cos m\,\lambda + h_n^m(t)\sin m\,\lambda\right] p_n^m(\cos\theta) \tag{5.74}$$

where

$$g_n^m = C_n^m A_n^m$$

$$h_n^m = S_n^m A_n^m$$

$$\bar{p}_n^m(\cos\theta) = P_n^m(\cos\theta)/A_n^m$$

The normalisation factors are defined by:

$$A_n^m = \sqrt{\frac{(n+m)!}{(n-m)!\ k}} \tag{5.75}$$

where

$$k = 1 \quad \text{if } m = 0$$
$$k = 2 \quad \text{if } m \neq 0$$

Interestingly, this differs from the equivalent definition in EGM96 (cf. (5.36)).

The WMM2005 documentation truncates the solution at $n = m = 36$. Coefficients are assumed to have a quadratic dependence on time from degree 1 to 8, a linear dependence from degree 9 to 12 and are constant thereafter. It is noted that the documentation only offers coefficients and first derivatives up to $n = m = 12$. For convenience, the time dependency relative to the 2005 epoch can be simplified as:

$$g_n^m(t) = g_n^m + \partial g_n^m(t - t_0) \tag{5.76}$$

$$h_n^m(t) = h_n^m + \partial h_n^m(t - t_0) \tag{5.77}$$

where the g, h, ∂g and ∂h terms are constants, defined in Table 5.2 up to $n = m = 6$.

By convention, the geomagnetic field vector B at any location is described by seven values:

X	North component
Y	East component
Z	Vertical component (+ve down)
F	Total magnitude
H	Horizontal magnitude
I	Inclination (or *magnetic dip*)
D	Declination (or *magnetic variation*)

Note that (unlike the gravity model) X, Y and Z do not represent Cartesian components: they are in fact spherical components, incorporating sign changes, as shown in Figure 5.9.

Under normal circumstances, the magnetic field vector would be evaluated as:

$$\mathbf{B} = -\nabla U = -\hat{\mathbf{r}}\frac{\partial U}{\partial r} - \hat{\boldsymbol{\theta}}\frac{1}{r}\frac{\partial U}{\partial \theta} - \hat{\boldsymbol{\lambda}}\frac{1}{r\sin\theta}\frac{\partial U}{\partial \lambda} \tag{5.78}$$

However, in this case, it has to be evaluated as follows:

$$\mathbf{B} = -\hat{\boldsymbol{\theta}}X + \hat{\boldsymbol{\lambda}}Y - \hat{\mathbf{r}}Z \tag{5.79}$$

Thus, the field vector is expressed as follows:

$$X = \frac{1}{r}\frac{\partial U}{\partial \theta} = \sum_{n=1}^{\infty}\sum_{m=0}^{n}\left(\frac{a}{r}\right)^{n+2}\left[g_n^m\cos m\lambda + h_n^m\sin m\lambda\right]\frac{\partial}{\partial \theta}p_n^m(\cos\theta) \tag{5.80}$$

$$Y = -\frac{1}{r\sin\theta}\frac{\partial U}{\partial \lambda} = \frac{1}{\sin\theta}\sum_{n=1}^{\infty}\sum_{m=0}^{n}\left(\frac{a}{r}\right)^{n+2}m\left[g_n^m\sin m\lambda - h_n^m\cos m\lambda\right]p_n^m(\cos\theta) \tag{5.81}$$

$$Z = \frac{\partial U}{\partial r} = -\sum_{n=1}^{\infty}\sum_{m=0}^{n}\left(\frac{a}{r}\right)^{n+2}(n+1)\left[g_n^m\cos m\lambda + h_n^m\sin m\lambda\right]p_n^m(\cos\theta) \tag{5.82}$$

The other magnetic field parameters are then derived as follows:

$$H = \sqrt{X^2 + Y^2} \tag{5.83}$$

$$F = \sqrt{X^2 + Y^2 + Z^2} \tag{5.84}$$

$$D = \tan^{-1}(Y/X) \tag{5.85}$$

$$I = \tan^{-1}(Z/H) \tag{5.86}$$

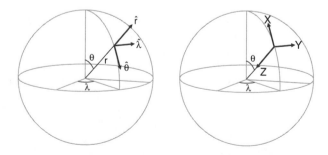

Figure 5.9 Spherical coordinates vs magnetic field components.

Table 5.2 WMM2005 coefficients (truncated at n=m=6).

Degree	Order	Gauss Coefficients		Secular Variation	
n	m	g_n^m	h_n^m	∂g_n^m	∂h_n^m
1	0	−29556.8		8.0	
1	1	−1671.7	5079.8	10.6	−20.9
2	0	−2340.6		−15.1	
2	1	3046.9	−2594.7	−7.8	−23.2
2	2	1657.0	−516.7	−0.8	−14.6
3	0	1335.4		0.4	
3	1	−2305.1	−199.9	−2.6	5.0
3	2	1246.7	269.3	−1.2	−7.0
3	3	674.0	−524.2	−6.5	−0.6
4	0	919.8		−2.5	
4	1	798.1	281.5	2.8	2.2
4	2	211.3	−226.0	−7.0	1.6
4	3	−379.4	145.8	6.2	5.8
4	4	100.0	−304.7	−3.8	0.1
5	0	−227.4		−2.8	
5	1	354.6	42.4	0.7	0.0
5	2	208.7	179.8	−3.2	1.7
5	3	−136.5	−123.0	−1.1	2.1
5	4	−168.3	−19.5	0.1	4.8
5	5	−14.1	103.6	−0.8	−1.1
6	0	73.2		−0.7	
6	1	69.7	−20.3	0.4	−0.6
6	2	76.7	54.7	−0.3	−1.9
6	3	−151.2	63.6	2.3	−0.4
6	4	−14.9	−63.4	−2.1	−0.5
6	5	14.6	−0.1	−0.6	−0.3
6	6	−86.3	50.4	1.4	0.7

Time variation in (X,Y,Z) is calculated by substituting Gauss coefficients and their secular variations. Time variation in the remaining values is calculated using:

$$\dot{H} = \frac{X\dot{X} + Y\dot{Y}}{H} \tag{5.87}$$

$$\dot{F} = \frac{X\dot{X} + Y\dot{Y} + Z\dot{Z}}{F} \tag{5.88}$$

$$\dot{D} = \frac{X\dot{Y} + Y\dot{X}}{H^2} \tag{5.89}$$

$$\dot{I} = \frac{H\dot{Z} + Z\dot{H}}{F^2} \tag{5.90}$$

5.3.4   *Approximate Dipole*

The WMM2005 solution for geomagnetic potential is given by (5.73):

$$U = a \sum_{n=1}^{\infty} \sum_{m=0}^{n} \left(\frac{a}{r}\right)^{n+1} \left[g_n^m \cos m\lambda + h_n^m \sin m\lambda \right] p_n^{-m}(\cos\theta)$$

This is a generalised *multipole* solution, as indicated by the presence of terms of degree $n > 1$. The best-fit dipole solution ($n=1$) is derived as:

$$U = a\left(\frac{a}{r}\right)^2 \left[g_1^0 p_1^{-0}(\cos\theta) + \left(g_1^1 \cos\lambda + h_1^1 \sin\lambda\right) p_1^{-1}(\cos\theta)\right] \qquad (5.91)$$

Note the absence of a h_1^0 term, which is consistent with Table 5.2.
    From the WMM2005 normalisation scheme, it is recognised that:

$$p_1^{-0}(\cos\theta) = P_1^0(\cos\theta) \equiv P_1(\cos\theta) \qquad (5.92)$$

$$p_1^{-1}(\cos\theta) = P_1^1(\cos\theta) \qquad (5.93)$$

Applying the definitions for Legendre polynomials and associated Legendre polynomials:

$$p_1^{-0}(\cos\theta) = \cos\theta \qquad (5.94)$$

$$p_1^{-1}(\cos\theta) = \sin\theta \qquad (5.95)$$

Thus, the dipole solution can be specified as follows:

$$U = a\left(\frac{a}{r}\right)^2 \left[g_1^0 \cos\theta + \left(g_1^1 \cos\lambda + h_1^1 \sin\lambda\right) \sin\theta\right] \qquad (5.96)$$

Using WMM2005:

$$g_1^0 = -29556.8$$
$$g_1^1 = -1671.7$$
$$h_1^1 = 5079.8$$

The magnetic field vector can be established as follows:

$$\mathbf{B} = -\hat{\mathbf{r}} \frac{\partial U}{\partial r} - \hat{\boldsymbol{\theta}} \frac{1}{r} \frac{\partial U}{\partial \theta} - \hat{\boldsymbol{\lambda}} \frac{1}{r \sin\theta} \frac{\partial U}{\partial \lambda}$$

$$= 2\left(\frac{a}{r}\right)^3 \left[g_1^0 \cos\theta + \left(g_1^1 \cos\lambda + h_1^1 \sin\lambda\right)\sin\theta\right]\hat{\mathbf{r}}$$

$$+ \left(\frac{a}{r}\right)^3 \left[g_1^0 \sin\theta - \left(g_1^1 \cos\lambda + h_1^1 \sin\lambda\right)\cos\theta\right]\hat{\boldsymbol{\theta}} \qquad (5.97)$$

$$+ \left(\frac{a}{r}\right)^3 \left(-g_1^1 \sin\lambda + h_1^1 \cos\lambda\right)\hat{\boldsymbol{\lambda}}$$

This can be re-written as:

$$\mathbf{B} \equiv B_r \hat{\mathbf{r}} + B_\theta \hat{\boldsymbol{\theta}} + B_\lambda \hat{\boldsymbol{\lambda}} \qquad (5.98)$$

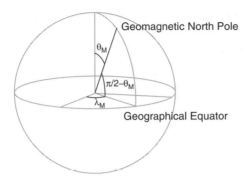

Figure 5.10 Geomagnetic North Pole.

Define the geomagnetic North Pole (along the dipole axis) to be located at longitude and (geocentric) co-latitude of (λ_M, θ_M), as shown in Figure 5.10. Recall that the magnetic field vector for a dipole is symmetry about its axis and that the tangential component is zero along the axis. Thus, in order to find the 'best-fit' dipole, it is necessary to find the value $\lambda = \lambda_M$ for which $B_\lambda = 0$ and then to find the value $\theta = \theta_M$ for which $B_\theta = 0$.

The vertical plane of symmetry relative to the dipole is established by setting $B_\lambda = 0$:

$$-g_1^1 \sin \lambda_M + h_1^1 \cos \lambda_M = 0$$

$$\tan \lambda_M = \frac{h_1^1}{g_1^1} \tag{5.99}$$

This implies that the Geomagnetic North Pole is at longitude 71.78 °W, which is equivalent to $\lambda_M = -71.78°$. From this, the plane azimuth can be expressed using Gauss coefficients:

$$\sin \lambda_M = \frac{h_1^1}{\sqrt{\left(g_1^1\right)^2 + \left(h_1^1\right)^2}} \tag{5.100}$$

$$\cos \lambda_M = \frac{g_1^1}{\sqrt{\left(g_1^1\right)^2 + \left(h_1^1\right)^2}} \tag{5.101}$$

Now, the dipole axis (within the plane of symmetry) is found by constraining the magnetic field to have a purely radial component, i.e. by setting the tangential component to zero. Thus, $B_\theta = 0$ and $\lambda = \lambda_M$.

Thus, the derivation proceeds as follows:

$$g_1^0 \sin \theta_M - \left(g_1^1 \cos \lambda_M + h_1^1 \sin \lambda_M \right) \cos \theta_M = 0$$

$$g_1^0 \sin \theta_M - \sqrt{\left(g_1^1\right)^2 + \left(h_1^1\right)^2} \cos \theta_M = 0$$

$$\tan \theta_M = \frac{\sqrt{\left(g_1^1\right)^2 + \left(h_1^1\right)^2}}{g_1^0} \tag{5.102}$$

This which implies that $\theta_M = 10.26°$, i.e. the Geomagnetic North Pole is at latitude 79.74 °N. From this, the dipole inclination can be expressed using Gauss coefficients:

$$\sin \theta_M = \frac{\sqrt{\left(g_1^1\right)^2 + \left(h_1^1\right)^2}}{\sqrt{\left(g_1^0\right)^2 + \left(g_1^1\right)^2 + \left(h_1^1\right)^2}} \tag{5.103}$$

$$\cos \theta_M = \frac{g_1^0}{\sqrt{\left(g_1^0\right)^2 + \left(g_1^1\right)^2 + \left(h_1^1\right)^2}} \tag{5.104}$$

Applying (5.98), (5.100) and (5.101), the magnetic field at the equator ($\lambda = \lambda_M$, $\theta = \pi/2$, $r = a$) is:

$$\mathbf{B} = -2\sqrt{\left(g_1^1\right)^2 + \left(h_1^1\right)^2}\,\hat{\mathbf{r}} + g_1^0\,\hat{\boldsymbol{\theta}} \tag{5.105}$$

Recall that the components of a dipolar field are defined by (5.67), (5.68) and (5.69):

$$\mathbf{B} = B_0 \left[2 \cos \theta\,\hat{\mathbf{r}} + \sin \theta\,\hat{\boldsymbol{\theta}} \right] \tag{5.106}$$

where θ is measured with respect to the dipole axis and B_0 (as defined by (5.70)) is the strength of the dipole at its equator, i.e. the geomagnetic equator. With reference to Figure 5.9, the Earth's equator corresponds with dipole co-latitude $\theta = \pi/2 - \theta_M$ and so the magnetic field vector (5.105) becomes:

$$\mathbf{B} = B_0 \left[2 \sin \theta_M\,\hat{\mathbf{r}} + \cos \theta_M\,\hat{\boldsymbol{\theta}} \right]$$

$$\mathbf{B} = \frac{B_0}{\sqrt{\left(g_1^0\right)^2 + \left(g_1^1\right)^2 + \left(h_1^1\right)^2}} \left[2\sqrt{\left(g_1^1\right)^2 + \left(h_1^1\right)^2}\,\hat{\mathbf{r}} + g_1^0\,\hat{\boldsymbol{\theta}} \right] \tag{5.107}$$

However, an alternative derivation gives the result in (5.105) and, by direct comparison, this will be true if and only if

$$B_0 = \sqrt{\left(g_1^0\right)^2 + \left(g_1^1\right)^2 + \left(h_1^1\right)^2} \tag{5.108}$$

Thus, using the values of Gauss coefficients associated with (5.96), the dipole strength on the geomagnetic equator is calculated as 30036.7 nT. Also, from (5.70),

$$B_0 \approx -\frac{\mu_0 m}{4\pi a^3} \tag{5.109}$$

where $\mu_0 = 4\tilde{\pi}10^{-7}\,\text{H.m}^{-1}$ is the permeability of free space and $a = 6371200.0\,\text{m}$ is the standard Earth magnetic reference radius (McLean, 2004, p. 25). From this, the dipole moment is evaluated to be $m \approx 7.79 \times 10^{22}\,\text{A.m}^2$.

5.4   Geopotential Computation

In order to pave the way for practical computation, it is necessary to review the mathematical treatment of spherical harmonics and the relevant properties of Legendre polynomials. When applied to specific geopotential models, this will enable the design of efficient algorithms.

5.4.1   *EGM96 and WMM2005 Spherical Harmonics*

Laplace's Equation has been solved using spherical harmonics, as shown in (5.19):

$$U = -\frac{1}{a} \sum_{n=0}^{\infty} \sum_{m=0}^{n} \left[\left(C_n^m \left(\frac{a}{r}\right)^{n+1} + \overline{C}_n^m \left(\frac{r}{a}\right)^n \right) \cos m\lambda + \left(S_n^m \left(\frac{a}{r}\right)^{n+1} + \overline{S}_n^m \left(\frac{r}{a}\right)^n \right) \sin m\lambda \right] P_n^m (\cos\theta)$$

Recall that this is summation of terms involving Legendre polynomials of *degree* n and *order* m. For both gravitational and magnetic models, this can be simplified as follows:

$$U = -\sum_{n=0}^{\infty} \sum_{m=0}^{n} \frac{a^n P_n^m (\cos\theta)}{r^{n+1}} \left[C_n^m \cos m\lambda + S_n^m \sin m\lambda \right]$$

By convention, this is re-expressed using semi-normalised Legendre polynomials.

For EGM96, the solution is recast in the particular form (cf. (5.35)):

$$U = -\frac{GM}{r} \left[1 + \sum_{n=2}^{\infty} \sum_{m=0}^{n} \frac{a^n p_n^{-m} (\cos\theta)}{r^n} \left[c_n^m \cos m\lambda + s_n^m \sin m\lambda \right] \right]$$

where the first few 's' and 'c' parameters are given in Table 5.1 (truncated at $n = m = 6$). Note that $a = 6378137$ m is the WGS equatorial radius (3.60).

For WMM2005, the solution is recast in the particular form (cf. (5.73)):

$$U = a \sum_{n=2}^{\infty} \sum_{m=0}^{n} \frac{a^{n+1} p_n^{-m} (\cos\theta)}{r^{n+1}} \left[g_n^m \cos m\lambda + h_n^m \sin m\lambda \right]$$

where the first few Gauss coefficients ((5.75) and (5.76)) are given in Table 5.2 (truncated at $n = m = 6$), each with its *secular variation*. Note that $a = 6371200.0$m is the Earth magnetic reference radius.

These relatively minor distinctions between the two model formulations are an inconvenience. The practical approach is to adopt a common solution of the Laplace Equation and readjust the coefficients accordingly. Thus, the method of computation adopted here will be, as follows:

$$U = -\frac{1}{a} \sum_{n=0}^{\infty} \sum_{m=0}^{n} \left(\frac{a}{r}\right)^{n+1} \left[C_n^m \cos m\lambda + S_n^m \sin m\lambda \right] P_n^m (\cos\theta) \qquad (5.110)$$

Correspondence between Legendre polynomials and semi-normalised Legendre polynomials has to be established such that:

$$C_n^m P_n^m (\cos\theta) = c_n^m p_n^m (\cos\theta) \qquad (5.111)$$

$$S_n^m P_n^m (\cos\theta) = s_n^m p_n^m (\cos\theta) \qquad (5.112)$$

where

$$p_n^{-m} (\cos\theta) = P_n^m (\cos\theta) / A_n^m \qquad (5.113)$$

For EGM96, comparing (5.34) and (5.35), the necessary readjustment of coefficients 'c' and 's' in order to match (5.108) is, as follows:

$$C_0^0 = \left(\frac{GM}{a}\right) \qquad C_1^0 = C_1^1 = S_0^0 = S_1^0 = S_1^1 = 0 \tag{5.114}$$

$$C_n^m = \left(\frac{GM}{a}\right) C_n^{-m}/A_n^m \qquad S_n^m = \left(\frac{GM}{a}\right) S_n^{-m}/A_n^m \tag{5.115}$$

For WMM2005, comparing (5.73) and (5.74), the necessary readjustment of coefficients 'g' and 'h' in order to match (5.108) is, as follows:

$$C_1^0 = S_0^0 = 0 \tag{5.116}$$

$$C_n^m = -a^2 g_n^m/A_n^m \qquad S_n^m = -a^2 h_n^m/A_n^m \tag{5.117}$$

5.4.2   *Recurrence Formulae*

The Legendre polynomial of *degree* n can be derived from (5.20). The associated Legendre polynomial of *degree* n and *order* m can be derived from (5.21). For consistency, it is appropriate to apply a common notation and refer to both types of polynomial as 'Legendre polynomials'. Thus, the relevant definitions become:

$$P_n^0(x) = \frac{1}{2^n n!} \frac{d^n}{dx^n} \left(x^2 - 1\right) \tag{5.118}$$

$$P_n^m(x) = \left(1 - x^2\right)^{m/2} \frac{d^m}{dx^m} P_n(x) \tag{5.119}$$

Without venturing into the mathematical details, it is sufficient to note that a variety of recurrence formulae exist that relate polynomials of different *degree* and *order* (e.g. Abramovitz & Stegun, 1965; Hildebrand, 1967; Arfken & Weber, 2001). Two commonly quoted formulae are, as follows:

$$(n - m)P_n^m(x) = (2n - 1) x P_{n-1}^m(x) - (n + m - 1)P_{n-2}^m(x) \tag{5.120}$$

$$\sqrt{1 - x^2} \, P_n^m(x) = (n - m + 1) x P_n^{m-1}(x) - (n + m - 1)P_{n-1}^{m-1}(x) \tag{5.121}$$

As stated previously, it is conventional to express Legendre polynomials using trigonometric functions and to parameterise each polynomial with respect to $x = \cos\theta$. Thus, applying (5.116) and (5.117), a matrix of results up to 5th degree and 5th order is shown in Table 5.3. From this, it is seen that the elements on the leading diagonal can be generated using a simple recurrence formula:

$$P_m^m = (2m - 1) \sin\theta \, P_{m-1}^{m-1} \tag{5.122}$$

Furthermore, the first off-diagonal element in each column is generated by:

$$P_{m+1}^m = (2m + 1) \cos\theta \, P_m^m \tag{5.123}$$

Table 5.3 Legendre polynomials (for n=1..4 and m=1..n)

P_n^m	m=0	m=1	m=2	m=3	m=4
n=0	1				
n=1	$\cos\theta$	$\sin\theta$			
n=2	$\frac{1}{2}\left(3\cos^2\theta-1\right)$	$3\cos\theta\sin\theta$	$3\sin^2\theta$		
n=3	$\frac{1}{2}\left(5\cos^3\theta-3\cos\theta\right)$	$\frac{3}{2}\left(5\cos^2\theta-1\right)\sin\theta$	$15\cos\theta\sin^2\theta$	$15\sin^3\theta$	
n=4	$\frac{1}{8}\left(35\cos^4\theta-30\cos^2\theta+3\right)$	$\frac{5}{2}\left(7\cos^3\theta-3\cos\theta\right)\sin\theta$	$\frac{15}{2}\left(7\cos^2\theta-1\right)\sin^2\theta$	$105\cos\theta\sin^3\theta$	$105\sin^4\theta$

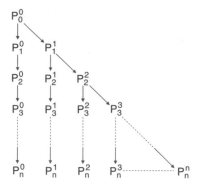

Figure 5.11 Sequence of evaluation of Legendre polynomials.

All subsequent elements (i.e. $n > m + 1$) are generated using a modified form of (5.117):

$$P_n^m = \frac{1}{n - m} \left[(2n - 1) \cos \theta \, P_{n-1}^m - (n + m - 1) P_{n-2}^m \right] \tag{5.124}$$

These three formulae provide a recursive scheme for evaluating Legendre polynomials, following the sequence shown in Figure 5.11.

The same basic sequence can also be applied to the EGM96 and WMM2005 scaling factors (cf. (5.36) and (5.37)). These are defined as:

$$A_0^0 = 1$$

$$A_n^m = \sqrt{\frac{1}{2} \frac{(n + m)!}{(n - m)!(2n + 1)}} \text{ for } n \neq 0 \text{ and } m \neq 0 \text{ (EGM96)}$$

$$A_n^m = \sqrt{\frac{1}{2} \frac{(n + m)!}{(n - m)!}} \text{ for } n \neq 0 \text{ and } m \neq 0 \text{ (WMM2005)}$$

The computational sequence for EGM96 scaling factors, for $m > 0$, is initialised as:

$$A_1^1 = \frac{\sqrt{3}}{3} \tag{5.125}$$

and the remaining values can be derived using the following recurrence formulae:

$$A_m^m = (2m - 1) \sqrt{\frac{2m}{2m + 1}} A_{m-1}^{m-1} \tag{5.126}$$

$$A_n^m = \sqrt{\left(\frac{n + m}{n - m} \right) \left(\frac{2n - 1}{2n + 1} \right)} A_{n-1}^m \tag{5.127}$$

The corresponding WMM2005 sequence is initialised as:

$$A_1^1 = 1 \tag{5.128}$$

and the remaining values can be derived using the following recurrence formulae:

$$A_m^m = \sqrt{2m(2m-1)}\, A_{m-1}^{m-1} \tag{5.129}$$

$$A_n^m = \sqrt{\left(\frac{n+m}{n-m}\right)}\, A_{n-1}^m \tag{5.130}$$

5.4.3   Cunningham's Method

Rather than plug numbers into (5.110) in order to compute a value (U) for the particular potential field, it makes more sense to utilise the recurrence formulae wherever possible. These are computationally more efficient, mainly because they avoid the explicit and repetitive use of trigonometric functions. The basic principle has been illustrated in Section 5.4.2 and, with some further elaboration, it is possible to obtain a complete scheme for geopotential computations.

To this end, a well-established method is due to Cunningham (1970) (which is summarised by Montenbruck and Gill, 2000). Although the original paper becomes mathematically intricate, the underlying principle is simple and the practical implementation is based on the recurrence formulae for Legendre polynomials. This was expressed in a form suitable for gravitational fields but the approach is completely general.

First, as discussed already, it is necessary to adjust the solution of Laplace's Equation given in (5.110), as follows:

$$U = -\frac{1}{a} \sum_{n=0}^{\infty} \sum_{m=0}^{n} \left(\frac{a}{r}\right)^{n+1} \left[C_n^m \cos\, m\lambda + S_n^m \sin\, m\lambda\right] P_n^m(\cos\theta)$$

Note 5.6

Cunningham (1970) sets up the solution of Laplace's Equation in a slightly different format from that used in this book. Introducing a few notational changes in order to remain consistent with the derivation so far, this can be written as follows:

$$U = \mathbf{real}\left\{ \sum_{n=0}^{\infty} \sum_{m=0}^{n} a^n \left[C_n^m - iS_n^m\right] V_n^m \right\}$$

where

$$V_n^m = \frac{P_n^m\,(\cos\, m\lambda + i\sin\, m\lambda)}{r^{n+1}}$$

and

$$i = \sqrt{-1}$$

(continued)

A general expression for high-order partial derivatives is derived as follows:

$$\frac{\partial^{\alpha+\beta+\gamma}}{\partial x^{\alpha}\partial y^{\beta}\partial z^{\gamma}}V_n^m = i^{\beta}\sum_{j=0}^{\alpha+\beta}\frac{(-1)^{\alpha+\gamma-j}}{2^{\alpha+\beta}}\frac{(n-m+\gamma+2j)\,!}{(n-m)\,!}C_{\alpha\beta k}V_{n+\alpha+\beta+\gamma}^{m+\alpha+\beta-2j}$$

where

$$C_{\alpha\beta j}=\sum_{k}(-1)^k(_{\alpha}C_{j-k})(_{\beta}C_k)$$

in which $\mathbf{max}(0,j-\alpha)\le k\le \mathbf{min}(\beta,j)$

Note that the symbol C, when used in combination with prefix and suffix, is the combinatorial quantity defined by:

$$_nC_r^n=\frac{n!}{r!(n-r)\,!}$$

Secondly, introduce a reformulation based on functions 'V' and 'W':

$$U=-\frac{1}{a}\sum_{n=0}^{\infty}\sum_{m=0}^{n}\left[C_n^m V_n^m + S_n^m W_n^m\right] \tag{5.131}$$

where

$$V_n^m=\left(\frac{a}{r}\right)^{n+1}P_n^m(\cos\theta)\cos m\lambda \tag{5.132}$$

$$W_n^m=\left(\frac{a}{r}\right)^{n+1}P_n^m(\cos\theta)\sin m\lambda \tag{5.133}$$

Thirdly, with reference to Figure 5.2, establish the cartesian coordinates of a general point in space (x,y,z) relative the centre of the Earth, as follows:

$$\begin{aligned}x&=r\sin\theta\cos\lambda\\y&=r\sin\theta\sin\lambda\\z&=r\cos\theta\end{aligned} \tag{5.134}$$

Fourthly, it should be recognised that recurrence formulae exist for trigonometric functions:

$$\begin{aligned}\sin m\lambda&=\sin(m-1)\,\lambda\cos\lambda+\cos(m-1)\,\lambda\sin\lambda\\\cos m\lambda&=\cos(m-1)\lambda\cos\lambda\ -\sin(m-1)\,\lambda\sin\lambda\end{aligned} \tag{5.135}$$

Now, the computation of geopotential can be cast in terms of cartesian coordinates (x,y,z). Applying the philosophy of Cunningham's method, the 'V' and 'W' functions satisfy the following recurrence formulae. The evaluation sequence is the same as for Legendre polynomials, as shown in Figure 5.11.

The diagonal elements are derived using (5.122), (5.132) and (5.133):

$$V_m^m=(2m-1)\left(\frac{a}{r^2}\right)\left[xV_{m-1}^{m-1}-yW_{m-1}^{m-1}\right] \tag{5.136}$$

$$W_m^m=(2m-1)\left(\frac{a}{r^2}\right)\left[xW_{m-1}^{m-1}+yV_{m-1}^{m-1}\right] \tag{5.137}$$

The first off-diagonal elements are derived using (5.123) and (5.132):

$$V_{m+1}^{m} = (2m + 1)\left(\frac{a}{r^2}\right)zV_m^m \tag{5.138}$$

$$W_{m+1}^{m} = (2m + 1)\left(\frac{a}{r^2}\right)zW_m^m \tag{5.139}$$

All subsequent elements are derived using (5.124), (5.132) and (5.133):

$$V_n^m - \frac{1}{n-m}\left(\frac{a}{r^2}\right)\left[(2n-1)\,zV_{n-1}^m - (n+m-1)\,V_{n-2}^m\right] \tag{5.140}$$

$$W_n^m = \frac{1}{n-m}\left(\frac{a}{r^2}\right)\left[(2n-1)\,zW_{n-1}^m - (n+m-1)\,W_{n-2}^m\right] \tag{5.141}$$

The sequence of calculations is initialised as follows:

$$V_0^0 = \left(\frac{a}{r}\right) \quad \text{and} \quad W_0^0 = 0 \tag{5.142}$$

Now, in order to obtain field vectors, it is necessary to perform the following operation:

$$\nabla U = \hat{\mathbf{x}}\frac{\partial U}{\partial x} + \hat{\mathbf{y}}\frac{\partial U}{\partial y} + \hat{\mathbf{z}}\frac{\partial U}{\partial z} \tag{5.143}$$

For EGM96, this is the gravitational acceleration ($\mathbf{g} = -\sigma U$) and, for WMM2005, this is the magnetic induction ($\mathbf{B} = -\sigma U$). These vectors are introduced in (5.14) and (5.61), respectively.

Applying Cunningham's derivation (cf. Note 5.6) in relation to (5.131) and (5.143), the following derivatives are obtained:

$$\frac{\partial U}{\partial x} = -\frac{1}{a}\sum_{n=0}^{\infty}\left[C_n^0 V_{n+1}^1 + \frac{1}{2}\sum_{m=1}^{n}\left[(C_n^m V_{n+1}^{m+1} + S_n^m W_{n+1}^{m+1}) - \frac{(n-m+2)!}{(n-m)!}(C_n^m V_{n+1}^{m-1} + S_n^m W_{n+1}^{m-1})\right]\right] \tag{5.144}$$

$$\frac{\partial U}{\partial y} = -\frac{1}{a}\sum_{n=0}^{\infty}\left[C_n^0 W_{n+1}^1 + \frac{1}{2}\sum_{m=1}^{n}\left[(C_n^m W_{n+1}^{m+1} - S_n^m V_{n+1}^{m+1}) + \frac{(n-m+2)!}{(n-m)!}(C_n^m W_{n+1}^{m-1} - S_n^m V_{n+1}^{m-1})\right]\right] \tag{5.145}$$

$$\frac{\partial U}{\partial z} = -\frac{1}{a}\sum_{n=0}^{\infty}\left[\sum_{m=0}^{n}(n-m+1)(C_n^m V_{n+1}^m + S_n^m W_{n+1}^m)\right] \tag{5.146}$$

The implementation of this method is outlined in Algorithm 5.1. The might be used, as follows:

```
>> g = calculateEGM96(100000,53,27)
g =
    x: -5.13587414723065
    y: -2.61735142789896
    z: -7.59931547962109
>> acc = sqrt(g.x^2 + g.y^2 + g.z^2)
acc =
  9.53820357887824
```

This gives the x, y and z components of gravitational acceleration at a geodetic latitude of 53 °N, longitude 27°E and altitude of 100km. This position can be expressed in cartesian coordinates:

$$x = 3.481038851070739e + 006$$
$$y = 1.773677885299388e + 006$$
$$z = 5.150407038863236e + 006$$

The radial distance from the centre of the Earth is:

$$r = 6.464538451385061e + 006$$

The simplest gravity model would give the following acceleration:

$$acc = GM/r^2 = 9.53811885113589 \text{m.s}^{-2}$$

Clearly, the variation between EGM96 and Newtonian gravity is slight (in this case, less than 0.0025 %) but this is a significant factor in predicting satellite orbits over very long periods.

```
function g = calculateEGM96(altitude,latitude,longitude)

% Constants
GM = 3.986004418e14 ;
a = 6378137.0 ;
c = 6356752.3 ;

% Position
cosLon = cosd(longitude) ;
sinLon = sind(longitude) ;
cosLat = cosd(latitude) ;
sinLat = sind(latitude) ;
cosColat = cosd(90-latitude) ;
sinColat = sind(90-latitude) ;

h = altitude ;
e = sqrt(1 - (c/a)^2) ;
n = a/sqrt(1 - (e*sinLat)^2) ;
x = (h + n)*cosLat*cosLon ;
y = (h + n)*cosLat*sinLon ;
z = (h + n*(1-e^2))*sinLat ;
r = sqrt(x^2 + y^2 + z^2) ;
```

```
% Resolution of gravity model
N = 6 ; M = 6 ;

% Offset for array indexing
L = 1 ;

% EGM96 Scaling factors
for im = 0:M
    if im==0
        A(L+im,L+im) = 1 ;
    else
        A(L+im,L+im) = ...
            (2*im-1) * sqrt(2*im/(2*im+1)) * A(L+im-1,L+im-1) ;
    end

    for in = im+1:N
        A(L+in,L+im) = ...
            sqrt((in+im)/(in-im)) * sqrt((2*in-1)/(2*in+1)) *A(L+in-1,L+im);
    end
end

% EGM96 coefficients
C(L+0,L+0) = 1 ;
C(L+1,L+0) = 0 ;
C(L+1,L+1) = 0 ;

S(L+0,L+0) = 0 ;
S(L+1,L+0) = 0 ;
S(L+1,L+1) = 0 ;

cc(L+2,L+0) = -0.484165371736e-3 ;
cc(L+2,L+1) = -0.186987635955e-9 ; ss(L+2,L+1) = 0.119528012031e-8 ;
cc(L+2,L+2) =  0.243914352398e-5 ; ss(L+2,L+2) = -0.140016683654e-5 ;
cc(L+3,L+0) =  0.957254173792e-6 ;
cc(L+3,L+1) =  0.202998882184e-5 ; ss(L+3,L+1) = 0.248513158716e-6 ;
cc(L+3,L+2) =  0.904627768605e-6 ; ss(L+3,L+2) = -0.619025944215e-6 ;
cc(L+3,L+3) =  0.721072657057e-6 ; ss(L+3,L+3) = 0.141435626958e-5 ;
cc(L+4,L+0) =  0.539873863789e-6 ;
cc(L+4,L+1) = -0.536321616971e-6 ; ss(L+4,L+1) = -0.473440265853e-6 ;
cc(L+4,L+2) =  0.350694105785e-6 ; ss(L+4,L+2) = 0.662671572540e-6 ;
cc(L+4,L+3) =  0.990771803829e-6 ; ss(L+4,L+3) = -0.200928369177e-6 ;
cc(L+4,L+4) = -0.188560802735e-6 ; ss(L+4,L+4) = 0.309953169333e-6 ;
cc(L+5,L+0) =  0.685323475630e-7 ;
cc(L+5,L+1) = -0.621012128528e-7 ; ss(L+5,L+1) = -0.944226127525e-7 ;
cc(L+5,L+2) =  0.652438297612e-6 ; ss(L+5,L+2) = -0.323349612668e-6 ;
cc(L+5,L+3) = -0.451955406071e-6 ; ss(L+5,L+3) = -0.214847190624e-6 ;
cc(L+5,L+4) = -0.295301647654e-6 ; ss(L+5,L+4) = 0.496658876769e-7 ;
cc(L+5,L+5) =  0.174971983203e-6 ; ss(L+5,L+5) = -0.669384278219e-6 ;
cc(L+6,L+0) = -0.149957994714e-6 ;
```

Continued

```
cc(L+6,L+1) = -0.760879384947e-7 ; ss(L+6,L+1) = 0.262890545501e-7 ;
cc(L+6,L+2) =  0.481732442832e-7 ; ss(L+6,L+2) = -0.373728201347e-6 ;
cc(L+6,L+3) =  0.571730990516e-7 ; ss(L+6,L+3) = 0.902694517163e-8 ;
cc(L+6,L+4) = -0.862142660109e-7 ; ss(L+6,L+4) = -0.471408154267e-6 ;
cc(L+6,L+5) = -0.267133325490e-6 ; ss(L+6,L+5) = -0.536488432483e-6 ;
cc(L+6,L+6) =  0.967616121092e-8 ; ss(L+6,L+6) = -0.237192006935e-6 ;

for im = 1:M
    for in = im:N
        C(L+in,L+im) = cc(L+in,L+im)/A(L+in,L+im) ;
        S(L+in,L+im) = ss(L+in,L+im)/A(L+in,L+im) ;
    end
end

C = C * GM/a ;
S = S * GM/a ;

C = C(L+0:L+N,L+0:L+M) ;
S = S(L+0:L+N,L+0:L+M) ;

% Cunningham functions
for im = 0:M
    if im==0
        V(L+0,L+0) = a/r ;
        W(L+0,L+0) = 0 ;
    else
        V(L+im,L+im) = (2*im-1)*(a/r^2)*(x*V(L+im-1,L+im-1)-y*W
                        (L+im-1,L+im-1)) ;
        W(L+im,L+im) = (2*im-1)*(a/r^2)*(x*W(L+im-1,L+im-1)+y*V
                        (L+im-1,L+im-1)) ;
    end

    if im+1<=N
        V(L+im+1,L+im) = (2*im+1)*(a/r^2)*z*V(L+im,L+im) ;
        W(L+im+1,L+im) = (2*im+1)*(a/r^2)*z*W(L+im,L+im) ;
    end

    for in = im+2:N
        V(L+in,L+im) = (a/r^2)*((2*in-1)*z*V(L+in-1,L+im) - ...
                        (in+im-1)*V(L+in-2,L+im))/(in-im) ;
        W(L+in,L+im) = (a/r^2)*((2*in-1)*z*W(L+in-1,L+im) - ...
                        (in+im-1)*W(L+in-2,L+im))/(in-im) ;
    end
end

% Gravity components
n = 0 ;
gx = C(L+n,L+0)*V(L+n+1,L+1) ;
gy = C(L+n,L+0)*W(L+n+1,L+1) ;
gz = C(L+n,L+0)*V(L+n+1,L+0) + S(L+n,L+0)*W(L+n+1,L+0) ;

for in = 1:N
    for im = 1:in
        f1 = (in-im+1) ;
```

```
        f2 = (in-im+2) ;
        if im+1<=M && in+1<=N
            gx = gx + (C(L+in,L+im)*V(L+in+1,L+im+1) + ...
                       S(L+in,L+im)*W(L+in+1,L+im+1))/2 ;
            gy = gy + (C(L+in,L+im)*W(L+in+1,L+im+1) - ...
                       S(L+in,L+im)*V(L+in+1,L+im+1))/2 ;
        end
        if in+1<=N
            gx = gx - (C(L+in,L+im)*V(L+in+1,L+im-1) + ...
                       S(L+in,L+im)*W(L+in+1,L+im-1))*f1*f2/2 ;
            gy = gy + (C(L+in,L+im)*W(L+in+1,L+im-1) - ...
                       S(L+in,L+im)*V(L+in+1,L+im-1))*f1*f2/2 ;
            gz = gz + (C(L+in,L+im)*V(L+in+1,L+im) +
                       S(L+in,L+im)*W(L+in+1,L+im))*f1 ;
        end
    end
end

g.x = -gx/a ;
g.y = -gy/a ;
g.z = -gz/a ;

return
```

Algorithm 5.1 'calculateEGM96'

5.5   Final Comment on Geopotential Models

To conclude this short discussion on geopotential modelling, it should be clear that the subject matter offers considerable sophistication in its analysis of the Earth. Aerospace applications use a variety of models and, while it may be desirable to have only one definition, practitioners need to be aware of the diversity. A review of gravitational models and magnetic models reveals significant differences in the way in which the mathematics are worked out and care is required when setting up calculations and interpreting the results. As with any modelling task, whatever is used in an actual implementation must be clearly declared and defined. In this case, the inherent complexity means that this principle is crucially important.

Chapter 6

Atmosphere

6.1　Overview

The atmosphere is a body of air that envelops the Earth. It extends up to extremely high altitudes, into what is normally thought of as space, although the vast bulk of the air mass is below 10 km. Thus, in terms of aviation (and, indeed, the entire biosphere), it is a relatively thin layer of air (when compared with the mean radius of the Earth, which is approximately 6370 km). Its overall structure is summarised in Figure 6.1. This emphasises three different ways of separating the atmosphere into layers, based on composition, temperature and radiation.

Below about 80 km, the composition of the atmosphere is an almost uniform mixture of nitrogen (~78 %), oxygen (~21 %) and argon (~1 %), with traces of other gases and variable amounts of water vapour and particulate matter. Of the other gases, carbon dioxide, ozone and methane are of particular significance in the debate on climate change. This is the so-called *homosphere*. Above about 80 km, the composition changes with altitude as the residual air tends to stratify according to the molecular mass of the component gases. Nitrogen is virtually absent above 200 km, above which the dominant components tend to be atomic oxygen (up to 1000 km), helium (up to 2000 km) and hydrogen (above 2000 km). This is the *heterosphere*.

The atmospheric layers that are defined by temperature are the *troposphere*, the *stratosphere*, the *mesosphere* and the *thermosphere*. The upper boundaries are delineated by the tropopause, stratopause, mesopause and thermopause, respectively. Note that the mesopause marks the top of the homosphere.

The troposphere contains about 90 % of the air mass and is characterised by a linear reduction in temperature with altitude, down to about −50 °C at the tropopause (at mid-latitudes). Given this very low temperature, almost no atmospheric water can exist at higher altitude. The thickness of the troposphere varies continually and seasonally. Typically, it is more than 20 km at the equator and less than 10 km at the poles. Note that most cloud formation occurs below 8 km.

The stratosphere is strongly influenced by ozone, which is created by a reaction between atomic and molecular oxygen. Atmospheric flow patterns cause a high concentration of ozone at around 25 km, which is known as the *ozone layer*. Because ozone absorbs ultraviolet radiation from the Sun, the temperature in the *upper* stratosphere (above the ozone layer) increases with altitude. In the lower stratosphere, temperature is virtually constant and circulation is limited, making this layer very stable. Particulate matter (e.g. carbon particles derived from the combustion

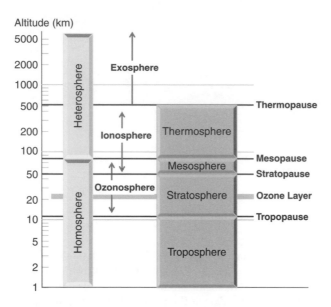

Figure 6.1 Overall structure of the atmosphere.

of petroleum products) that enters this layer is likely to remain there for prolonged periods (which is thought to be the key factor in global dimming).

The mesosphere is characterised by rapid temperature decrease with altitude, which is largely due to the role of carbon dioxide absorbing infrared radiation from the surface of the Earth (which is thought to be a key factor in global warming). It is also part of the process of ozone creation that operates in the stratosphere.

The thermosphere is a region of rapid temperature increase with altitude, leading to kinetic temperatures[1] of about $1500\,K$ at 300 km altitude. This is the result of absorbing ultraviolet radiation from the Sun and so this varies on a daily (or *diurnal*) cycle as well as on cycles of sunspot activity. Energy cannot be reradiated at this level and so must follow a slow conduction path to lower altitudes (below 100 km altitude) from where it can be radiated at night. Note that the atmosphere at these extreme altitudes is so rarified that, while the kinetic temperature is high, the specific heat capacity is practically zero. In other words, the external environment is cold!

Considering radiation, the designated atmospheric layers are the *ionosphere*, the *ozonospere* and the *exosphere*. These are of peripheral interest to this book and so will not be described in detail. In the ionosphere, cosmic and solar radiation ionises gas molecules and also causes the dissociation of molecular oxygen into atomic oxygen. This layer has high electrical conductivity (which affects radio communication) and can be energised by the solar wind (which creates aurorae in the skies around polar latitudes). The ozonosphere is a layer that corresponds with the stratosphere and the mesosphere, in which ultraviolet radiation dissociates any stray water vapour from lower layers into hydrogen and oxygen atoms. The process of ozone creation then takes place, as discussed already. The exosphere is essentially everything outside and it extends indefinitely. This contains the magnetosphere, which marks the extent of the geomagnetic field (as discussed at length in Chapter 4).

---

[1] Kinetic temperature is a measure of the energy state of individual atoms and molecules.

6.2   Standard Atmosphere Models

The development of standard models of the atmosphere has been underway since the 1920s, motivated by the need to provide a common basis for calibrating aircraft instruments and for analysing aircraft performance. In essence a standard atmosphere is a steady-state model that is averaged over a full year. It considers the air mass as a single entity that rotates with the Earth and comprises a homogeneous mixture of constituent gases (neglecting the effects of water vapour, particulate matter and solar activity). In this way, regional, diurnal and seasonal fluctuations are removed and so too is any dependency on latitude.

ESDU[2] Data Item 7702 1b offers a useful summary of developments, leading up to the two principal derivations that apply to aviation, namely the ISO[3] Standard Atmosphere[4] (ISO 2533:1975) and the US Standard Atmosphere (COESA, 1976). Early American and European efforts were harmonised in 1952 when the International Civil Aviation Organization (ICAO) adopted a standard atmosphere for altitudes up to 20 km. Various extensions followed as a result of experimental data gathered from high-altitude aircraft and, especially, rockets and satellites. Models are readily available for atmospheric properties up to 1000 km but, clearly, aviation rarely exceeds 20 km. Over this range all derivations of standard atmosphere are practically identical.[5] This includes the two mentioned already together with others, such as those adopted by ICAO (1993) and the World Meteorological Organization.

ISO 2533:1975 covers altitudes up to 80 km. The *International Standard Atmosphere* (ISA) refers strictly to the portion of this model up to 50 km, with the reminder being designated as an Interim Standard Atmosphere. In this context, the term 'interim' is not really appropriate as no serious change is envisaged in the foreseeable future. In any event, as already suggested, this is of no interest to aviation.

A useful extension to the ISA is ISO 5878:1982, which defines a set of reference models for aerospace use. This presents information on the temporal and spatial variations in atmospheric properties at levels between the surface and 80 km. This is achieved via reference atmospheres for tropical, subtropical, mid-latitude, subarctic and arctic regions. These are *nonstandard* or *off-standard* atmospheres that are used for more specific design investigations. This concept will be developed later in the chapter, with particular reference to ESDU Data Items 78008c and 78012b.

6.3   ISA Constants and Relationships

The International Standard Atmosphere is underpinned by the following constants:

Mean Earth Radius	$r_0 = 6356766 \, \text{m}$	(6.1)
Gravitational Acceleration	$g_0 = 9.80665 \, \text{m.s}^{-2}$	(6.2)
Speed of Sound	$a_0 = 340.294 \, \text{m.s}^{-1}$	(6.3)
Pressure	$p_0 = 1.01325 \times 10^{-5} \, \text{Pa}$	(6.4)
Temperature	$T_0 = 288.15 \, \text{K}$	(6.5)

---

[2] Engineering Sciences Data Unit (ESDU).
[3] International Standardisation Organisation (ISO).
[4] The coordination body within ISO is TC6 (*Technical Committee, Aircraft and Space Vehicles*) SC6 (*Sub-Committee, Standard Atmosphere*).
[5] A common definition exists for the altitude range from −2 km up to 32 km.

Density	$\rho_0 = 1.225 \text{ kg.m}^{-3}$	(6.6)
Molecular Mass (Dry Air)	$m_a = 28.96442 \text{ kg.kmol}^{-1}$	(6.7)
Thermal Conductivity	$k_0 = 2.5343 \times 10^{-2} \text{ W.m}^{-1}.\text{K}^{-1}$	(6.8)
Dynamic Viscosity	$\mu_0 = 1.7894 \times 10^{-5} \text{ N.s.m}^{-2}$	(6.9)
Kinematic Viscosity	$\nu_0 = 1.4607 \times 10^{-5} \text{ m}^2.\text{s}^{-1}$	(6.10)
Universal Gas Constant	$R_0 = 8314.32 \text{ J.kmol}^{-1}.\text{K}^{-1}$	(6.11)
Sutherland's Constants	$S = 110.4 \text{ K}$	(6.12)
	$\beta_s = 1.458 \times 10^{-6} \text{ N.s.m}^{-2}.\text{K}^{-0.5}$	(6.13)
Specific Heat Capacity	$C_p = 1005 \text{ J.kg}^{-1}.\text{K}^{-1}$	(6.14)

The parameter given in (6.14), namely C_P, is the specific heat capacity of a gas at constant pressure. An equivalent parameter, C_V, is defined at constant volume. The ratio of these parameters for dry air is defined as follows:

$$\gamma = C_p/C_v = 1.4 \tag{6.15}$$

The concentration of atmospheric gases can be approximated as given in Table 6.1. Bearing in mind that summary data of this type can vary in precision, the end result is a mean molecular mass of dry air that is usually quoted between 28.96 and 28.97 kg/kmol. In this case, the numbers conveniently add up to the relevant ISA constant although the detail is unimportant. In addition to the gas components included in this calculation, note that there are smaller concentrations of other gases (such as helium and methane) and, recognising that the ISA is based on dry air, the presence of water vapour is ignored.

As stated earlier, a standard atmosphere treats the air mass as a single entity that comprises a homogeneous mixture of constituent gases. For practical purposes, the same mixture exists up to 80 km and accordingly the value of m_a is constant. Thus the value of the gas constant $R = R_0/m_a$ applies to the whole atmosphere model:

$$R = \frac{R_0}{m_a} = \frac{8314.32}{28.96442} = 287.05287 \text{ J.Kg}^{-1}.\text{K}^{-1} \tag{6.16}$$

For the reasons discussed in Section 6.1, the atmospheric composition changes above 80 km. Although not of direct relevance to aviation, it is informative to see the impact of that change. This is defined in the US Standard Atmosphere and is depicted in Figure 6.2.

The speed of sound (a) in a gas is defined as follows:

$$a = \sqrt{\gamma R T} \tag{6.17}$$

where R is the gas constant, T is the gas temperature and γ is the ratio of specific heat capacities C_P and C_V. Note that the specific heat capacity is the amount of energy required in order to

Table 6.1 Approximate composition of dry air.

Gas	Molecular Mass	Fraction	Mass Fraction
N_2	28.013	78.083%	21.87339
O_2	31.999	20.946%	6.70251
Ar	39.948	0.933%	0.37272
CO_2	44.010	0.035%	0.01540
Ne	20.180	0.002%	0.00040
Molecular Mass (Dry Air)			**28.96442**

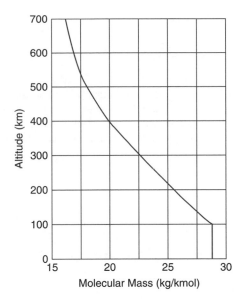

Figure 6.2 Variation of molecular mass with altitude.

effect a 1 K temperature rise per kilogram. In this case, this refers to a gas and the subscript 'P' denotes gas at constant pressure and the subscript 'V' denotes gas at constant volume. The ratio $\gamma = C_P/C_V$ has a value of 1.4 for dry air, as given in (6.15). Thus, the speed of sound in dry air can be derived as follows:

$$a = 20.046796\sqrt{T} \tag{6.18}$$

The relationship between pressure (p) and temperature (T) is:

$$p = \rho RT \tag{6.19}$$

where R is the gas constant and ρ is the gas density.

Thermal conductivity (k) is expressed as a function of temperature:

$$k = 2.648151 \times 10^{-3}\,\frac{T^{3/2}}{T + 245.4 \times 10^{(-12/T)}} \tag{6.20}$$

Dynamic viscosity (μ) is also expressed as a function of temperature:

$$\mu = \beta_s\frac{T^{3/2}}{T + S} \tag{6.21}$$

where S and β_s are Sutherland's empirical constants, given in (6.12) and (6.13).

Substituting numerical values, this becomes:

$$\mu = 1.458 \times 10^{-6}\,\frac{T^{3/2}}{T + 110.4} \tag{6.22}$$

From this, kinematic viscosity (v) is derived as:

$$v = \frac{\mu}{\rho} \tag{6.23}$$

where ρ is the gas density.

As well as using actual values of pressure (p), temperature (T) and density (ρ), it is often appropriate to use nondimensional parameters that relate actual values to reference values. In the context of atmospheric models, this is established using the ISA constants that are given in (6.4) to (6.6). Thus, the definitions of relative pressure (δ), relative temperature (θ) and relative density (σ) are given by:

$$\delta = \frac{p}{p_0} \tag{6.24}$$

$$\theta = \frac{T}{T_0} \tag{6.25}$$

$$\sigma = \frac{\rho}{\rho_0} \tag{6.26}$$

Note that, applying these parameters, Equations (6.17) and (6.19) can be recast as follows:

$$a = a_0 \sqrt{\theta} \tag{6.27}$$

$$\delta = \sigma\theta \tag{6.28}$$

6.4   Geopotential Altitude

6.4.1   *Standard Definition*

The variation of atmospheric properties with altitude is based on the principle of *hydrostatic equilibrium*, i.e. the air is stationary with respect to the Earth. This postulates a thin block of air with thickness dh and with unit horizontal area. Pressure p acts from below and pressure $p + dp$ acting from above. The difference in these pressures is equivalent to the reduction in weight of the vertical column of air that results from subtracting that block of air. Thus hydrostatic equilibrium is established in the form:

$$p = p + dp + \rho g \, dh$$

This simplifies (very simply!) to the standard form:

$$dp = -\rho g \, dh \tag{6.29}$$

This is known as the *hydrostatic equation*.

The value of gravitational acceleration (g) varies with altitude. As defined in (4.24), this obeys the inverse square law:

$$g = \frac{GM}{r^2} \tag{6.30}$$

As conventionally defined, r is the radial distance from the centre of mass and GM is the product of the gravitational constant and the mass of the Earth. Recall that WGS84 assigns a numerical value of GM $= 3.986004418 \times 10^{14}\,\mathrm{m^3 s^{-2}}$, as stated in (4.25). Assuming a mean radius of $r_0 = 6356766\,\mathrm{m}$ (as given in (6.1)), this gives a familiar-looking nominal value for gravitational acceleration of about $9.86\mathrm{m.s^{-1}}$. This is greater than the corresponding ISA constant, which will be discussed later.

In order to avoid dealing with variations in gravity with altitude, (6.26) can be rewritten in an equivalent form, namely:

$$dp = -\rho g\ dh = -\rho g_0\ dH \qquad (6.31)$$

Now the hydrostatic equation is based on the value of gravitational acceleration at sea-level and, as such, the altitude scale must be adjusted. The equivalence is expressed as:

$$dH = \frac{g}{g_0}\,dh \qquad (6.32)$$

where h is the *geometric* altitude and H is the so-called *geopotential* altitude. At sea-level, (6.27) becomes:

$$g_0 = \frac{GM}{r_0^2} \qquad (6.33)$$

At altitudes above sea-level, recognising that $r = r_0 + h$, the following integration can be performed:

$$H = \int_0^h \frac{g}{g_0}\,dh = \int_0^h \left(\frac{r_0}{r}\right)dh = \int_0^h \left(\frac{r_0}{r_0+h}\right)^2 dh = -\left[\frac{r_0^2}{r_0+h}\right]_0^h$$

The result is a simple relationship between geometric and geopotential altitudes:

$$H = \frac{r_0\,h}{r_0+h} \qquad (6.34)$$

The corresponding inverse relationship is, as follows:

$$h = \frac{r_0\,H}{r_0-H} \qquad (6.35)$$

6.4.2   *Generalised Definition*

The standard definition of geopotential altitude is based on ISA constants, which means that it is strictly valid at a single latitude north or south of the equator. A generalised definition needs to refer to geospatial and gravitational models from which those constants derive. In previous chapters, for example, this has been addressed via the WGS84 ellipsoid (cf. Section 6.2.1) and its associated gravity formula (cf. Section 4.2.9). In the current context (as defined in EDSU Data Items 77022b and 79018, and elsewhere), this is addressed via an 'International Ellipsoid of Reference' and a gravity formula that is called the Lambert Equation.

The parameters of the reference ellipsoid are, as follows:

$$\text{Earth Rotation Rate}\quad \omega = 7.29212 \times 10^{-5}\,\mathrm{rad.s^{-1}} \qquad (6.36)$$

$$\text{Equatorial Radius} \quad a = 6378178\,\text{m} \tag{6.37}$$

$$\text{Polar Radius} \quad c = 6356798\,\text{m} \tag{6.38}$$

The Lambert Equation specifies gravitational acceleration at mean sea level as a function of geodetic latitude[6] (φ):

$$g_0(\varphi) = 9.80616\left(1 - 0.0026373\cos 2\varphi + 0.0000059\cos^2 2\varphi\right) \tag{6.39}$$

Setting this quantity equal to ISA gravity as given in (6.2), this implies that ISA constants are valid for the nominal latitude of $\varphi = 45.5425°$.

The inverse square law for gravitation is stated in (6.30). The effect of altitude can be approximated by the following ratio:

$$g(\varphi) = g_0(\varphi)\left(\frac{r_0}{r_0 + h}\right)^2 \tag{6.40}$$

This specifies gravitational acceleration at altitude (h) above mean sea level as a function of geodetic latitude (φ). In fact, it measures altitude radially outwards from the centre of the Earth rather than orthogonal to the reference ellipsoid. A more accurate method relates to (2.76) and (2.77) but, for practical purposes, the eccentricity of the Earth is sufficiently small as to cause no great problem (as already discussed in Section 2.6.4).

Incorporating the centrifugal effects of Earth rotation (cf. Section 4.2.3), (6.39) is modified:

$$g(\varphi) = \left[g_0(\varphi) + \omega_0^2 r_0(\varphi)\cos^2\varphi\right]\left(\frac{r_0(\varphi)}{r_0(\varphi) + h}\right)^2 - \omega_0^2(r_0(\varphi) + h)\cos^2\varphi \tag{6.41}$$

However, rather than propagate the additional algebra, it is preferable to revert to the simpler structure of (6.40) and introduce a fictitious Earth radius (r_e). Thus,

$$g(\varphi) = g_0(\varphi)\left(\frac{r_e(\varphi)}{r_e(\varphi) + h}\right)^2 \tag{6.42}$$

where values of this fictitious radius are given by:

$$r_e(\varphi) = \frac{2g_0(\varphi)}{3.085462 \times 10^{-6} + 2.27 \times 10^{-9}\cos 2\varphi - 2 \times 10^{-12}\cos 4\varphi} \tag{6.43}$$

Note that, at the nominal latitude for ISA constants (discussed above), the numerical value is $6356766\,\text{m}$ at $\varphi = 45.5425°$, which accounts for (6.1).

From consideration of hydrostratic equilibrium, the general version of (6.31) is:

$$dp = -\rho g(\varphi)\,dh = -\rho g_0(\varphi)\,dH$$

---

[6] Note that geodetic latitude is denoted here by φ where, in Chapter 2, it was denoted by μ.

Accordingly, the equivalence with sea-level values (as originally stated in (6.32)) now becomes:

$$dH = \frac{g(\varphi)}{g_0(\varphi)}\left(\frac{g_0(\varphi)}{g_0}\right)dh \tag{6.44}$$

This integrates to give the generalised relationship between geometric and geopotential altitudes (cf. (6.34)), namely:

$$H = \frac{r_e(\varphi)\,h}{r_e(\varphi)+h}\left(\frac{g_0(\varphi)}{g_0}\right) \tag{6.45}$$

Remember that the gravitational acceleration at mean sea level $g_0(\varphi)$ is given by (6.39).

Recalling the discussion of geospatial geometry in Chapter 2, the radial distance from the centre of the Earth to any point on its surface is given by:

$$r_e(\varphi) = \sqrt{\frac{a^4 + c^4\tan^2\varphi}{a^2 + c^2\tan^2\varphi}} \tag{6.46}$$

or, equivalently:

$$r_e(\varphi) = \sqrt{\frac{a^4\cos^2\varphi + c^4\sin^2\varphi}{a^2\cos^2\varphi + c^2\sin^2\varphi}} \tag{6.47}$$

where the parameters where the parameters a and c are defined by (6.37) and (6.38) and where the notation from Chapter 2 has been modified to avoid a clash with ISA parameters.

6.5   Vertical Structure of the Atmosphere

Atmosphere models provide parametric data as functions of geopotential altitude. The vertical structure is based on a temperature profile that is appropriate for a given representation of the atmosphere. This takes the form of a multilayered model with a linear temperature variation within each layer. What distinguishes a particular model is the number layers, where the boundaries occur between layers and what temperature gradients are adopted within each layer. The general temperature-altitude relationship is defined as:

$$T = T_n + L_n(H - H_n) \tag{6.48}$$

This applies to the n-th layer with a base altitude H_n, a base temperature T_n (defined at H_n) and a linear gradient L_n (above H_n). Also, by implication, there is a base pressure p_n (defined at H_n).

The fundamental relationships governing pressure are given in (6.28) and (6.19):

$$dp = -\rho g_0\,dH$$

$$\rho = \frac{p}{RT}$$

These can be combined in order to give:

$$\frac{dp}{p} = -\frac{g_0}{R} \frac{1}{T} dH \tag{6.49}$$

Therefore,

$$\int \frac{dp}{p} = -\frac{g_0}{R} \int \frac{1}{T} dH \tag{6.50}$$

Applying the temperature profile from (6.48), there are two cases to be considered, namely $L_n = 0$ and $L_n \neq 0$.

When $L_n = 0$, the integral is trivial:

$$\log_e p - \log_e p_n = -\frac{g_0}{RT_n} \int_{H_n}^{H} dH = -\frac{g_0}{RT_n} (H - H_n)$$

$$\log_e \left(\frac{p}{p_n} \right) = -\frac{g_0}{RT_n} (H - H_n)$$

$$\frac{p}{p_n} = \exp\left(-\frac{g_0}{RT_n} (H - H_n) \right) \tag{6.51}$$

Alternatively, altitude can be expressed as a function of pressure ratio:

$$H - H_n = -\frac{RT_n}{g_0} \log_e \left(\frac{p}{p_n} \right) \tag{6.52}$$

When $L_n \neq 0$, the integral is nearly as trivial. From (6.33), it is seen that:

$$dT = L_n dH \tag{6.52}$$

Thus,

$$\log_e p - \log_e p_n = -\frac{g_0}{RL_n} \int_{T_n}^{T} \frac{dT}{T} = -\frac{g_0}{RL_n} (\log_e T - \log_e T_n)$$

$$\log_e \left(\frac{p}{p_n} \right) = -\frac{g_0}{RL_n} \log_e \left(\frac{T}{T_n} \right)$$

$$\frac{p}{p_n} = \left(\frac{T}{T_n} \right)^{-g_0/RL_{ni}} \tag{6.53}$$

Again, as an alternative, altitude can be expressed as a function of pressure ratio:

$$H - H_n = \frac{T - T_n}{L_n} = \frac{T_n}{L_n}\left(\frac{T}{T_n} - 1\right)$$

$$H - H_n = \frac{T_n}{L_n}\left[\left(\frac{p}{p_n}\right)^{-RL_n/g_0} - 1\right] \tag{6.54}$$

Using these expressions, the ISA vertical structure can be developed, using the values of H_n, T_n and L_n given in Table 6.2. The associated variations in pressure and temperature are shown in Figure 6.3. The underlying data for that figure was computed using Algorithm 6.1.

Table 6.2 Data for the International Standard Atmosphere.

n	H_n	T_n	L_n	p_n	g_0/RL_n	g_0/RT_n
0	0	288.15	-6.5×10^{-3}	1.013250×10^5	-5.255880	n/a
1	11000	216.65	0	2.263204×10^4	n/a	1.576885×10^{-4}
2	20000	216.65	1.0×10^{-3}	5.474879×10^3	34.16322	n/a
3	32000	228.65	2.8×10^{-3}	8.680160×10^2	12.20115	n/a
4	47000	270.65	0	1.109058×10^2	n/a	1.262266×10^{-4}
5	51000	270.65	-2.8×10^{-3}	6.693853×10^1	-12.20115	n/a
6	71000	214.65	-2.0×10^{-3}	3.956392	-17.08161	n/a
7	80000	196.65	n/a	8.862722×10^{-1}	n/a	n/a

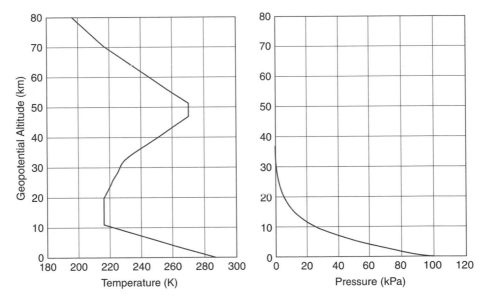

Figure 6.3 ISA temperature and pressure profiles.

Table 6.3 Pressure envelopes based on atmospheric measurement.

H (km)	High Pressure (mbar)					Low Pressure (mbar)				
	Max	1%	5%	10%	20%	Min	1%	5%	10%	20%
0	1084	–	–	–	–	870	–	–	–	–
1	930	920	918	917	916	842	847	856	861	868
2	821	817	816	815	814	736	742	748	752	757
4	643	642	641	640	639	548	550	558	565	569
6	501	499	497	496	495	406	408	413	418	422
8	385	384	383	382	381	296	299	303	306	309
10	294	293	292	291	290	215	218	221	223	225
12	226	226	224	223	222	154	157	158	160	162
14	168	167	166	165	164	111	111	113	115	117
16	123	123	122	121	120	79	79	80	82	84
18	88	88	87	86	85	56	56	57	58	59
20	65	65	64	63	62	40	41	42	43	45
22	45	45	44	44	43	28	29	30	31	32
24	35	34	33	32	32	20	21	24	27	28
26	26	25	25	24	24	14	15	18	20	21
28	20	19	19	18	18	10	11	13	15	16
30	15	15	14	14	13	7	9	10	11	12

Table 6.4 Temperature envelopes based on atmospheric measurement.

H (km)	High Temperature (°C)					Low Temperature (°C)				
	Max	1%	5%	10%	20%	Min	1%	5%	10%	20%
0	58	49	46	45	–	−68	−61	−57	−54	−51
1	41	40	39	38	34	−54	−53	−51	−50	−49
2	32	30	29	28	27	−47	−41	−36	−34	−31
4	19	17	14	13	12	−53	−48	−46	−44	−40
6	8	6	4	3	0	−61	−57	−56	−54	−51
8	−4	−5	−6	−9	−11	−68	−66	−65	−64	−61
10	−13	−13	−17	−19	−20	−75	−74	−72	−70	−65
12	−22	−22	−24	−30	−31	−80	−73	−72	−70	−67
14	−30	−30	−35	−36	−40	−77	−75	−74	−73	−70
16	−35	−37	−39	−39	−40	−87	−86	−84	−83	−82
18	−35	−37	−38	−39	−40	−88	−86	−84	−82	−81
20	−31	−32	−38	−39	−40	−87	−86	−83	−81	−71
22	−29	−30	−38	−38	−39	−85	−84	−83	−82	−82
24	−33	−33	−38	−39	−39	−86	−85	−83	−82	−81
26	−27	−28	−37	−37	−38	−84	−84	−83	−81	−80
28	−22	−23	−28	−33	−36	−84	−83	−81	−79	−78
30	−17	−18	−23	−28	−33	−85	−83	−81	−79	−76

Stacking layers of the atmospheric model together, the pressure and temperature at any altitude can be obtained as follows:

$$\frac{T}{T_0} = \left(\frac{T}{T_n}\right)\left(\frac{T_n}{T_{n-1}}\right)\left(\frac{T_{n-1}}{T_{n-2}}\right)\cdots\left(\frac{T_2}{T_1}\right)\left(\frac{T_1}{T_0}\right) \tag{6.55}$$

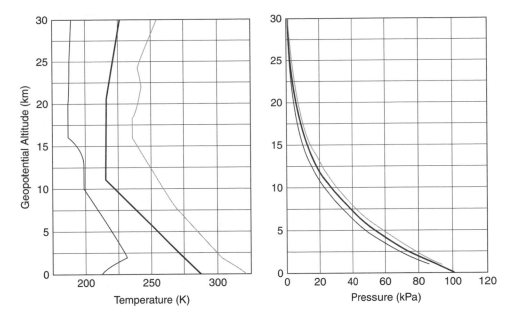

Figure 6.4 ISA profiles with 99 % likelihood envelopes.

$$\frac{P}{P_0} = \left(\frac{P}{P_n}\right)\left(\frac{P_n}{P_{n-1}}\right)\left(\frac{P_{n-1}}{P_{n-2}}\right)\cdots\left(\frac{P_2}{P_1}\right)\left(\frac{P_1}{P_0}\right) \tag{6.56}$$

This information can be set in an appropriate context by comparing it with data presented in MIL-HDBK-310. This provides a wide range of environmental data intended to guide the development of military products. Amongst other items, there are pressure and temperature envelopes based on measurement, as given in Tables 6.3 and 6.4. These do not represent coherent profiles but are simply maximum and minimum values at each altitude, together with related values for frequencies of occurrence of 1 %, 5 %, 10 % and 20 %. This is illustrated in Figure 6.4, where the ISA pressure and temperature profiles are contained within envelopes with 99 % likelihood of occurrence (i.e. between the 1 % values at the high and low extremes).

```
function (pressure,temperature) = InternationalStandardAtmosphere
  (altitude)
H = (0,11,20,32,47,51,71,80)*1000 ;
L = (-6.5,0.0,1.0,2.8,0.0,-2.8,-2.0)/1000 ;

isa.p0 = 101325 ;
isa.T0 = 288.15 ;
isa.g0 = 9.80665 ;
isa.R = 287.05287 ;

altitude = sort(altitude) ;
altitude(altitude<min(H)|altitude>max(H)) = [] ;
```

Continued

```
(P,T) = calculateIsaProfile(isa,H,L);
n = findLayer(altitude,H);
(pressure,temperature) = calculateProfile(isa,altitude,n,H,L,P,T);

function (P,T) = calculateIsaProfile(isa,H,L)
P(1) = isa.p0;
T(1) = isa.T0;
for iLayer = 2:length(H)
 gradient = L(iLayer-1);
 dH = H(iLayer) - H(iLayer-1);
 T(iLayer) = T(iLayer-1) + dH*gradient;

 switch gradient
 case 0
 exponent = -isa.g0*dH/(isa.R*T(iLayer-1));
 P(iLayer) = P(iLayer-1)*exp(exponent);
 otherwise
 exponent = -isa.g0/(isa.R*gradient);
 ratio = T(iLayer)/T(iLayer-1);
 P(iLayer) = P(iLayer-1)*ratio^exponent;
 end
end

function (pressure,temperature)=calculateProfile(isa,altitude,
  n,H,L,P,T)
for iAltitude = 1:length(altitude)
 iLayer = n(iAltitude);
 gradient = L(iLayer);
 dH = altitude(iAltitude) - H(iLayer);
 temperature(iAltitude) = T(iLayer) + dH*gradient;

 switch gradient
 case 0
 exponent = -isa.g0*dH/(isa.R*T(iLayer));
 pressure(iAltitude)=P(iLayer)*exp(exponent);
 otherwise
 exponent=-isa.g0/(isa.R*gradient);
 ratio = temperature(iAltitude)/T(iLayer);
 pressure(iAltitude)=P(iLayer)*ratio^exponent;
 end
end

function n = findLayer(altitude,H)
for iAltitude = 1:length(altitude)
 nLayer = max(find(altitude(iAltitude)>=H));
 switch nLayer
 case length(H)
 n(iAltitude) = nLayer - 1 ;
 otherwise
 n(iAltitude) = nLayer ;
 end
end
```

Algorithm 6.1 'InternationalStandardAtmosphere'.

6.6   Pressure Altitude

By definition, the *pressure altitude* of the point in any atmosphere is the geopotential altitude in the ISA that gives the same pressure. The ISA relationship between pressure and geopotential altitude is used as the calibration law for altimeters, which means that an altimeter will display pressure altitude if it is set to display zero when $p = p_0$, where p_0 is the ISA sea-level pressure defined in (6.4).

Combining (6.19) and (6.31), the relationship between pressure and geopotential altitude can be written as:

$$dp = -\frac{pg_0}{RT}\, dH \qquad (6.57)$$

This can be rewritten specifically for ISA:

$$dp = -\frac{pg_0}{RT_{ISA}}\, dH_P \qquad (6.58)$$

where pressure altitude (H_P) can be used because the temperature variation follows the ISA profile (as denoted by the subscript 'ISA'). Therefore the variation of geopotential altitude with pressure altitude is given by:

$$\frac{dH}{dH_P} = \frac{T}{T_{ISA}} \qquad (6.59)$$

The ISA temperature profile is defined by (6.48) and is redefined here, as follows:

$$T_{ISA} = T_n + L_n(H_P - H_{Pn}) \qquad (6.60)$$

where H_P is pressure altitude and H_{Pn} is the pressure altitude that corresponds with the base of the n-th layer in the atmosphere model. Any other temperature at any altitude can be produced simply by adding an increment (ΔT):

$$T = T_{ISA} + \Delta T \qquad (6.61)$$

Thus, (6.59) becomes:

$$\frac{dH}{dH_P} = \frac{L_n(H_P - H_{Pn}) + T_n + \Delta T}{L_n(H_P - H_{Pn}) + T_n} \qquad (6.62)$$

There are two cases to be considered, namely $L_n = 0$ and $L_n \neq 0$.

When $L_n = 0$, the integral is trivial:

$$\int_{H_n}^{H} dH = \frac{T_n + \Delta T}{T_n} \int_{H_{Pn}}^{H_P} dH_P$$

$$H - H_n = \left(1 + \frac{\Delta T}{T_n}\right)(H_P - H_{Pn})$$

$$= H_P - H_{Pn} + \left(\frac{\Delta T}{T_n}\right)(H_P - H_{Pn})$$

The relevant ISA relationship between geopotential altitude and pressure ratio was given by (6.52). By definition, this is equally valid for pressure altitude:

$$H_P - H_{Pn} = -\frac{RT_n}{g_0} \log_e \left(\frac{p}{p_n}\right) \tag{6.63}$$

Therefore, in the case when $L_n = 0$, the final expression that links geopotential altitude in any atmosphere with its equivalent pressure altitude is:

$$H - H_n = H_P - H_{Pn} - \Delta T \frac{R}{g_0} \log_e \left(\frac{p}{p_n}\right) \tag{6.64}$$

When $L_n \neq 0$, the integral is slightly more demanding:

$$\int_{H_n}^{H} dH = \int_{H_{Pn}}^{H_P} \left[1 + \frac{\Delta T}{L_n(H_P - H_{Pn}) + T_n}\right] dH_P$$

$$H - H_n = H_P - H_{Pn} + \frac{\Delta T}{L_n} \log_e \left[\frac{L_n(H_P - H_{Pn}) + T_n}{T_n}\right]$$

$$= H_P - H_{Pn} + \frac{\Delta T}{L_n} \log_e \left[1 + \frac{L_n}{T_n}(H_P - H_{Pn})\right]$$

The relevant ISA relationship between geopotential altitude and pressure ratio was given by (6.54). By definition, this is equally valid for pressure altitude:

$$H_P - H_{Pn} = \frac{T_n}{L_n}\left[\left(\frac{p}{p_n}\right)^{-RL_n/g_0} - 1\right] \tag{6.65}$$

A small amount of manipulation gives:

$$1 + \frac{L_n}{T_n}(H_P - H_{Pn}) = \left(\frac{p}{p_n}\right)^{-RL_n/g_0}$$

Therefore, in the case when $L_n \neq 0$, the final expression that links geopotential altitude in any atmosphere with its equivalent pressure altitude is:

$$H - H_n = H_P - H_{Pn} - \Delta T \frac{R}{g_0} \log_e \left(\frac{p}{p_n}\right) \tag{6.66}$$

Noting that (6.60) and (6.62) are identical, it is possible to write a single equation that defines the difference between geopotential altitude and pressure altitude:

$$H - H_P = H_n - H_{Pn} - \Delta T \frac{R}{g_0} \log_e \left(\frac{p}{p_n}\right) \tag{6.67}$$

This applies within the n-th layer of the atmosphere model. Lower layers are governed by equations of the form:

$$H_i - H_{Pi} = H_j - H_{Pj} - \Delta T \frac{R}{g_0} \log_e \left(\frac{p_i}{p_j} \right) \tag{6.68}$$

where $i = j+1$. Aggregating (6.67) and (6.68) across all layers in the atmosphere model, the following result is obtained:

$$H - H_P = H_0 - H_{P0} - \Delta T \frac{R}{g_0} \left[\log_e \left(\frac{p}{p_n} \right) + \log_e \left(\frac{p_n}{p_{n-1}} \right) \right.$$
$$\left. + \ldots \log_e \left(\frac{p_2}{p_1} \right) + \log_e \left(\frac{p_1}{p_0} \right) \right]$$

$$H - H_P = H_0 - H_{P0} - \Delta T \frac{R}{g_0} \log_e \left(\frac{p}{p_0} \right) \tag{6.69}$$

Thus, the difference between *geopotential altitude* and *pressure altitude* at any altitude above sea-level is determined as the difference of these quantities at sea level decremented by a function of temperature increment (ΔT) (from (6.61)) and pressure ratio (δ) (from (6.24)). In appropriate notation, this becomes:

$$H - H_P = \Delta H_0 - \Delta T \frac{R}{g_0} \log_e \delta \tag{6.70}$$

where ΔH_0 is the geopotential altitude that corresponds with zero pressure altitude. By convention, sea-level pressure is standard in all reference atmospheres and so $H = H_0 = H_{P0} = 0$, i.e. $\Delta H_0 = 0$. Using the constants defined in Section 6.1, (6.70) reduces to the following practical formula:

$$H - H_P = \Delta H_0 - 29.271247 \Delta T \ln \delta \tag{6.71}$$

6.7   Reference Atmospheres

As discussed, a standard atmosphere model is constructed as a global average of atmospheric parameters over a full year. Variations due to regional, diurnal and seasonal factors are removed, as well as those due to latitude. Thus, a single model is available for performance analysis and instrument calibration. As has been seen already, the International Standard Atmosphere is valid for a geodetic latitude of $45.5425°$.

Nonstandard *reference atmospheres* provide a wider context for aircraft design and operation, by characterising typical profiles that are associated with climatic zones and with seasonal variations are different latitudes. Further information can be found in ESDU Data Items 78008c and 78012b. In addition, a wide range of climatic and atmospheric data is presented in MIL-HDBK-310.

For current purposes it is appropriate to summarise the temperature profiles of four commonly quoted atmospheres in relation to the ISA profile, namely (1) Tropical Maximum, (2) Temperate and Arctic Maximum, (3) Tropical and Temperate Minimum and (4) Arctic Minimum. This is given in the form of a graphical sketch in Figure 6.5, with the underlying data contained in Table 6.5. Note that these data are calibrated against pressure altitude and only extend up to 20 km. Attention is drawn to the different number of layers in each

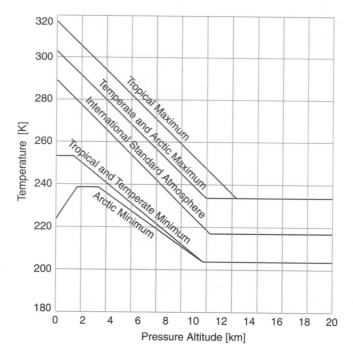

Figure 6.5 Temperature profiles for reference atmospheres.

Table 6.5 Data for reference atmospheres.

Atmosphere	n	H_{Pn}	T_n	L_n
International	0	0	288.15	−0.0065
Standard	1	11000	216.65	0
Atmosphere (ISA)	2	20000	216.65	n/a
Tropical	0	0	318.15	−0.0065
Maximum	1	13077	233.15	0
	2	20000	233.15	n/a
Temperate and	0	0	303.15	−0.0065
Arctic	1	10769	233.15	0
Maximum	2	20000	233.15	n/a
Tropical and	0	0	253.15	0
Temperate	1	1219	253.15	−0.0052917
Minimum	2	10668	203.15	0
	3	20000	203.15	n/a
Arctic	0	0	223.15	0.0098425
Minimum	1	1524	238.15	0
	2	3048	238.15	−0.0045932
	3	10668	203.15	0
	4	20000	203.15	n/a

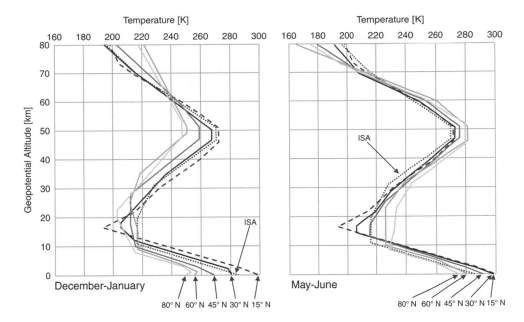

Figure 6.6 Mean temperature profiles at various latitudes.

atmosphere model, with specific emphasis on the low temperature near ground level in the Arctic Minimum climate.

6.8   Seasonal Variation

In order to illustrate the extent of seasonal variation, further information is sketched graphically in Figure 6.6. This shows the difference between midwinter climate (averaged over December and January) and midsummer climate (averaged over June and July) for a selection of northerly latitudes. For purposes of comparison, the ISA profile is included together with the mean annual profile for latitude 15 °N. Note that these data are calibrated against geopotential altitude and extend up to 80 km. Attention is drawn to the high number of layers in these atmosphere model, as opposed to the reference atmospheres giving global or climatic averages.

6.9   Climatic Regions

Standard atmospheres and reference atmospheres have the virtue of offering a small number of well-characterised *average* environments that can be used for engineering analysis and calibration. In order to set the context for this type of modelling, this section will give a very brief introduction to the climatic classification that defines the regional surface environments on the Earth. Two schemes will be considered, namely MIL-HDBK-310 and Köppen-Geiger.

6.9.1   *MIL-HDBK-310 Classification*

MIL-HDBK-310 is intended to give guidance on a wide of range of environmental factors that are relevant to military product design. In terms of surface environment, it defines five regional types of climate. Four of these pertain to land areas and are based on temperature and humidity, as these are important factors for equipment design. The four types of climatic region are designated Basic, Hot, Cold and Severe Cold.

The *Basic Region* is defined by 1 % values[7] of temperature of $-31.7\,°C$ and $43.3\,°C$. It relates to mid-latitudes, which would normally by thought of as having temperate or moderate climates, as well as humid tropics, which are warm throughout the whole year (without extremes of very high temperature). This is divided into five sub-regions with distinctive diurnal cycles, namely:

- Basic/Hot
- Basic/Cold
- Basic/Constant High Humidity
- Basic/Variable High Humidity
- Basic/Cold-Wet

Most attention is focused on the Basic Region because this type of climate applies is 'roughly coincident with the more densely populated, industrialised, and agriculturally productive areas of the world; therefore, most of the land areas with the highest probability of combat operations are within its limits' (MIL-HDBK-310, Section 6.2.1).

The Hot Region has a 1 % value of high temperature at $49\,°C$ in its hottest places. This is subdivided into Hot/Dry and Hot/Humid diurnal cycles, which notionally coincide with deserts and hot tropics, respectively. The Cold Region has a 1 % value of low temperature at $-45.6\,°C$ in its coldest places. This is usually associated with high latitudes and high mountains. The Severe Cold Region has a 20 % value of low temperature at $-51\,°C$ and, effectively, represents polar climate.

Diurnal variations of temperature (in °C) and relative humidity (in percent) are summarised in Tables 6.6 and 6.7, respectively, for each type of climate region, with subdivisions as appropriate. The time-scale is given in hours, relating to local solar time. A diurnal cycle is not defined for the Severe Cold region because temperature and humidity remain almost constant.

It is clear that this information addresses a different class of problem from the atmosphere models discussed previously. This only applies to the surface environment and explicitly includes water vapour and daily fluctuations. In terms of aircraft performance, this would be appropriate for take-off and landing trials in a range of conditions (with particular emphasis on extreme conditions). Also, it would be appropriate for considering thermal performance of systems (especially engines) that are operating from start-up under different ambient or soak conditions.

6.9.2   *Köppen-Geiger Classification*

The Köppen classification is the most widely used framework for classifying climatic regions. It is based on the work of Wladimir Köppen from 1900 onwards, which was

---

[7] A '1 % value' relates to a frequency of occurrence of 1 %. In other words, values that are less severe will be expected to occur on 99 % of occasions.

Table 6.6 Diurnal temperature cycles for MIL-HDBK-310 climatic regions.

Time Hrs	Basic region					Hot region		Cold region
	Hot	Cold	Constant high humidity	Variable high humidity	Cold-wet	Dry	Humid	
01	33	−31	24	27	−3	35	30	−46
02	32	−32	24	26	−3	34	31	−46
03	32	−32	24	26	−4	34	31	−46
04	31	−32	24	26	−4	33	31	−46
05	30	−32	24	26	−4	33	31	−46
06	30	−32	24	26	−4	32	32	−46
07	31	−30	24	27	−3	33	34	−45
08	34	−28	24	29	−2	35	36	−44
09	37	−26	24	31	−1	38	37	−43
10	39	−24	24	32	0	41	38	−41
11	41	−22	24	33	1	43	39	−39
12	42	−21	24	34	1	44	40	−37
13	43	−21	24	34	2	47	41	−37
14	43	−21	24	35	2	48	41	−37
15	43	−21	24	35	2	48	41	−37
16	43	−22	24	34	1	49	41	−38
17	43	−24	24	33	1	48	39	−39
18	42	−25	24	32	0	48	37	−39
19	40	−26	24	31	−1	46	36	−41
20	38	−27	24	29	−2	42	34	−42
21	36	−28	24	28	−2	41	33	−43
22	35	−29	24	28	−3	39	32	−44
23	34	−30	24	27	−3	38	32	−44
24	33	−31	24	27	−3	37	31	−45

Table 6.7 Diurnal humidity cycles for MIL-HDBK-310 climatic regions.

Time Hrs	Basic region					Hot region		Cold region
	Hot	Cold	Constant high humidity	Variable high humidity	Cold-wet	Dry	Humid	
01	36	N	100	100	100	6	88	N
02	38	E	100	100	100	7	88	E
03	41	A	100	100	100	7	88	A
04	44	R	100	100	100	8	88	R
05	44		100	100	100	8	88	
06	44	S	100	100	100	8	88	S
07	41	A	98	94	100	8	83	A
08	34	T	97	88	100	6	78	T
09	29	U	95	82	100	6	73	U
10	24	R	95	80	100	5	70	R
11	21	A	95	77	100	4	66	A
12	18	T	95	75	95	4	63	T

Table 6.7 (Continued).

Time Hrs	Basic region					Hot region		Cold region
	Hot	Cold	Constant high humidity	Variable high humidity	Cold-wet	Dry	Humid	
13	16	I	95	75	95	3	60	I
14	15	O	95	74	95	3	60	O
15	14	N	95	74	95	3	60	N
16	14		95	77	95	3	60	
17	14		95	79	100	3	64	
18	14		95	82	100	3	69	
19	17		97	86	100	3	74	
20	20		98	91	100	4	79	
21	22		100	95	100	5	85	
22	25		100	97	100	6	86	
23	28		100	98	100	6	87	
24	33		100	100	100	6	88	

probably motivated by patterns and delineations in vegetation. It is intended to reflect long-term mean climatic conditions and thus it gives an appropriate context for this book. Note that, for present purposes, it is an aside. The detail of this framework is less important than the recognition that a formal framework exists and is under continual development. This is of interest given the debate on climate change, especially with respect to the impact of aviation. It is also of interest when trying to make sense of the MIL-HDBK-310 climatic regions.

The good overview of climatic regions (and the associated issues of data and methodology) is presented by Peel *et al.* (2007), who describe their work on an updated world map for the Köppen-Geiger classification. It has the additional benefit of high-quality colour maps that allow the climatic distribution to be assimilated readily. The classification scheme (i.e. symbols and criteria) is defined in Table 6.8.

In Table 6.8, the following nomenclature is applied:

$$
\begin{aligned}
\text{MAP} &= \text{mean annual precipitation} \\
\text{MAT} &= \text{mean annual temperature} \\
T_{hot} &= \text{temperature of the hottest month} \\
T_{cold} &= \text{temperature of the coldest month} \\
T_{mon10} &= \text{number of months in which temperature is above } 10\ ^{\circ}\text{C} \\
P_{dry} &= \text{precipitation in driest month} \\
P_{sdry} &= \text{precipitation in driest month of summer} \\
P_{wdry} &= \text{precipitation in driest month of winter} \\
P_{swet} &= \text{precipitation in wettest month in summer} \\
P_{wwet} &= \text{precipitation in wettest month in winter} \\
P_{threshold} &= 2\text{xMAT (if 70\% of MAP occurs in winter)} \\
&\quad or\ 2\text{xMAT}+28\ \text{(if 70\% of MAP occurs in summer)} \\
&\quad or\ 2\text{xMAT}+14\ \text{(otherwise).}
\end{aligned}
$$

Note that Summer (winter) is defined as the warmer (cooler) six-month period of ONDJFM and AMJJAS.

Table 6.8 Köppen climate symbols and criteria.

Symbols			Description	Criteria
1st	**2nd**	**3rd**		
A			**Tropical**	$T_{cold} \geq 18$
	f		- Rainforest	$P_{dry} \geq 60$
	m		- Monsoon	$Not(Af)$ & $P_{dry} \geq 100 - MAP/25$
	w		- Savannah	$Not(Af)$ & $P_{dry} < 100 - MAP/25$
B			**Arid**	$MAP < 10 \times Pthreshold$
	W		- Desert	$MAP < 5 \times Pthreshold$
	S		- Steppe	$MAP \geq 5 \times Pthreshold$
		h	-- Hot	$MAT \geq 18$
		k	-- Cold	$MAT < 18$
C			**Temperate**	$T_{hot} > 10$ & $0 < T_{cold} < 180$
	s		- Dry Summer	$P_{sdry} < 40$ & $P_{sdry} < P_{wwet}/3$
	w		- Dry Winter	$P_{wdry} < P_{swet}/10$
	f		- Without Dry Season	$Not(Cs\ or\ Cw)$
		a	-- Hot Summer	$T_{hot} \geq 22$
		b	-- Warm Summer	$Not(a)$ & $T_{mon10} \geq 4$
		c	-- Cold Summer	$Not(a\ or\ b)$ & $1 \leq T_{mon10} < 4$
D			**Cold**	$T_{hot} > 10$ & $T_{cold} \leq 0$
	s		- Dry Summer	$P_{sdry} < 40$ & $P_{sdry} < P_{wwet}/3$
	w		- Dry Winter	$P_{wdry} < P_{swet}/10$
	f		- Without Dry Season	$Not(Ds\ or\ Dw)$
		a	-- Hot Summer	$T_{hot} \geq 22$
		b	-- Warm Summer	$Not(a)$ & $T_{mon10} \geq 4$
		c	-- Cold Summer	$Not(a, b\ or\ d)$
		d	-- Very Cold Winter	$Not(a\ or\ b)$ & $T_{cold} < -38$
E			**Polar**	$T_{hot} < 10$
	T		- Tundra	$T_{hot} > 0$
	F		- Frost	$T_{hot} \leq 0$

As can be seen in the applicable criteria, this framework is based on averages of temperature and precipitation. The five major climate types can be summarised in everyday terms, as follows[8]:

A: Moist tropical climate with permanently high temperatures and very high rainfall
B: Dry climate with little rainfall and large diurnal variations in temperature
C: Humid mid-latitude climate in which land/water differences are significant and which have warm, dry summers and cool, wet winters
D: Continental climate in the interior of very large land areas, with modest rainfall and large seasonal variations in temperature
E: Very cold climate in which ice and tundra are permanent

The percentage distribution of major climate types for the world and for continents is presented in Table 6.9, as discussed by Peel *et al.* (2007). Here, the boundary between Asia and Europe is

---

[8] www.blueplanetbiomes.org/climate.htm.

Table 6.9 Distribution of climate types.

Climate Type	A	B	C	D	E
Description	Tropical	Arid	Temperate	Cold	Polar
World	19.0%	30.2%	13.4%	24.6%	12.8%
Africa	31.0%	57.2%	11.8%		
Asia	16.3%	23.9%	12.3%	43.8%	3.8%
N America	5.9%	15.3%	13.4%	54.5%	11.0%
S America	60.1%	15.0%	24.1%		0.8%
Europe		36.3%	17.0%	44.4%	2.3%
Australia	8.3%	77.8%	13.9%		

Table 6.10 Proposed equivalence of climate types.

MIL-HDBK-310 climate type	Köppen-Geiger climate type	Description
HOT	BWh	Arid-Desert-Hot
	Cwa	Temperate-Dry Winter-Hot Summer
COLD	Dsc	Cold-Dry Summer-Cold Summer
	Dwc	Cold-Dry Winter-Cold Summer
	Dfc	Cold-Dry without Season-Cold Summer
	Dsd	Cold-Dry Summer-Very Cold Winter
VERY COLD	Dwd	Cold-Dry Winter-Very Cold Winter
	Dfd	Cold-Dry without Season-Very Cold Winter
BASIC	Remaining 22 types	Refer to Table 6.8

drawn nominally as a north–south line through the Ural Mountains down to the Arabian Sea. Also, North America is taken to include Central America and the West Indies.

It is observed that there are thirty possible climate types available in the full classification. The most extensive is BWh (Arid-Desert-Hot), covering 14.2 % of the land surface, followed by Aw (Tropical-Savannah), covering 11.5 %. At the other extreme, Cwc (Temperate-Dry Winter-Cold Summer) occurs over just 0.002 % of the land surface and Csc (Temperate-Dry Summer-Cold Summer) does not occur at all.

It is possible to identify a general mapping of climate types between Köppen-Geiger and MIL-HDBK-310, accepting the process is necessarily imprecise because different criteria are applied in each case. (Recall that the former classification combines temperature and precipitation while the latter combines temperature and humidity.) This is proposed in Table 6.10.

6.10   Air Density

The relationship between pressure (p) and temperature (T) has already been given in (6.19), with dependency on gas constant (R) and air density (ρ). Thus, air density can be written as:

$$\rho = \frac{p}{RT} \tag{6.72}$$

This is valid in all cases, for any atmosphere model. Following on from the discussion of pressure altitude in Section 6.5, this can be rewritten explicitly for ISA conditions, as follows:

$$\rho_{ISA} = \frac{p_{ISA}}{RT_{ISA}} \qquad (6.73)$$

For any other atmosphere, the air density can be calibrated against pressure altitude, thereby allowing direct comparison with the ISA. Recalling (6.61), it is known that temperature variation from ISA conditions is established by introducing an increment (ΔT):

$$T = T_{ISA} + \Delta T$$

Therefore, (6.72) becomes:

$$\rho = \frac{p}{R(T_{ISA} + \Delta T)} \qquad (6.74)$$

Picking two points with the same pressure altitude, such that $p = p_{ISA}$, it is clear that:

$$\rho = \frac{RT_{ISA}}{R(T_{ISA} + \Delta T)} \rho_{ISA}$$

$$\rho = \frac{1}{1 + \dfrac{\Delta T}{T_{ISA}}} \rho_{ISA} \qquad (6.75)$$

For any other atmosphere, the air density at any point can be compared with the ISA via pressure altitude.

6.11   Water Vapour

Standard atmospheres and reference atmospheres all assume dry air, i.e. air without any water vapour. Environmental specifications and guidance documents, such as MIL-HDBK-310, most definitely do include water vapour. This raises the need to be able to account for the effect of water vapour in atmospheric calculations.

6.11.1   *Gas Constant*

The gas constant (R) of a homogeneous mixture of gases is derived as:

$$R = \frac{\sum R_i M_i}{\sum M_i} \qquad (6.76)$$

where R_i is the gas constant for a particular component of the gas mixture and M_i is the mass of that component. The gas law (6.19) is applied to each component as:

$$p_i = \rho_i R_i T \qquad (6.77)$$

or, alternatively,

$$p_i = \frac{M_i}{V} R_i T \qquad (6.78)$$

This relates pressure (p_i) and temperature (T) for a mass of gas (M_i) occupying a fixed volume (V). In this context, the pressure of a particular component of a gas mixture is called the *partial pressure*.

Knowing that the gas constant is defined as

$$R_i = \frac{R_0}{m_i} \tag{6.79}$$

where R_0 is the *universal* gas constant and m_i is the molecular mass, the partial pressure can be rewritten as:

$$p_i = \frac{R_0 T}{V}\left(\frac{M_i}{m_i}\right) = \frac{R_0 T}{V}n_i \tag{6.80}$$

where n_i is the number of moles of gas. This demonstrates Avogadro's Law, which states that equal volumes of any gas contain equal numbers of atoms or molecules, under the assumption of equal temperature.

For a mixture of dry air and water vapour, this hybrid gas constant is derived in the following way:

$$R = \frac{R_a M_a + R_v M_v}{M_a + M_v} \tag{6.81}$$

where subscript 'a' denotes dry air and 'v' denotes water vapour.

The molecular masses of dry air and water vapour are:

$$m_a = 28.96442 \text{ kg.kmol}^{-1} \text{ (as def ined in (6.7))} \tag{6.82}$$

$$m_v = 18.01594 \text{ kg.kmol}^{-1} \tag{6.83}$$

Recall that the universal gas constant is:

$$R_0 = 8314.32 \text{ J.kmol}^{-1}.\text{K}^{-1} \text{ (as defined in(6.11))}$$

Thus, the associated gas constants (from (6.79)) are given by:

$$R_a = 3287.05287 \text{ J.kg}^{-1}.\text{K}^{-1} \text{ (as derived in (6.16))} \tag{6.84}$$

$$R_v = 461.49798 \text{ J.kg}^{-1}.\text{K}^{-1} \tag{6.85}$$

It is often preferable to express (6.81) in terms of the partial pressures of dry air and water vapour. Given that partial pressure is defined by (6.78), the ratio of partial pressures in this case is defined by:

$$\frac{p_v}{p_a} = \frac{M_v}{M_a}\left(\frac{R_v}{R_a}\right) \tag{6.86}$$

Applying (6.79), this becomes:

$$\frac{p_v}{p_a} = \frac{M_v}{M_a}\left(\frac{m_a}{m_v}\right) \tag{6.87}$$

Equivalently, the mixture ratio is:

$$\frac{M_v}{M_a} = \frac{p_v}{p_a}\left(\frac{m_v}{m_a}\right) \tag{6.88}$$

With these equivalences in mind, the gas constant for the air/vapour mixture (cf. (6.81)) is:

$$R = \frac{R_a M_a + R_v M_v}{M_a + M_v} = R_a \frac{1 + \dfrac{M_v}{M_a}\left(\dfrac{R_v}{R_a}\right)}{1 + \dfrac{M_v}{M_a}}$$

$$R = R_a \frac{1 + \dfrac{p_v}{p_a}}{1 + \dfrac{p_v}{p_a}\left(\dfrac{m_v}{m_a}\right)} \tag{6.89}$$

Rearranging this expression, recognising that the total pressure $p = p_a + p_v$, the gas constant can be written in an alternative and frequently quoted form:

$$R = R_a \frac{p_a + p_v}{p_a + p_v\left(\dfrac{m_v}{m_a}\right)} = R_a \frac{p}{p - p_v + p_v\left(\dfrac{m_v}{m_a}\right)}$$

$$R = R_a \frac{p}{p - p_v\left(1 - \dfrac{m_v}{m_a}\right)} \tag{6.90}$$

Using the numerical values quoted in (6.82) and (6.83), this simplifies as follows:

$$R = \frac{287.05287\ p}{p - 0.377998\ p_v} \tag{6.91}$$

A method for calculating the partial pressure of water vapour is covered in the next section.

6.11.2   *Humidity*

Equations for the determination of humidity are given by Parish and Putnum (1977). This is based on vapour pressure (which is the partial pressure of water vapour in moist air) and saturated vapour pressure (at which two phases of water coexist in equilibrium). These pressures are *with respect to water (or ice)* when the air mass is over a plane of water (or ice) at the same temperature. In the case of saturated vapour pressure, any reduction in temperature will cause the formation of dew (or frost). Equivalently the *dew point* is the temperature at which the actual vapour pressure is equal to the saturated vapour pressure. If the dew point is at or below freezing, it is referred to as the frost point.

Absolute (or volumetric) humidity (AH) is the mass of water vapour per unit volume of air at a given temperature and pressure. With reference to the gas law given in (6.78), this is defined as follows:

$$AH = \frac{p_v}{R_v T} \tag{6.92}$$

Relative humidity (RH) is defined as the ratio of vapour pressure (p_v) to saturated vapour pressure (p_s):

$$RH = \frac{p_v}{p_s} \tag{6.93}$$

By recommendation, vapour pressure is evaluated with respect to water or ice (as appropriate) but saturated vapour pressure is always evaluated with respect to water.

The transformation from *water* to vapour is called *evaporation*; the transformation from *ice* to vapour is called *sublimation*. Both transformations require the addition of heat, or so-called latent heat (E), which is defined by:

$$\frac{dE}{dT} = C_{Pv} - C \tag{6.94}$$

where C_{Pv} is the specific heat capacity of water vapour at constant pressure and C is the specific heat capacity of water (or ice). As these parameters are almost constant for normal atmospheric pressures and temperatures, (6.94) becomes:

$$E = (C_P - C)(T - T_0) + E_0 \tag{6.95}$$

The Claperon-Clausius Equation relates pressure and temperature when two phases of matter are in equilibrium:

$$\left(\frac{1}{p_s}\right)\frac{dp_s}{dT} = \frac{E}{R_v T^2} \tag{6.96}$$

Thus,

$$\left(\frac{1}{p_s}\right)\frac{dp_s}{dT} = \frac{(C_{Pv} - C)\ (T - T_0)\ + E_0}{R_v T^2} = \frac{(C_{Pv} - C)}{R_v}\left(\frac{1}{T}\right)$$
$$+\frac{E_0 - T_0(C_{Pv} - C)}{R_v}\left(\frac{1}{T}\right)^2$$

This integrates to give:

$$\log_e p_s = \frac{(C_{Pv} - C)}{R_v} \log_e T - \frac{E_0 - T_0(C_{Pv} - C)}{R_v}\left(\frac{1}{T}\right) + K \tag{6.97}$$

where K is the constant of integration. In general form, this can be rewritten as:

$$\log_e p_s = a \log_e T + b\left(\frac{1}{T}\right) + c \tag{6.98}$$

or equivalently:

$$p_s = T^a \exp\left[b\left(\frac{1}{T}\right) + c\right] \tag{6.99}$$

Now, by definition, the dew point is the temperature at which the actual vapour pressure is equal to the saturated vapour pressure. Therefore, (6.99) can be applied directly for vapour pressure (p_v) and drew point (D):

$$p_v = D^a \exp\left[b\left(\frac{1}{D}\right) + c\right] \tag{6.100}$$

Relevant constants in this context are:

Reference Temperature:	$T_0 = 273.15\,\text{K}$	(6.101)
Specific Heat Capacity:		
Ice at 0°C	$Ci = 2060\,\text{J.kg}^{-1}.\text{K}^{-1}$	(6.102)
Water at 0°C	$C_w = 4185\,\text{J.kg}^{-1}.\text{k}^{-1}$	(6.103)
Water Vapour	$C_{Pv} = 1911\,\text{J.kg}^{-1}.-1^{-1}$	(6.104)
Vapour Pressure:		
Ice at 0°C	$p_{I0} = 6.107\,\text{mbar}$	(6.105)
Water at 0° C	$p_{w0} = 6.11\,\text{mbar}$	(6.106)
Latent Heat:		
Sublimation	$E_{i0} = 2834000\,\text{J.kg}^{-1}$	(6.107)
Evaporation	$E_{w0} = 2500000\,\text{J.kg}^{-1}$	(6.108)

Note that pressure is measured in millibars.[9] Constants that appear in (6.98) and (6.99) need to be evaluated with respect to ice and water, using separate instances in (6.97):

$$\log_e p_{i0} = \frac{(C_{Pv} - C_i)}{R_v} \log_e T_0 - \frac{F_{i0} - T_0(C_{Pv} - C_i)}{R_v}\left(\frac{1}{T_0}\right) + K_i$$

$$\log_e p_{w0} = \frac{(C_{Pv} - C_w)}{R_v} \log_e T_0 - \frac{E_{w0} - T_0(C_{Pv} - C_w)}{R_v}\left(\frac{1}{T_0}\right) + K_w$$

---

[9] 1 bar = 100000 Pa (as distinct from 1 *atmosphere* = 101325 Pa).

These expressions relate to separate instances of (6.98):

$$\log_e p_s = a_i \log_e T + b_i \left(\frac{1}{T}\right) + c_i \tag{6.109}$$

$$\log_e p_s = a_w \log_e T + b_w \left(\frac{1}{T}\right) + c_w \tag{6.110}$$

where

$$a_i = -0.322862$$
$$b_i = -6229.0616$$
$$c_i = 26.425244$$

and

$$a_w = -4.927432$$
$$b_w = -6763.0699$$
$$c_w = 54.212470$$

In accordance with recommendation, saturated vapour is evaluated as:

$$p_s = T^{a_w} \exp\left[b_w \left(\frac{1}{T}\right) + c_w\right] \tag{6.111}$$

Actual vapour pressure is evaluated as:

$$p_v = D^{a_w} \exp\left[b_w \left(\frac{1}{D}\right) + c_w\right] \quad \text{when } D > 273.15 \text{ K} \tag{6.112}$$

$$p_v = D^{a_i} \exp\left[b_i \left(\frac{1}{D}\right) + c_i\right] \quad \text{when } D \le 273.15 \text{ K} \tag{6.113}$$

Note 6.1

For purposes of calculation, Equations (6.111) to (6.113) are not particularly well-conditioned in the sense that the first part of the evaluation (T^a) typically gives numbers of the order 10^{-13} while the second part typically gives numbers of the order 10^{12}. This can be resolved by using a normalised temperature, as defined in (6.25) and extended here to cover the dew point temperature:

$$\theta = \frac{T}{T_0} \quad \text{and} \quad \Delta = \frac{D}{T_0}$$

Also, there is often a preference for common logarithms in calculation rather than natural logarithms, and therefore terms of the form e^x (or $\exp(x)$) will have to be substituted by terms of the form 10^x.

Thus, the saturated vapour pressure (cf. (6.111)) is evaluated as:

$$p_s = \theta^{a_w} 10^{\left[b_w \left(\frac{1}{\theta}\right) + c_w\right]}$$

Actual vapour pressure (cf. (6.112) and (6.113)) is evaluated as:

$$p_v = \Delta^{a_w} 10^{\left[b_w \left(\frac{1}{\Delta} \right) + c_w \right]} \quad \text{when } \Delta > 1$$

$$p_v = \Delta^{a_i} 10^{\left[b_i \left(\frac{1}{\Delta} \right) + c_i \right]} \quad \text{when } \Delta \leq 1$$

In this context, the relevant constants become:

$$a_i = -0.322862$$
$$b_i = -9.903888$$
$$c_i = 10.689717$$

and

$$a_w = -4.927432$$
$$b_w = -10.752935$$
$$c_w = 11.538976$$

Note that these equations give vapour pressure measured in millibars, which means that the corresponding partial pressure for dry air (in the absence of any other component) is given by:

$$p_a = 1000 - p_v \tag{6.114}$$

Thus, in this case, the mixture ratio (6.88) can be expressed as:

$$\frac{M_v}{M_a} = \left(\frac{m_v}{m_a} \right) \frac{p_v}{1000 - p_v} = 0.622002 \frac{p_v}{1000 - p_v} \tag{6.115}$$

To illustrate, Figure 6.7 shows the variation with temperature of (1) the saturated vapour pressure (in millibars) and (2) the vapour/air mixture ratio (in grams of vapour per kilogram of dry air). The variation of dew point (or frost point if it occurs at of below 0°C) is shown in Figure 6.8.

The calculation of vapour pressure (for Figure 6.9) is undertaken using Algorithm 6.2. This code implements the set of equations specified in Note 6.1, via a generic computation with specialisation depending on whether the vapour pressure is with respect to water or ice. Temperature input is specified in degrees Celsius and all necessary conversion and normalisation is conducted internally. This algorithm transforms ambient temperature into saturated vapour pressure (recalling that this is always calculated with respect to water) and transforms the dew point temperature (if known) into actual vapour pressure.

Given that different equations apply to *water* and *ice*, an additional constant [d] is introduced in order to ensure that the switch-over occurs numerically at 0 °C. Using the original constants (a,b,c) to their given resolution, the switch actually occurs at 0.04 °C. Two iterations (using the residual error on each occasion) provided the necessary correction and, hence, the parameter 'ice.d' is composed of two numbers summed together. While this is an *ad hoc* procedure and as such might be open to criticism, it is always important (vitally important) that any transfer between different methods of calculation should occur at the intended transition, with as small a discontinuity as possible.

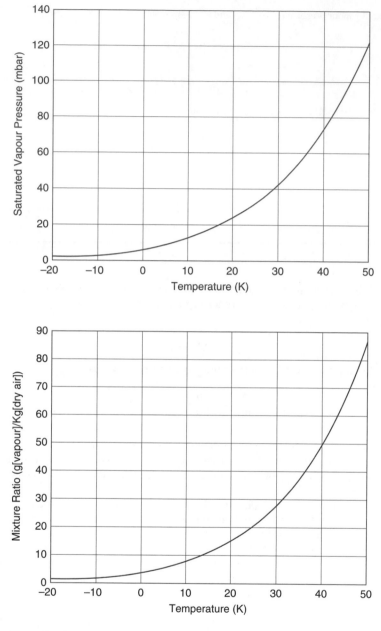

Figure 6.7 Vapour pressure and mixture ratio for saturated air.

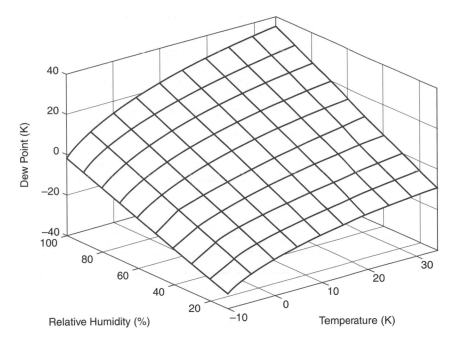

Figure 6.8 Dew point (frost point) variation vs ambient temperature and relative humidity.

```
function (p_value,p_deriv) = calculateVapourPressue(T,phase)

kelvin = @(celsius) celsius + 273.15 ;
normalise = @(T) T./273.15 ;

t = normalise(kelvin(T)) ;

ice.a = -0.322862 ;
ice.b = -9.903888 ;
ice.c = 10.689717 ;
ice.d = 0.00298185522617 + 2.664535259100376e-015 ;

water.a = -4.927432 ;
water.b = -10.752935 ;
water.c = 11.538976 ;
water.d = 0.0 ;

switch phase
 case 'water'
 constant = water ;
 case 'ice'
 constant = ice ;
 otherwise
 error('Phase must be ''water'' or ''ice''') ;
end
```

Continued

```
exponent = constant.c + constant.b./t ;
p_value = t.^constant.a .* 10.^exponent + constant.d ;

factor = constant.a - log(10)*constant.b./t ;
p_deriv = t^(constant.a-1) * 10.^exponent * factor ;
p_deriv = normalise(p_deriv) ;
```

Algorithm 6.2 'calculateVapourPressure'.

Environmental data often specifies ambient temperature and relative humidity. This requires a method for determining the dew point (or frost point). An example implementation is given in Algorithm 6.3. Current temperature is used to determine the saturated vapour pressure (via Algorithm 6.2) and the actual vapour pressure is determined from the relative humidity (as defined in (6.93)). This pressure is then used to find the dew point, based initially on the assumption that the vapour pressure is with respect to water; if the result is below 0 °C, the computation is repeated such that the vapour pressure is with respect to ice.

```
function D = calculateDewPoint(T,H)

fraction = @(percent) percent/100 ;
between = @(number,lower,upper) min(upper,max(lower,number)) ;

H = fraction(between(H,0,100)) ;

saturationPressure = calculateVapourPressure(T,'water') ;
vapourPressure = H * saturationPressure ;
D = calculateTemperature(vapourPressure,'water') ;
if isempty(D) || D<=0
 D = calculateTemperature(vapourPressure,'ice') ;
End
```

Algorithm 6.3 'calculateDewPoint'.

The determination of dew point is a temperature calculation that is implemented in Algorithm 6.4. This solves (6.112) and (6.113) (using the corresponding equations given in Note 6.1) by applying an iterative *gradient search* technique (refer to Note 6.2) because there is no convenient explicit formula for temperature as a function of pressure. The technique is highly effective and convergence should be rapid given the form of numerical relationship, as shown in Figure 6.7. The algorithm is simple but needs to be protected from circumstances that may prevent convergence, almost certainly involving very low vapour pressures. Thus, checks are made for zero gradient (to prevent division by zero) and for an unreasonably large number of iterations (to prevent an infinite loop). In situations where a result cannot be found, an empty result is returned. Thus, the computation always terminates in a controlled manner.

```
function T = calculateTemperature(vapourPressure,phase)

T = 100 ;

maxError = 1e-6 ;
nIteration = 0 ;
maxIterations = 50 ;
```

```
iterating = 1 ;
while iterating
  nIteration = nIteration + 1 ;
  (y,m) = calculateVapourPressure(T,phase) ;
  dy = y - vapourPressure ;

  if m==0 || nIteration>maxIterations
  T = [] ;
  iterating = 0 ;
  elseif abs(dy)<=maxError
  iterating = 0 ;
  else
  dT = dy/m ;
  T = T - dT ;
  end
end
```

Algorithm 6.4 'calculateTemperature'.

Note 6.2

The gradient search technique is one approach to solving nonlinear equations of the form $y = f(x)$. For a given value of y, several attempts are made to find a value of x that satisfies the equation. Because this is a numerical procedure, solutions are approximate except in special cases. Thus, the final value of x will be such that

$$| y - f(x) | < \varepsilon$$

i.e. the magnitude of the error between y and $f(x)$ is less than a small number, ε.

With reference to the diagram, the initial guess is $x = x_0$. The initial error is

$$\Delta y_0 - f(x_0) - y$$

The initial gradient is determined as m_0 where

$$m_0 = \frac{\Delta y_0}{\Delta x_0}$$

The associated change in x is calculated as

$$\Delta x_0 = \frac{\Delta y_0}{m_0}$$

Thus the next value of x is chosen to be:

$$x_1 = x_0 - \Delta x_0$$

Looking at the diagram, this steps closer the solution (where the line $y = $ input intersects with the line $y = f(x)$). The process is repeated until convergence has been achieved. As can be seen, convergence is rapid in this particular application.

(continued)

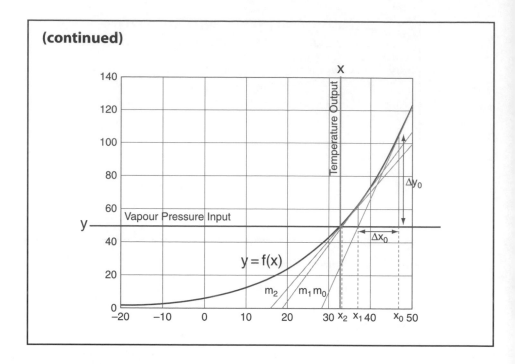

6.12   Weather Systems

In contrast to 'average' atmospheric models, weather systems provide a time-varying environment for aircraft flight, both in terms of the dynamics of the air at any given location at any given time and in terms of the progress along a flight path between locations. So the requirement is to provide time-varying weather data covering a three-dimensional (or 3-D) space, thereby yielding a so-called four-dimensional (or 4-D) model. Specific data can then be interpolated along a given flight path, with time variation calibrated against path length.

In this book, the real interest lies with the types of weather information that are relevant to aviation and some of the considerations that relate to data availability. A readily accessible source of aviation weather information is provided by the Aviation Weather Center,[10] which is part of the NOAA[11] National Weather Service.[12] Data downloads are available via the Aviation Digital Data Service (ADDS).[13]

A number of example charts are presented in this section. Figure 6.9 shows wind speeds and streamlines at the surface, 12,000 ft, 24,000 ft and 36,000 ft. What is immediately striking in this case is that surface winds do not follow the trends of winds aloft and, also, that average wind speeds increase with altitude. Note the presence of winds in excess of

---

[10] aviationweather.gov.
[11] National Oceanic and Atmospheric Administration.
[12] www.nws.noaa.gov.
[13] adds.aviationweather.gov.

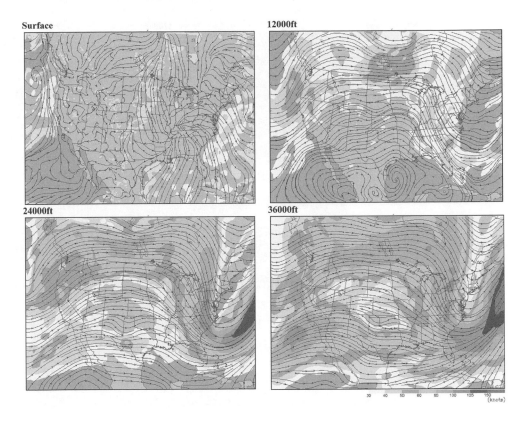

Figure 6.9 Example wind speed and streamline charts.

140 kts. Figure 6.10 shows the temperatures corresponding with those same altitudes. Note the distinction at surface level between land and sea temperatures. Above ground, given the wide application of the International Standard Atmosphere (ISA) (or, equivalently, the ICAO[14] Standard Atmosphere), it is convenient to quote temperatures with respect to the respective ISA datum values (i.e. −8.3 °C at 12,000 ft, −31.6 °C at 24,000 ft, −56.5 °C at 36,000 ft).

In this book, the real interest lies with the types of weather information that are relevant to aviation and some of the considerations that relate to data availability. A readily accessible source of aviation weather information is provided by the Aviation Weather Center[15], which is part of the NOAA[16] National Weather Service.[17] Data downloads are available via the Aviation Digital Data Service (ADDS).[18]

---

[14] International Civil Aviation Organisation.
[15] aviationweather.gov.
[16] National Oceanic and Atmospheric Administration.
[17] www.nws.noaa.gov.
[18] adds.aviationweather.gov.

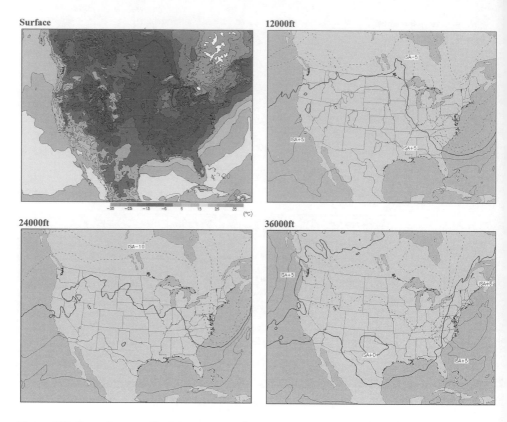

Figure 6.10 Example atmospheric temperature charts.

```
DATA BASED ON 030600Z
VALID 031200Z FOR USE 0800-1500Z. TEMPS NEG ABV 24000

FT 3000 6000 9000 12000 18000 24000 30000 34000 39000
ABR 2149 2438+12 2532+06 2527-02 2833-20 2940-32 293949 294458 284468
```

Figure 6.11 Typical wind/temperature text download format.

The ADDS offers text download for winds and temperatures in the form of a so-called NCEP Winds Aloft Forecast. A example of which is given in Figure 6.11. This is for low-level (up to 39,000 ft); a high-level forecast is available up to 53,000 ft. The header is decoded as follows:

DATA BASED ON 030600Z

Data based on computer forecasts generated for Day 3 of the month at 0600 UTC (cf. Section 3.3.3).

VALID 031200Z FOR USE 0800-1500Z. TEMPS NEG ABV 24000

Forecast is valid for Day 3 of the month at 1200 UTC and data can be used between 0800 and 1500 UTC. Temperatures are negative above 24,000 ft.

FT 3000 6000 9000 12000 18000 24000 30000 34000 39000

'FT' denotes the forecast location and the numbers denote altitudes.

The main body of data then follows. The forecast location is specified by its call sign and can be found by searching a database of NOAA weather stations.[19] The numerical data is given a number string for each altitude. For example, '2220 + 04' means that the wind forecast is 220° at 20 kts and the temperature forecast is +4 °C. As stated in the header, temperatures are negative above 24,000 ft and so the number sign is not used for higher altitudes.

Wind direction is specified in 10° increments. Thus, when wind speed is less than 100 kt, wind direction is encoded as a number between 01 and 36 (in increments of 10°) and wind speed is encoded between 00 and 99. When wind speed exceeds this range, wind direction is encoded between 51 and 86 and wind speed is encoded between 00 and 99. Thus, in order to obtain actual values, subtract 50 from the first pair of digits and add 100 to the second pair of digits. For example, a data item '770249' for 34,000 ft signifies a wind forecast of 270° at 102 kt and a temperature forecast of −49 °C.

Actual wind speeds are encoded between 5 kt and 199 kt; slower winds are treated as 'zero' and faster winds are treated as '199 knots or greater'. The code sequence becomes '9900' (meaning winds light and variable) and, for example, '2599' (meaning 250° at over 199 kt).

The example given in Figure 6.13 produces data for Aberdeen Regional Airport, Aberdeen, South Dakota. This is decoded to produce the results in Table 6.11. For reference, data on this particular forecast station is presented in Table 6.12. From a modelling perspective, this allows the data to be associated with an altitude scale projecting upwards from a geographical location.

Table 6.11 Wind/temperature forecast for Aberdeen Regional Airport, South Dakota (valid for 1200UTC, 3 January 2008).

Altitude (ft)	Wind Direction (deg)	Wind Speed (kt)	Temperature (°C)
3000	210	49	not defined
6000	240	38	12
9000	250	32	6
12000	250	27	−2
18000	280	33	20
24000	290	40	32
30000	290	39	49
34000	290	44	58
39000	280	44	68

---

[19] lwf.ncdc.noaa.gov/oa/climate/stationlocator.html.

Table 6.12 NOAA Weather Station Data for Aberdeen Regional Airport, SD.

Aberdeen Rgnl Airport
Aberdeen, SD, United States

Type	ASOS-NWS COOP WSO-N ASOS ABC NEXRAD
Call Sign/ICS	ABR/KABR
WBAN	14929
COOP ID	390020
Climate Division	SD-03 – Northeast
WMI ID	72659
In Service	01 Aug 1934 to Present
Elevation	395.3m (1297') above s/l
Lat/Lon	45°27'N / 98°25'W
County	Brown

Data on surface pressure distribution is available from weather charts in the form of isobars with centres of High and Low pressure regions, together with weather fronts. The average surface pressure is nominally equal to the ISA constant given in (6.4), which is usually expressed as 1013 mbar in this context. The obvious practical issue is that an aircraft flying along a path at constant pressure altitude will actually be flying along a path of varying geometric (or geopotential) altitude. Equivalently, an altimeter would not display zero when at mean sea level.

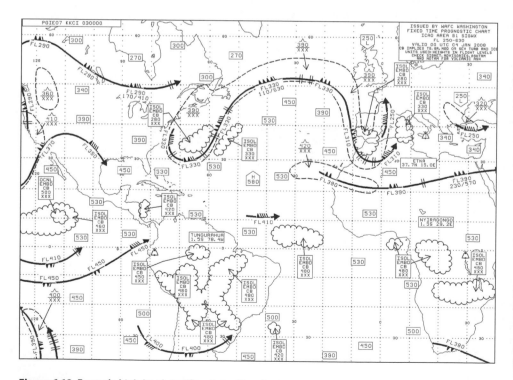

Figure 6.12 Example high-level significant weather chart.

The problem of atmospheric pressure variation is handled by allowing the datum pressure for the altimeter to be selected, depending on local or regional conditions. Two common settings are known as QNII and QFE. The QNH value is the pressure at mean sea level for a particular region and is required in order to establish height relative to topographical features identified on aeronautical charts. The QFE value is the pressure at a given field elevation and is required in order that the altimeter will read height above ground level during take-off and landing. Altimeters are set to the ISA datum above a threshold altitude (which is 3000 ft in the UK) and, in this setting, an altitude is expressed as a Flight Level (FL). Note that flight levels are resolved to 100 ft increments, such that 36,000 ft would be referred to as FL360.

Figure 6.12 shows a High-Level Significant Weather (SWH) chart. This type of information applies to flight levels between FL250 and FL630 above and is intended for the en-route sections of long-range flights. In addition, there are Medium-level Significant Weather (SWM) charts that cover the range FL100 to FL450. Collectively these are known as Significant Weather (SIGWX) charts and are issued by the World Area Forecast Centres (WAFCs) in London and Washington. The underlying data format is the Binary Universal

Table 6.13 BUFR features for SWH charts.

Feature	Designator
Jet-stream	JET
Clear Air Turbulence	TURB
Cloud	CLOUD
Tropopause Height	TROP
Tropical Cyclone and Sandstorms	STORM
Volcanic Eruption	VOLCANO
Radiation Event	RADIATION

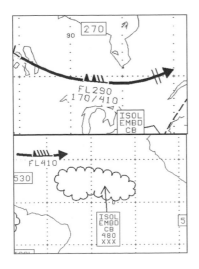

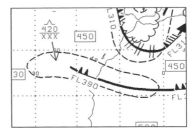

Figure 6.13 Main graphical features for SWH charts (JET,TURB,CLOUD).

Form for the Representation of meteorological data (BUFR). This is neutral data standard for SIGWX data that is defined in WMO[20] Manual 306 Part B. The features contained in SWH data are summarised in Table 6.13 and the main drawing features are shown in Figure 6.13.

A jet-stream is defined by a directed path, wind symbols and change bars. By ICAO definition this starts and stops at 80 kt. Wind symbols show direction via a line segment and speed via one or more fleches. Fleches are of three types, namely a triangle (denoting a 50 kt increment), a line (10 kt) and a short line (5 kt). The level of the maximum jet speed is shown, together wih the vertical depth (given by the lower and upper levels of the 80 kt isotachs). Thus, in the left pane of Figure 6.13, a 120 kt jet-stream is located at FL290, with 80 kt isotachs at FL170 and FL410. Change bars are orthogonal to the jet-stream, each denoting a 20 kt speed change.

Clear Air Turbulence (CAT) is represented by a dashed area, together with an index number or summary data. In the centre pane of Figure 6.13, the CAT area is the lozenge-shaped feature and the summary data (with an arrow pointing into the area) consists of a chevron (indicating moderate turbulence) together with a ceiling at FL420 and a base below FL250. Cloud is represented by a scalloped area, marked out by latitude and longitude, with an associated information box. In the right pane of Figure 6.13, the cloud area is annotated as ISOL/EMBED CB (isolated/embedded cumulo-nimbus) with a ceiling at FL480 and a base below FL250. Tropopause heights are shown by three-digit numbers enclosed within rectangular boxes, expressing height as a flight level.

A further set of descriptions, definitions and illustrations are available for SWM charts. More information is provided in a guidance document entitled 'Representing WAFS Significant Weather (SIGWX) Data in BUFR' Version 4.1,[21] issued by the WAFCs under joint patronage of the NOAA and UK Meteorological Office.

---

[20] World Meteorological Organisation.
[21] Accessible via the Help page associated with aviationweather.gov/products/swh/.

Appendix A

Introduction to MATLAB

A.1 MATLAB

MATLAB is a tool for computation and visualisation with a wide range of scientific and technical applications. It has its own programming language and an extensive function library, making it a useful tool for processing information, trying out calculations and for developing software prototypes. It is convenient to use because of the facilities for importing, exporting and plotting data. The MATLAB product family incorporates a wide range of *toolboxes* that provide ready-made functions relevant to major subject areas and it is common practice for users to create toolboxes that meet their own particular requirements.

Above all else, MATLAB is a tool for solving numerical problems. The user interface is intended to be quick and easy to use, employing notation that is more flexible than other programming languages and provides a degree of automation in defining data. The philosophy is clearly one of allowing the user to concentrate effort on the application rather than the tool. From the outset, it is seen that MATLAB is a good place to perform quick calculations, as it provides access to powerful numerical procedures as well as handling the mundane tasks of data manipulation and plotting. After years of MATLAB development, the programming language, the toolboxes and the graphics capabilities have become quite sophisticated. While these innovations may not satisfy purists or specialists, it is widely accepted that MATLAB represents a good compromise between the needs of quick calculation and structured programming.

The aim of this appendix is to provide general guidance on MATLAB that would be of interest and value to a new user. To this end, the focus of attention will be on how to run a MATLAB session, where to find relevant information and how to undertake simple tasks, such as performing simple calculations and plotting data. Hopefully this will provide sufficient background to enable individual research into more specialised aspects of the MATLAB product family.

Computational Modelling and Simulation of Aircraft and the Environment   D.J. Diston
© 2009 John Wiley & Sons, Ltd

A.2   The MATLAB Product Family

The MATLAB product family is developed and supplied by The Mathworks Inc.,[1] which is a company focused on technical computing and related applications. At the time of writing, it about 1,000,000 customers, which enables it to maintain momentum in product development. An important consideration in the writing of this book is that MATLAB is used in all the major aerospace companies and research facilities, as well as being available in 3500 universities worldwide. So, quite aside from its technical merits, it made sense to adopt MATLAB for computation because it has become a de facto standard, widely used and widely available.

MATLAB is a high-level for technical computing, supported by facilities for user interaction, data handling, numerical calculation and so on. It enables the development of algorithms and applications, together with bespoke interfaces for data visualisation, user control and file input-output. A large collection of functions are provided, in the following categories:

- Desktop Tools and Development Environment
- Mathematics
- Data Analysis
- Programming and Data Types
- File I/O
- Graphics
- 3-D Visualisation
- Graphical User Interfaces
- External Interfaces

An extensive library of user-contributed code is available for open exchange via MATLAB Central,[2] which is a useful place to look in order to find examples of how other people have attempted to solve application problems. It also offers good pointers and tips on programming style and how to use some of the more specialised features of MATLAB.

Note also that more than 900 books[3] have already been published that make explicit use of MATLAB and SIMULINK. As well as providing authoritative sources of information, these give a good insight into the practical application of MATLAB and related products in many and varied application-specific contexts.

A.3   Getting Started

There are two methods for starting MATLAB. Predominantly, it will be started from a windows environment by double-clicking a desktop icon. Alternatively, from a command line interface (such as a Linux xterm), it will probably be started by typing MATLAB. The details depend on the operating system and the administrative preferences, neither of which are of any great interest here.

Once started, MATLAB opens a window for user interaction. This can be split into frames, as configured via the 'Desktop' menu (selected from the menu bar). The 'Command Window' frame provides the primary user interface. A command prompt appears as a double

---

[1] www.mathworks.com.
[2] www.mathworks.com/matlabcentral.
[3] www.mathworks.com/support/books.

chevron '>>' at the start of each new command line; selecting the Command Window will initially set the cursor at the prompt and the user is then free to type in a string of characters from the keyboard. On hitting the 'Return' key, MATLAB will then attempt to interpret that string as an instruction to do something, conforming to its language specification. All commands are added to the Command History, which therefore provides a complete log of all the instructions given to MATLAB via the command line.

Arrow keys, as well as PageUp and PageDown, can be used to move around the Command Window and, by trial and error, it quickly becomes obvious how this works. Usefully, 'Up' and 'Down' keys enable the command history to be traversed from bottom to top and back down again. By typing just a few characters, the 'Up' and 'Down' keys will then jump to successive lines in the command history that start with those characters.

(*Note: If you ever get lost, hitting the 'Esc' key will return you to the bottom of the command history, on a new line. Alternatively, start typing . . . anything!*)

The Command Window is also the place where, by default, the results of any operation (including warnings and errors) will be displayed. Scroll bars appear when the amount of information to be displayed exceeds the display area bounded by the command window. The scrollable area is large but not unlimited; when displaying a lot of data, the display will truncate such that only the last few hundred lines can be recovered. In practice, this is no real problem and, in any case, dumping large amounts of data on to a screen is not a particularly effective means of extracting information.

A.4   Getting Help

MATLAB contains an extensive set of documentation that describes how to use its many features. Typing 'help' will bring up a complete list of categories for MATLAB plus any other products from the MATLAB family that happen to be installed on the computer. The first few lines are as follows:

```
>> help
Help topics

matlab\general - General purpose commands.
matlab\ops - Operators and special characters.
matlab\lang - Programming language constructs.
matlab\elmat - Elementary matrices and matrix manipulation.
matlab\elfun - Elementary math functions.
matlab\specfun - Specialized math functions.
matlab\matfun - Matrix functions - numerical linear algebra.
matlab\datafun - Data analysis and Fourier transforms.
matlab\polyfun - Interpolation and polynomials.
matlab\funfun - Function functions and ODE solvers.
matlab\sparfun - Sparse matrices.
matlab\scribe - Annotation and Plot Editing.
matlab\graph2d - Two dimensional graphs.
```

. . . and so on.

Narrowing the focus of the search produces a list of functions. For instance, the matlab\lang category is expanded as follows:

```
>> help lang
Programming language constructs.

Control flow.
if - Conditionally execute statements.
else - Execute statement if previous IF condition failed.
elseif - Execute if previous IF failed and condition is true.
end - Terminate scope of control statements.
for - Repeat statements for a specific number of times.
while - Repeat statements an indefinite number of times.
```

. . . and so on.

Then it is a simple matter to focus in on the function of interest, e.g. 'if'.

```
>> help if
IF Conditionally execute statements.
The general form of the IF statement is

IF expression
statements
ELSEIF expression
statements
ELSE
statements
END
```

. . . and so on. The text summarises what the format of the command, what it does, an example of its usage and a list of related functions.

As stated, MATLAB has its own programming language, which enables the development of quite sophisticated computation and visualisation procedures. The commands entered from the keyboard are language statements. MATLAB programs are collections of statements saved in so-called M-files (i.e. text files with extension 'm'). The help text for a function is contained as a set of comments (i.e. the first unbroken block of text, with each line commencing with the character %) at the start of the associated M-file. Thus, the command 'help if' tells MATLAB to find the file **if.m** and to copy the first block of comments to the Command Window. The location of the M-file can be found using a function called 'which'.

```
>> which if
C:\MATLAB701\toolbox\matlab\lang\if.m
```

Note that it is possible to have several M-files with the same name, located in different places. This can be checked by using the flag '-all':

```
>> which if -all
C:\MATLAB701\toolbox\matlab\lang\if.m
```

Thus, because only one instance of **if.m** has been found, it must be unique. But, of course, that search is necessarily limited to places that MATLAB knows about. The important question of 'where?' will be covered in the next section.

So far, the help information is fine as a means of confirming the details of a known function or browsing out of general interest. Looking for a way of doing something new is much less straightforward. A function called 'lookfor' is available that will search for a specified character string in the help text of all M-files that MATLAB can find. This relies on the authors writing sensible comments in their M-files but, nonetheless, this is a useful facility when all else fails.

```
>> lookfor('tangent')
ACOT Inverse cotangent.
ACOTD Inverse cotangent, result in degrees.
ACOTH Inverse hyperbolic cotangent.
ATAN Inverse tangent.
ATAN2 Four quadrant inverse tangent.
ATAND Inverse tangent, result in degrees.
ATANH Inverse hyperbolic tangent.
COT Cotangent.
COTD Cotangent of argument in degrees.
COTH Hyperbolic cotangent.
TAN Tangent.
TAND Tangent of argument in degrees.
TANH Hyperbolic tangent.

>> help ACOTD
ACOTD Inverse cotangent, result in degrees.
ACOTD(X) is the inverse cotangent, expressed in degrees,
of the elements of X.

See also cotd, acot.

Reference page in Help browser
doc acotd
```

For a better overview of MATLAB capabilities, many users prefer to use the Help browser. As seen in the example above, there is a reference page on the function of interest. Simply by typing 'doc', a web-style browser appears that has the advantage of having all source pages available in one place, together with a Help Navigator. Following the many links is often a good way of tripping over useful information on MATLAB features that would be difficult or time-consuming to find by other means.

Before moving on, an additional piece of helpful information is the MATLAB version currently in use. For projects that are likely to have a long lifespan, this can be extremely important as the MATLAB specification (i.e. what it does and how it does it) may change from version to version. Despite the best intentions of tool developers, there is no absolute guarantee of compatibility between versions; backward compatibility is a common problem but eventually forward compatibility can become a problem as certain features become obsolete and die.

There are two main functions that can be used for this purpose, namely 'version' and 'ver'.

```
>> version
ans =7.0.1.24704 (R14) Service Pack 1
```

(*Note: 'ans =' is the standard way that MATLAB introduces the result of a function call, i.e. the answer.*)

```
>> ver
-----------------------------------------------------------------
MATLAB Version 7.0.1.24704 (R14) Service Pack 1
MATLAB License Number: ******
Operating System: Microsoft Windows XP Version 5.1 (Build 2600: Service
Pack 2)
JAVA VM Version: Java 1.4.2 04 with Sun Microsystems Inc. Jave HotSpot (TM)
Client VM
-----------------------------------------------------------------
MATLAB Version 7.0.1 (R14SP1)
Simulink Version 6.1 (R14SP1)
Control System Toolbox Version 6.1 (R14SP1)
Mapping Toolbox Version 2.0.3 (R14SP1)
Stateflow Version 6.1 (R14SP1)
```

(*Note: The license number in this example has been obscured.*)

A.5   Where?

Once started, MATLAB opens a window for user interaction, as explained previously. It also sets a predefined working directory, as will be explained here. MATLAB is typically started by double-clicking a MATLAB icon on the desktop display. Under Windows XP, for example, icon properties are accessed by right-clicking the icon and selecting 'Properties' from the pop-up menu. Looking at MATLAB icon properties, on the 'Shortcut' tab, the 'Target' will be identified as C:\MATLAB701\bin\win32\matlab.exe (or something similar) and the 'Start in' directory is the initial working directory.

The whole issue of what to deposit in files, how to organise them and how to tell MATLAB where to find them is rather involved and is largely a matter of preference, except insofar as MATLAB preassigns locations to its components. Files will be used extensively to hold MATLAB programs and to hold data. Files located in the current working directory take precedence over files anywhere else. That is why the working directory is so important.

The working directory can be changed at any time during the session by typing something like:

```
>> cd 'C:\Documents and Settings\misterX\My Documents\workspace'
```

The current working directory can be confirmed as follows:

```
>> pwd
ans =
C:\Documents and Settings\misterX\My Documents\workspace
```

where 'pwd' stands for 'print working directory'.

Apart from the working directory, MATLAB can access files on its search path. This is obtained by typing 'path', as shown below:

```
>> path

MATLABPATH

C:\MATLAB701\toolbox\matlab\general
C:\MATLAB701\toolbox\matlab\ops
C:\MATLAB701\toolbox\matlab\lang
C:\MATLAB701\toolbox\matlab\elmat
C:\MATLAB701\toolbox\matlab\elfun
C:\MATLAB701\toolbox\matlab\specfun
C:\MATLAB701\toolbox\matlab\matfun
C:\MATLAB701\toolbox\matlab\datafun
```

. . . and so on.

The MATLAB search path defines the precedence of all MATLAB-related files, after searching in the working directory (also known as the *local* directory). The search proceeds from the top of the path list to the bottom and stops when the required file has been located. The path can be modified in various ways, a few of which are shown below:

```
>> addpath 'C:\Documents and Settings\mrX\My Documents\projectX'
```

which adds the new path to the beginning of the search path; or

```
>> addpath 'C:\Documents and Settings\mrX\My Documents\projectX' -end
```

which adds the new path to the end of the search path; or by using the MATLAB 'pathtool' and then performing modifications interactively.

A.6   Numbers: Variables and Literals

MATLAB is primarily a tool that handles and manipulates numbers, which are held either as *variables* or *literals*. Variables exist physically in designated memory locations although MATLAB hides these locations from users; virtually, they exist as named items in the MATLAB *workspace* and the content of the workspace can be displayed using the function 'who'. Literals exist as actual numerical values, such as 0, 1.0 or $-2.5e5$.

Numbers are configured as arrays of arbitrary size, shape and dimension, i.e. these attributes do not have to be predefined. An array can be a scalar, a matrix (which is a two-dimensional array) or a multidimensional array (although these will not be addressed here). A few simple examples are shown below, together with equivalent MATLAB coding:

$$x = 1 \quad \mathbf{r} = (1 \quad 2 \quad 3 \quad 4) \quad \mathbf{c} = \begin{pmatrix} 1 \\ 2 \\ 3 \\ 4 \end{pmatrix} \quad \mathbf{A} = \begin{pmatrix} 1 & 2 \\ 3 & 4 \end{pmatrix}$$

```
>> x = 1      % scalar
x =
1
>> r = [1 2 3 4] % row vector
r =
1 2 3 4
>> r = [1, 2, 3, 4] % ...another way of creating a row vector

r =
1 2 3 4
>> c = [1; 2; 3; 4] % column vector
c =
1
2
3
4
>> A = [ 1 2; 3 4 ] % 2-by-2 matrix
A =
1 2
3 4
```

Variable names start with a letter of the alphabet (not a number or a punctuation mark) and can have up to 63 characters (combining alphanumeric characters and underscores). Names in MATLAB are case sensitive, meaning that upper case and lower case letters are interpreted as different letters, which is not necessarily so in other programming languages.

The *equals* sign is an assignment, i.e. what is calculated on the right-hand side of a statement is placed in the memory location that corresponds with the variable name on the left-hand side of the statement. Thus, the statement

```
>> x = x + 1
```

does not imply that one is equal to zero! It actually takes whatever value is currently held in x, increments it by one and overwrites x with the new value.

Square brackets are used to contain an array specification. Commas are used to combine numbers in rows (a process called *horizontal concatenation*) and semicolons are used to stack numbers in columns (called *vertical concatenation*).

Generally, array elements are indexed using natural numbers (which are *counting* numbers starting at 1). A specific element can be accessed by specifying the variable name followed by an indexed position enclosed in parentheses. For a (one-dimensional) vector, a single number is sufficient to locate the element; for a (two-dimensional) matrix, two numbers are needed (written as an ordered pair), the first locating the row and the second locating the column.

```
>> rowVector = [1 2 3 4];
>> rowVector(2)
ans =
2
>> matrix = [1 2; 3 4];
>> matrix(2,1)
ans =
3
```

Notice that the display of results can be suppressed by placing a semicolon at the end of a statement. This also acts as a separator, enabling more than one statement to appear on a command line. Another separator is a comma; this enables multiple statements but without suppressing the display. Combinations are possible but commas, semicolons and brackets must be used in the correct order (a way of organising written text that is called *syntax*).

Arrays can be constructed in a variety of other ways, as illustrated below:

```
>> row1 = [1 2];
>> row2 = [3 4];
>> matrix = [ row1; row2 ] % Vertical concatenation
matrix =
1 2
3 4
>> column1 = [1; 3];
>> column2 = [2; 4];
>> matrix = [ column1, column2 ] % Horizontal concatenation
matrix =
1 2
3 4
```

Note that the concatenation rules are completely generic but they require the numerical objects have the right size and shape to be combined in this way.

As an additional refinement, vectors can be created as number sequences, as follows:

```
>> rowVector = [1:4]
rowVector =
1 2 3 4
>> rowVector = 1:4
rowVector =
1 2 3 4
>> anotherRowVector = 1:2:4
anotherRowVector =
1 3 5 7
```

A number sequence can be specified between upper and lower limits in the form *lo:hi*, resulting in a sequence $(lo, lo + 1, lo + 2, \ldots)$ up to but not exceeding *hi*. A more general form *first:inc:last* also specifies the increment to be used between successive numbers in the sequence. The first number in the sequence is *first*. If the increment *inc* is positive, the last number must not be greater than *last*; otherwise it must be less than *last*.

Note that, by default, a sequence is created as a row vector. In order to create a column vector, the row vector is simply transposed. This operation is programmed using a single inverted comma following a number sequence or, equivalently, its associated variable name.

```
>> columnVector = [1:4]'
columnVector =
1
2
3
4
```

For completeness, MATLAB permits the existence of empty arrays:

```
>> sequence = 1:2:-1
sequence =
Empty matrix: 1-by-0
>> z = []
z =
[]
>> length(z)
ans =
0
>> size(z)
ans =
0 0
```

The two variables, sequence and z, are empty by construction. One is a number sequence that logically cannot exist and the other is specified as an empty array (as indicated by square brackets with no content). Empty arrays have zero length or, equivalently, zero size (i.e. no rows and no columns). The functions 'length' and 'size' will reveal this information. Note that any operation that involves an empty array will generate an empty array.

More interestingly, because MATLAB uses the IEEE 754-1985[4] number standard for binary floating-point arithmetic, numbers can be divided by zero without causing catastrophic failure.

```
>> a = 1 / 0
Warning: Divide by zero.
a =
Inf
>> n = 0 / 0
Warning: Divide by zero.
n =
NaN
```

where '/' denotes division.

Dividing a nonzero number by zero generates infinity (Inf) and for consistency note that $1/$Inf is zero, Inf+1 is Inf and so on. Dividing zero by zero is *Not a Number* (NaN) and any operation that involves NaN will generate the result NaN.

The above example generated a warning message. All warnings and errors have message identifiers associated with them of the form 'component:mnemonic'; in this case the message identifier would be MATLAB:divideByZero. These allow control to be exercised over what is presented to the user by disabling specific errors or warnings. For example, the warning for division by zero can be suppressed as follows:

```
>> warning off MATLAB:divideByZero
>> a = 1 / 0
a =
Inf
```

. . . and it can be restored via a corresponding 'warning on MATLAB:divideByZero' command. Note that all warnings can be suppressed by typing 'warning off' although blanket action of this type is not recommended. Note that error messages cannot be suppressed because, by definition, they signify events that conflict with the MATLAB language specification and so preclude continuation to a safe or predictable result.

---

[4] ISBN 0-7381-1165-1 (shop.ieee.org/ieeestore/Product.aspx?product_no=SS10116).

A.7   Arithmetic

MATLAB arithmetic is performed using unary and binary operators, which are applied respectively to single numbers and pairs of numbers. In this context, numbers are referred to as *operands*. The standard arithmetic operators are summarised as follows:

Operator	Usage	Equivalent Function	Description	
+	+x	uplus(x)	Unary Plus	
−	−x	uminus(x)	Unary Minus	
+	x + y	plus(x,y)	Addition	$x + y$
−	x − y	minus(x,y)	Subtraction	$x - y$
*	x * y	mtimes(x,y)	Matrix Multiplication	xy
/	x / y	mrdivide(x,y)	Matrix Right Division	xy^{-1}
\	x \ y	mldivide(x,y)	Matrix Left Division	$x^{-1}y$
^	x ^ y	mpower(x,y)	Matrix Power	x^y
.'	x.'	transpose(x)	Transpose	
'	x'	ctranspose(x)	Complex Conjugate Transpose	

Arithmetic operators are designed for different combinations of numerical objects and MATLAB will sort out the appropriate operators for a particular combination and will ensure that the numerical objects are compatible (i.e. have the correct type, size and shape). Thus, if A and B are matrices, then MATLAB makes sure that the expression A + B makes sense before attempting to calculate a result; in this context, both objects must have the same size and shape (i.e. number of rows and columns) or one object must be a scalar.

After some basic experimentation, it should be clear how these operators work. For illustration, a few calculations are performed below, together with their MATLAB equivalents:

$$\mathbf{A} = \begin{pmatrix} 1 & 2 \\ 3 & 4 \end{pmatrix} \quad \mathbf{B} = \begin{pmatrix} 5 & 7 \\ -2 & 0 \end{pmatrix}$$

```
>> A = [1 2; 3 4];
>> B = [5 7; -2 0];
```

$$\mathbf{A}' = \begin{pmatrix} 7 & 15 \\ 10 & 22 \end{pmatrix}$$

```
>> A'
ans =
7   15
10   22
```

$$\mathbf{A} + \mathbf{B} = \begin{pmatrix} 6 & 9 \\ 1 & 4 \end{pmatrix} \quad \mathbf{A}^2 = \begin{pmatrix} 7 & 10 \\ 15 & 22 \end{pmatrix}$$

```
>> A+B
ans =
   6   9
   1   4
>> A^2
ans =
   7   10
   15   22
   10   22
```

$$\mathbf{AB} = \begin{pmatrix} 1 & 7 \\ 7 & 21 \end{pmatrix} \quad \mathbf{AB}^{-1} = \begin{pmatrix} 0.2857 & 0.2143 \\ 0.5714 & -0.0714 \end{pmatrix} \quad \mathbf{A}^{-1}\mathbf{B} = \begin{pmatrix} -12 & -14 \\ 8.5 & 10.5 \end{pmatrix}$$

```
>> A*B
ans =
   1   7
   7   21
>> A/B
ans =
   0.2857   0.2143
   0.5714  -0.0714
>> A\B
ans =
  -12.0000  -14.0000
    8.5000   10.5000
```

Operator *precedence* dictates the order in which operators are applied in an expression. Concentrating here on arithmetic operators only, this is summarised as follows:

1. Transpose (.'), Complex Conjugate Transpose ('), Matrix power (^)
2. Unary Plus (+), Unary Minus (-)
3. Matrix Multiplication (*), Matrix Right Division (/), Matrix Left Division (\)
4. Addition (+), Subtraction (-)

Where adjacent operators have the same precedence, the order of execution proceeds from left to right in the arithmetic expression. More information is given in Section 12.6.3.

Parentheses provide a way of grouping numbers and variables together in order to force the order of calculation. For example, $2 + 3*5$ is the same as $2 + (3*5)$ but is not the same as $(2 + 3)*5$. Parentheses can be nested inside other parentheses, effectively creating a hierarchy of arithmetic expressions; following common sense, the innermost expressions are evaluated first and then the calculation works its way outwards. Thus, an expression like

$$1 + 2^*(1 - 3^*(1 + 4^*(1 - 5^*2)))$$

would be evaluated in the following sequence:

$$1 + 2^*(1 - 3^*(1 + 4^*(-9)))$$
$$1 + 2^*(1 - 3^*(-35))$$
$$1 + 2^*(106)$$
$$213$$

Fortunately, programming languages look after all of this so that users do not have to waste time partitioning calculations into bite-size chunks.

Note that brackets are also a means of grouping numbers and variables together but care is needed because, strictly, these symbols are intended to create arrays. For example, the expression $(2+3)^*5$ is the same as $(2+3)^*5$ but is not the same as $(2+3)^*5$ ... one gives a scalar and the other gives a row vector of length 2 (because of the space between '2' and '+3'). However, $(2+3)^*5$ is the same as $(2+3)^*5$ ... because of the additional space between '+' and '3'. So, use brackets and parentheses for the purposes that they were designed for!

Equivalent functions (anticipating the introduction to functions in Section A.10) are specified in order to place arithmetic symbols ('+', '*' and so on) into the standard context of any other packaged calculation. Thus, an addition could be coded in one of two ways:

```
>> x = a + b ;
>> x = plus(a,b) ;
```

Here, the addition is being specified as an algebraic operation and then repeated by means of a function call (with two numbers a and b defined as inputs). If help is required on the use of arithmetic operators, type 'help' followed by the name of the equivalent function.

In addition to standard arithmetic operators, it is necessary to process data arrays element-by-element in many numerical applications; for this purpose, MATLAB provides a set of array operators, which are summarised as follows:

Operator	Usage	Equivalent Function	Description	
.*	x .* y	times(x,y)	Multiplication	$x_{ij}y_{ij}$
./	x ./ y	rdivide(x,y)	Right Division	$x_{ij}y_{ij}^{-1}$
.\	x .\ y	ldivide(x,y)	Left Division	$x_{ij}^{-1}y_{ij}$
.^	x .^ y	power(x,y)	Power	$x_{ij}^{(y_{ij})}$

Again, for information on operator precedence, refer to Section 12.6.3.

In common with its treatment of standard operators, MATLAB will ensure that numerical objects are compatible before attempting to apply array operators. Thus, if A and B are arrays, then MATLAB makes sure that the expression $A+B$ makes sense before attempting to calculate a result; in this context, both objects have to have the same size and shape or one object must be a scalar.

Note that an m-by-n array is written explicitly as

$$A = \begin{pmatrix} a_{11} & a_{12} & a_{13} & . & a_{1n} \\ a_{21} & a_{22} & a_{21} & . & a_{1n} \\ a_{31} & a_{32} & a_{33} & . & a_{1n} \\ . & . & . & . & . \\ a_{m1} & a_{m2} & a_{m3} & . & a_{mn} \end{pmatrix}$$

with each element (a_{ij}) indexed by its row (i) and column (j). With this is mind, it is important to remember that array operators are specified for a pair or arrays, say **A** and **B**, but the actual calculation is performed on array elements, typified as a_{ij} and b_{ij}, picked from corresponding indexed locations. For illustration, a few examples are given below:

```
>> A = (1 2; 3 4];
>> B = (5 7; -2 0];
>> A*B
ans =
   4   7
  10  17
>> A.*B
ans =
   2   6
   3   8
>> A.^B
ans =
  1.0000  128.0000
  0.1111    1.0000
>> A./B
Warning: Divide by zero.
ans =
  0.2000  0.2857
 -1.5000  Inf
>> A.\B
ans =
  5.0000  3.5000
 -0.6667  0
```

A.8   Logic

In addition to numbers, MATLAB also provides *logical* variables and literals (which can be TRUE or FALSE). MATLAB logicals can take values 0 or 1 but note that these are not numerical values: they are logical values and are treated as a distinct data type. Generally, logical values are generated by logic operations but what are commonly referred to a logic operations actually fall into two categories, one driven by *relational* operators and the other by *logical* operators.

Relational operators are defined as follows (with equivalent MATLAB functions):

Operator	Usage	Equivalent Function	Description
==	x==y	eq(x,y)	Is Equal
~=	x~=y	ne(x,y)	Is Not Equal
<	x<y	lt(x,y)	Is Less Than
>	x>y	gt(x,y)	Is Greater Than
<=	x<=y	le(x,y)	Is Less Than or Equal
>=	x>=y	ge(x,y)	Is Greater Than or Equal

Logical operators are defined as follows (with equivalent MATLAB functions):

Operator	Usage	Equivalent Function	Description
&&	x&&y		Short-circuit Logical AND
||	x||y		Short-circuit Logical OR
&	x&y	and(x,y)	Element-wise Logical AND
|	x|y	or(x,y)	Element-wise Logical OR
~	~x	not(x)	Logical NOT
xor		xor(x,y)	Exclusive OR
any		any(x)	Vector contains nonzero element
all		all(x)	All vector elements are nonzero

For information on operator precedence, refer to Section 12.6.3.

Relational operators require two operands. Logical AND and OR operators require two operands, while the remaining logical operators only require one. The MATLAB help information will clarify issues of usage but two comments are worth making here.

First, operators and functions (where equivalence exists) can be used interchangeably, e.g.

'a==b' is the same as 'eq(a,b)' and ' ~x' is the same as 'not (x)'.

Secondly, 'AND' is TRUE if both operands are TRUE while 'OR' is TRUE is either of the operands is TRUE[2]. The distinction between element-wise and short-circuit operations can lead to confusion. Element-wise operations are comprehensive in the sense that both operands are evaluated (element-by-element assuming that the operands are arrays) and the result is obtained by logical combination. Short-circuit operations require scalar operands and evaluate the second operand only if the result is not fully determined by the first operand; they have the advantage of offering simpler conditional tests but users need to take care because the logic is evaluated in a cascade and may be incomplete for a particular application.

A very simple example, as follows, asks whether the number held in variable 'a' is greater than that held in variable 'b':

```
>> a = 10 ;
>> b = 1 ;
>> a>b
ans =
1
```

which, in this case, returns the answer '1' meaning TRUE. However, it is emphasised again that these are not numerical values: they are logical values.

An alternative implementation serves to emphasise this distinction:

```
>> a = 10 ;
>> b = 1 ;
>> TRUE = logical(1) ;
>> FALSE = logical(0) ;
>> a_bigger_than_b = (a>b)==TRUE
a_bigger_than_b =
1
```

The form 'logical()' here references a function (cf.Section A.10) that creates logical values from numerical values; using this function, number '0' is mapped on to logical '0' while all others numbers are mapped on to logical '1' (. . . so be careful!).

In fact, this can be simplified using the built-in MATLAB functions 'true' and 'false':

```
>> a = 10 ;
>> b = 1 ;
>> a_bigger_than_b = (a>b)==true
a_bigger_than_b =
1
```

Without going into detail, the other main possibilities that exist with respect to logical values are arithmetic combinations and array indexing. Ordinary arithmetic is defined for numbers only and most programming languages (in their basic form) do not define arithmetic operations between numeric and logical variables: MATLAB does and, provided that users understand what is being combined, this can be a useful and compact way of programming compound logic. Ordinary array indexing, as explained already, uses natural numbers. MATLAB also provides the mechanism for using logical values such that elements of an array can be accessed via a congruent array of logical values, such that the numeric element is selected if the logical element is '1'. This works directly for vectors; two-dimensional arrays are processed column-by-column, resulting in a column vector of selected numbers. Note that higher-dimensional arrays are not discussed here.

A.9   M-Files and Functions

In order to save typing in commands over and over again when performing repetitive tasks, the required statements can be written to a text file (without the command prompts), which is then saved with a name that uses the extension 'm'. This is an M-file in its most basic form, called a script file. Thus, in stead of manually typing in all the statements separately, it is much easier just to type in the name of the M-file (without the extension). That will execute all the statements contained in that file.

A rather silly M-file can be written that assumes the existence of a variable x, increments its vales by two, doubles it and then subtracts fives – all as separate steps.

File: **runSillyTask.m**

```
y = x + 2 ;
z = 2*y ;
result = z - 5 ;
```

This could be invoked as follows:

```
>> x = 3 ;
>> runSillyTask ;
>> z
z =
5
```

Although perfectly valid, this is not good programming. The M-file has to presuppose the existence of x and it creates additional variables, which then clutter up the workspace unless they are explicitly removed (type 'help clear' for more information). In fact, it has to presuppose that x contains valid data for the numerical task. Also, the user has to know that the result is called z before being able to examine its value.

It is far better to use *functions* wherever possible for stand-alone tasks. This means changing the M-file so that it starts with a function declaration that states the function name together with its input and output arguments. Note that MATLAB completely ignores the function name inside the M-file and uses the M-file name.

File: **sillyFunction.m**

```
function result = sillyFunction(inputValue)
y = inputValue + 2 ;
z = 2*y ;
result = z - 5 ;
```

This could be invoked as follows:

```
>> x = 3 ;
>> z = sillyFunction(x)
z =
5
```

This is much tidier all round. It is clear what is being done, what the input is and what the output is. Also, additional variables are now defined exclusively for the function when it is invoked and they are not visible outside; when the function is complete, control is passed back to the place from where the function was called and temporary variables are cleared.

As a general indication of how functions can be organised, it is worth implementing the individual steps using trivially simple functions that add, subtract and multiply numbers.

File: **sillyFunction2.m**

```
function result = sillyFunction(inputValue)
y = add(inputValue,2) ;
z = multiplyByTwo(y,2) ;
result = subtract(z,5) ;

function result = add(inputValue,increment)
result = inputValue + increment;

function result = multiply(inputValue,scale)
result = inputValue * scale ;

function result = subtract(inputValue,decrement)
result = inputValue - decrement ;
```

Although the example is completely artificial, this method of functional decomposition is extremely useful for complex programs. In this configuration, the first function is called the *primary* function and the new functions are *subfunctions* that are available for use only within the M-file and cannot be seen from outside.

It is quite common to presuppose the validity of input variables for the numerical task at hand, as has been done in the examples here. This is acceptable for reasonable small tasks and for well-understood tasks. For more complex procedures that draw data from many different sources, validation is highly desirable. Given the flexibility that MATLAB provides, there are no built-in constraints on data type, dimension, size or shape.

A revision to **sillyFunction.m** could be made such that inputValue is checked in order to ensure that is a number. This involves a conditional statement that allows the numerical calculation to go ahead if the input is valid; otherwise it returns a null result.

File: **lessSillyFunction.m**

```
function result = lessSillyFunction(inputValue)
if isnumeric(inputValue)
  y = inputValue + 2 ;
  z = 2*y ;
  result = z - 5 ;
else
  result = [] ;
end
```

It so happens that the 'if-else-end' construction was one of the examples shown in Section A.3 on Getting Help. This is a universal method (in all programming languages) of allowing decisions to be made about what to do next.

A.10 Built-in Functions

MATLAB provides its own set of *built-in functions*, of which there are a large number. These are prepackaged with the tool and implemented internally, i.e. not via M-files (although M-files will exist as a place to deposit the help text for those functions).

The most commonly used built-in functions are likely to be

```
sin, cos, tan, asin, acos, atan, atan2, exp, log, log10, sqrt
```

where these functions have conventional interpretations (as can be confirmed using help).

Some of these are illustrated in the following example:

```
>> m1 = 10^(log10(2.5)*3.5)
m1 =
24.7053
>> m2 = exp^(log(2.5)*3.5)
m2 =
24.7053
>> angle = 30 ; sine = sin(angle*pi/180) ; cosine = cos(angle*pi/180) ;
>> calculatedAngle = atan2(sine,cosine)*180/pi
calculatedAngle =
30.0000
```

A.11 Constants

It is stressed that numbers are held as *variables* and, in MATLAB, there is no concept of a *constant* except for a few built-in constants like π, which is obtained by typing pi, and $\sqrt{-1}$, which is obtained by typing i or j. As with most programming languages, a constant is simply a variable that is not changed.

Note that if these predefined constants have been overwritten for any reason, their original values can be restored via the following statements:

```
>> pi = 4*atan(1) ;
>> pi = imag(log(-1)) ;
>> i = sqrt(-1) ;
>> j = sqrt(-1) ;
```

A.12 Creating Graphs

A major feature of MATLAB usage is data visualisation. At its most basic level, this means plotting graphs and the process is best introduced via an example problem.

> **Problem Statement:**
> Calculate values of x^2 for values of x in the range 0 to 10 and plot the result.

which is a simple problem!

Because MATLAB is a numerical tool, x has to be defined as a vector of numbers ranging from 0 to 10. Since the result has to be plotted, it is necessary to ensure that the finished plot appears continuous (i.e. it does not *look* as if it make up from straight line segments although in fact it will be). A reasonable x-scale for a high resolution graph would be about 400 pixels and it would make sense to use linear (or equal) spacing. This can be achieved as follows:

```
>> x = linspace(0,10,400);
```

This creates a row vector of length 400. In order to check its contents, it is sufficient to display the first few elements:

```
>> x(1:5)
ans =
0 0.0251 0.0501 0.0752 0.1003
```

Because the number format has been truncated to four decimal places, the spacing does not look equal although the actual numbers are equally spaced. However, the spacing is not particularly intuitive because an inclusive number range (including the end values) has been split using an awkward subdivision. It would have been better to choose a subdivision into 401 parts in order to achieve a spacing of 0.025:

```
>> x = linspace(0,10,401);
>> x(1:5)
ans =
0 0.0250 0.0500 0.0750 0.1000
```

In order to calculate 'x^2' it is necessary to square each element of the vector x:

```
>> y = x.^2;
>> y(1:5)
ans =
0 0.0250 0.0500 0.0750 0.1000
```

Thus, there are now two vectors, containing values of coordinates x and y respectively. The plot function will create a graphical representation. in order to introduce some formality, the

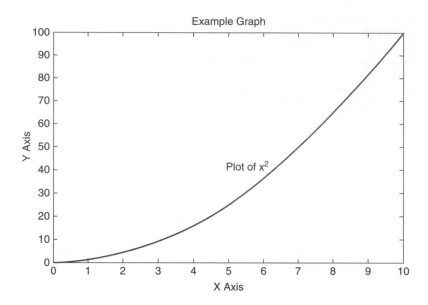

Figure A. 1 Example of graph-plotting in MATLAB.

graph can be given a title and axis labels, plus text annotation. A typical method is shown below:

```
>> plot(x,y);
>> title('Example Graph');
>> xlabel('X Axis');
>> ylabel('Y Axis');
>> text(5,40,texlabel('Plot of x^2'));
```

The end result is shown in Figure A.1. Character strings are specified within single inverted commas (unlike many programming languages, which use conventional speech marks). The text statement positions a written statement at the (x,y) coordinates specified in graph axes and the text can be a literal character string or a TeX character sequence, which in this case is generated by the function texlabel. TeX sequences have the advantage of being able to process transliterated Greek variables (e.g. 'delta') into actual Greek letters (e.g. δ).

A.13   Summary of Appendix A

MATLAB is started typically by double-clicking on a desktop icon. It opens a window for user interaction and sets a predefined working directory.

User interaction is primarily via a Command Window that accepts input from the keyboard and displays the results of any operation (including warnings and errors).

Extensive help is available in the form of explanatory text displayed in the Command Window and source pages displayed in a web-style browser.

The working directory can be changed at any time and so too can MATLAB's search path, thereby enabling MATLAB to find information that is held in files.

Numerical values are held as variables in the MATLAB workspace and these can be manipulated using conventional arithmetic operators, supported by MATLAB language statements.

Logical values can be derived from relational operators and logical operators and these can be used in combination with numerical data for decision-making in MATLAB functions.

Data can be visualised using MATLAB's graph-plotting capabilities.

Statements can be collected together into a *script* or, more usefully, a *function*. Either form is stored as an M-file and can be invoked by typing the filename, thereby removing the need for the user to type and retype statements when performing repetitive tasks. Illustrations have been presented in this lecture for manipulating single variables within a single-pass process. More sophisticated capabilities will be discussed in subsequent lectures.

Eventually, MATLAB can be closed (a) by typing exit or quit, (b) by selecting 'Exit MATLAB' from the 'File' menu or (c) by closing the MATLAB window via the window manager provided by the particular operating system.

Finally, note that there are two M-files that are reserved for starting and finishing a MATLAB session. These are **start.m** and **finish.m**, neither of which are predefined and both of which are optional. On starting MATLAB, if **start.m** exists in the initial working directory, MATLAB will read it and run the commands that it contains. Typically, this is done in order to set project-specific data and to update MATLAB's search path. On closing MATLAB, if **finish.m** exists in the final working directory or on the search path, MATLAB will read it and run the commands that it contains. Typically, this will undo any project-specific provisions.

(*Note: According to the* help *information,* quit *should be the only method of closing MATLAB that invokes **finish.m** but, by experimentation with MATLAB version 7 (R14SP1), all methods are seen to have this effect.*)

(*Technical note: Integrity in MATLAB computations is achieved via the LAPACK*[5] *and BLAS*[6] *numerical libraries.*)

---

[5] www.netlib.org/lapack.

[6] www.netlib.org/blas.

Appendix B

Data and Functions

B.1   Types of Data

Appendix A concentrated on simple operations performed on numbers, which were actually double-precision floating-point numbers (each having a 64-bit binary representation). This is an example of a data type, otherwise referred to as a *class*. By default, numbers in MATLAB belong to a class called *double*. In the course of discussion (cf. Section A.8), it was appropriate to convert numerical values to logical values, which created data of a new class called *logical*.

In general, when in doubt, the class of a particular item of data can be established using the MATLAB functions 'class' and 'isa', as shown in the following example:

```
>> class(1)
ans =
double
>> isa(1,'double')
ans =
1
```

Conversion between data types (provided there is a meaningful way of re-expressing data values from one class to another) can be illustrated using *double* and *logical* values:

```
>> true = logical(1) ;
>> one = double(true) ;
```

What is happening here is that values are being cast in a different data type. MATLAB provides a function called 'cast' that enables this conversion to be made highlighted explicitly:

```
>> true = cast(1,'logical') ;
>> one = cast(true,'double') ;
```

in which a new class of character string data (called *char*) is introduced in order to specify the name of the target classes, namely 'logical' and 'double'.

It is worth recalling the distinction drawn in Section A.6 between *variables* or *literals*. This distinction is applied to any class such that (a) variables exist as named objects that have actual values assigned to them and (b) literals exist as actual values.

```
>> x = f(x,y,z) ; % function operating on variables x, y and z
>> x = g(1,2,3) ; % another function operating on literals 1, 2 and 3
```

In order to do useful things with a programming language, a range of data types are needed, to allow for appropriate and convenient packaging of information. MATLAB provides many but, for present purposes, the main ones are, as follows:

- **double**
- **logical**
- **char**
- **struct**
- **cell**
- **function_handle**

There is also a capability for users to create their own classes (albeit with a few limitations), which will be considered towards the end of this chapter.

Strictly speaking, an extra category should be introduced into the discussion, namely

- **graphics_handle**

This is a number that is returned from a graphical operation (such as would be involved in plotting data, as shown in Section A.12). Although MATLAB creates this number as a *double*, it actually references a dataset containing all the properties of a given graphical object and provides the means for manipulating graphics and interfaces. Arguably graphics handles should have their own data type but, in stead, MATLAB provides a function 'ishandle' in order to establish whether or not a *double* is actually a graphics handle.

Because the distinction between application-specific data and graphical data (i.e. *content* and *presentation*) is fundamentally important in the design of computational schemes, the two categories will be discussed separately in this book. This chapter will deal with pure *data*, leading into a discussion of functional architectures that allow the flow of computation to be managed. The next chapter will concentrate on the specialised techniques that are necessary in order to design and implement graphics and interfaces.

B.2   Data Type Descriptions

B.2.1   *'double'*

Double-precision floating-point numbers form the default number representation in MATLAB. Variables that are created by assigning them a numerical value will be classified as *double*. Literal formats can be selected from the following options:

- integer value, e.g. 1
- real value, e.g. 1.234
- real value plus exponent, e.g. 1.234e6 (which is equivalent to 1.234×10^6)

```
>> x = 1.0e6 ;
>> class(x)
ans =
double
```

Arrays of *doubles* can be assembled as shown in Section A.6. Arrays can have one, two or more dimensions. Individual elements of an array can be accessed by appropriate indexing.

```
>> A = [1 2 3; 4 5 6; 7 8 9] ;    % 3x3 array
>> A(2,1)                         % Array element (row 2, column 1)
ans =
 4
>> A(1:2,[1,3])                   % Sub-array (rows 1&2, columns 1&3)
ans =
 1 3
 4 6
```

This method of indexing can be used in order to overwrite elements of an existing array. Additionally, it can be used in order to create a new array with selected values.

```
>> A = [1 2; 3 4] ;
>> A(2,1) = 100
A =
 1 2
 100 4
>> B(2,2) = 1
B =
 0 0
 0 1
```

B.2.2   *'logical'*

Logical variables are the result of logical operations performed on data, i.e. tests that return a result that is true or false.

```
>> x = 4 ; y = 2 ;
>> x_greater_than_y = x>y
x_greater_than_y =
 1
>> class(x_greater_than_y)
ans =
logical
```

Similar tests can be performed on arrays, giving a result that is a logical array. The logical array itself can be used in lieu of indexing as a means of data selection. Just in case there is any confusion, the equivalent indices can be generated via the MATLAB function 'find'.

```
>> x = [0 2 4 6] ; y = 2 ;
>> x_greater_than_y = x>y
x_greater_than_y =
 0 0 1 1
>> bigger_x = x(x_greater_than_y)
ans =
 4 6
>> indexList = find(x_greater_than_y)
ans =
 3 4
```

B.2.3   *'char'*

Character strings are interpreted as text objects, i.e. sets of characters that do not necessarily have any mathematical or numerical significance. These are classified as *char* data and are specified in MATLAB is characters enclosed within apostrophes.

```
>> greeting = 'hello'
greeting =
hello
>> class(greeting)
ans =
char
```

Note that if the text is intended to contain an apostrophe then, in order to distinguish it from the text delimiters, it is written as a double apostrophe.

```
>> sentence = 'I''m having fun.'
sentence =
I'm having fun.
```

A simple character string like this is actually a one-dimensional array of *char* objects (i.e. a character array) and, as such, is subject to the same construction rules and indexing rules as any other array.

```
>> sentence = 'I''m having fun.';
>> sentence(5:8) =
 hav
```

A set of functions for string manipulation is provided within MATLAB and more information can be found by typing 'help strfun'.

B.2.4   'cell'

Specific types of data are given there own class designation, which up to now has involved numerical, logical and textual objects. This is the way in which information is encoded within a programming language and the basic principle is that a data type is tailored to the particular method of expressing information. More generally, however, it may be sufficient to define place-holders for data in much the same way as a spreadsheet does. MATLAB provides a facility for this, using data that is organised in cells. Accordingly, this is classified as *cell* data.

A *cell* array does nothing more than contain data, regardless of its class. This type of object is created by enclosing a piece of data within braces (i.e. curly brackets)

```
>> c = {1} ;
c =
 [1]
>> class(c)
ans =
cell
>> d = {1, 'Hello'; 23, []}
d =
 [ 1] 'Hello'
 [23] []
```

Note that, from this example, a *cell* array can hold anything! The rules for creating and accessing *cell* arrays are almost the same as for arrays of anything else. If array indices are enclosed in parentheses, this will access the specified cells in the array, effectively creating another *cell* array. If array indices are enclosed in braces, this will access the cell contents.

```
>> d(1,1:2)
ans =
 [1] 'Hello'
>> d{1,1:2}
ans =
 1
ans =
Hello
```

When accessing cell contents in this way, there are limits on what can be done because cell contents can be of arbitrary class and therefore will not necessarily permit the creation of new arrays using the data contents.

```
>> newCellArray = d(1,1:2)
newCellArray =
 [1] 'Hello'
>> newContentArray = d{1,1:2}
??? Illegal right hand side in assignment. Too many elements.
```

which means that MATLAB does not like what it is being asked to do!

The use of cell arrays seems to take some getting used to, partly because of the subtly different notation and partly because of the initial confusion about how to implement a set of data containers in this way. It should always be remembered that this serves the same basic purpose as a spreadsheet and, conceptually, it is laid out in the same way.

B.2.5   *'struct'*

Data structures represent a major departure from simple items of data grouped together in arrays. These are a powerful weapon in the programmer's arsenal as they allow data to be grouped together by name. In this context, variables are classified as *struct* and are constructed with 'fields', as would be appropriate in an inventory.

```
>> x.number = 1 ;
>> x.array = [1 2; 3 4] ;
>> x.text = 'This is a data structure' ;
>> x
x =
 number: 1
 array: [2x2 double]
 text: 'This is a data structure'
```

In examples like this, take care that a variable does not already exist as something other than a *struct*. MATLAB provides considerably greater freedom than many programming languages in creating data but, even here, converting or replacing an unstructured object with a structure is not good programming practice. Note that, if a variable were empty (cf. Section A.6), redefining it as a structure would be perfectly acceptable.

Additional fields can be introduced and the depth of the structure can be changed by embedding structures within structures. Thus, structures offer the opportunity of organising information in a hierarchy, whereas arrays offer an indexed map (which shows essentially flat, anonymous data).

```
>> x.data = 0.0001 ;
>> x.structure.a = 1 ;
>> x.structure.b = 2 ;
>> x
x =
  number: 1
   array: [2x2 double]
   text: 'This is a data structure'
   data: 0.0001
  structure: [1x1 struct]
>> x.structure
ans =
 a: 1
 b: 2
```

On being presented with a *struct* variable, its class and content can be established as follows:

```
>> class(x)
ans =
struct
>> nameList = fieldnames(x)
nameList =
 'number'
 'array'
 'text'
 'data'
 'structure'
>> isfield(x,'data')
ans =
 1
```

The function 'fieldnames' produces a *cell* array containing the top-level fields of the particular structure. Each field name is a character string, i.e. a *char* object. The function 'isfield' performs a check that a nominated field exists in the structure and returns a *logical*.

Up to now, examples have been based on structures that are created and accessed using literal field names. As an alternative, the field names can be specified as a *char* variable, in which case the field name is enclosed within parentheses.

```
>> field = 'number';
>> x.(field)
ans =
 1
```

Structures can contain arrays and can be created as arrays; so there is a distinction between an array of structures and a structure of arrays. In common with all arrays, data are created and assigned to a specific position in the array and are accessed with appropriate indices.

```
>> x(1).number = 1 ;
>> x(1).array = [1 2; 3 4] ;
>> x(2).number = 24 ;
>> x
x =
1x2 struct array with fields:
 number
 array
>> x(2)
ans =
 number: 24
 array: []
```

Note that all structures in a *struct* array have the same fields and that fields that have no values assigned to them will contain an empty array. It should be mentioned that, when processing an array of data that involves structures, it is computational quicker to process a structure of arrays than it is to process an array of structures – typically a lot quicker!

As a final note in this section, there is a similar distinction between a cell array that contain structures and a structure that contains cell arrays. Initially, this can cause confusion!

```
>> x.cell = {1, 'number'};
>> y.a = 1 ; y.b = 2 ;
>> z.a = -1 ; z.b = -2 ;
>> cellArray = {y, z} ;
```

B.2.6   *'function_handle'*

The basic principles of MATLAB functions were introduced in Sections A.9 and A.10. Functions are invaluable as a means of packaging sequences of computation and allowing convenient and uncluttered management of the MATLAB workspace.

An arbitrary example of a function might be defined as follows:

```
% File: addOne.m
function outputValue = addOne(inputValue)
outputValue = inputValue + 1 ;
```

This might be used as follows:

```
>> one = 1 ;
>> two = addOne(one)
two =
  2
```

Although rather pointless, the example clearly shows the principles of how a function is defined and used. The function definition is contained in a file somewhere and, in order for it to be invoked, MATLAB has to find it on its search path (cf. Section A.5). It also has to ensure that input/output variables are created and linked to equivalent variables used in the calling statement. In short, MATLAB needs to know where the function is located and how to interface with it.

The example shows a direct function call. However, it is possible to call a function indirectly by means of a so-called *function_handle*. This acts as a reference to an actual function, which stores the information required for its execution. Once the handle exists, it can then be used in lieu of the actual function name.

```
>> f = @addOne ;
>> class(f)
ans =
function_handle
>> two = f(1)
two =
  2
```

The facility opens up various options for managing MATLAB code and for simplifying the MATLAB path for specific applications. For instance, large numbers of functions could easily be grouped into a small number of M-files serving as function libraries, each returning a *struct* containing *function_handle* objects.

```
% File: trigLibrary.m
function f = trigLibrary
f.sine = @sin ;
f.cosine = @cos ;
f.tangent = @tan ;
```

This type of library might be incorporated within an application, as follows:

```
% File: application.m
function output = application(input)
trig = trigLibrary ;
output = trig.sin(trig.cos(input)) ;
```

This neatly and explicitly collects together functions of a particular category. In fact, if implemented like this, it also provides a *struct* of *function_handle* objects that can be passed to other functions as an input.

One convenient usage of function handles is in the definition of generic functions or function templates (cf. Section B.2.3.2), i.e. a structure of computation that allows the functional implementation to be determined at run-time. In addition to data input, this requires the function to be given a *function_handle* before it can actually calculate anything.

```
% File: trig.m
function output = trig(f,degrees)
radians = degrees * pi/180 ;

output = f(radians) ;
```

```
>> trig(@sin,45)
ans =
  0.8509
```

In addition to named functions, it is possible to establish one-line calculations and associate them with function handles. This has the virtue of collecting together sets of trivial tasks and organising them for ease of use, without clogging up the main body of code or incurring the overhead of creating separate functions.

```
% File: trig.m
function output = trig(f,angle)
radians = @(degrees) degrees*pi/180 ;

output = f(radians(angle)) ;
```

In this formulation, the conversion from degrees to radians is referenced to a so-called anonymous function, i.e. a *function_handle* that is associated with input variables and a line of calculation...but no name. The example offers a satisfactory illustration of this principle although in a rather trivial context with a far-from-ideal implementation.

A note of clarification is warranted for functions that do not require input values. Functions of this kind are used to create 'universal' data for an application that is, by definition, not dependent the application context. However, if such a function is invoked by a *function_handle* then it must be followed by an empty input list in order to distinguish between an instruction to calculate something (i.e. execute the function) and an instruction just to look at something (i.e. show the value of the particular function handle).

```
% File: gravity.m
function g = gravity
g = 9.80665 ;
```

```
>> g = @gravity ;
>> g
g =
 @gravity
>> g ()
ans =
 9.80665
```

There are many variations on this theme and it is difficult to generalise as to the suitability of function handles in different applications. Depending on the design philosophy, they are certainly useful in managing function libraries (especially those associated with numerous small components). They are also useful in establishing an explicit dependency on functions at the start of a large program and then propagating that dependency throughout that program. Largely this is a matter of preference on the part of the developer.

B.3 Program Structure

Programs consist of machine-readable code and conform to a strict language specification, which defines the rules for their constructing specific task elements (like calculations). Programs consist of a set of language statements and, in the absence of any instruction to the contrary, their execution will proceed sequentially from the start of the program to the end. This is not adequate because, inevitably, parts of a calculation will vary in different circumstances and may have to be repeated. Accordingly, different program segments will be required together with some means for controlling their execution. In MATLAB, programs are implemented as functions and, recalling the earlier discussions, it is essential to know where these functions reside and how to interface with them. This is even more important given that, unlike most other programming languages, MATLAB allows variables to be created and used without being preceded by a formal declaration. Also, MATLAB provides the novel facility for handling variable numbers of inputs and outputs for a function and the not-so-novel facility for introducing *global* data and *persistent* data. Thus, it is necessary to devote some discussion to the subject of program structure.

B.3.1 Syntax

Programs follow a language specification and the most important part is the *syntax*, which defines how symbols and keywords can be used in the program construction. This is a straightforward matter of establishing a valid grammatical structure and is totally separate from the meaning of the symbols or keywords (which is called *semantics*). In this section, it is intended to show the architectural elements of MATLAB functions with reference to a basic set of syntax rules as they apply to MATLAB functions. No way is this a complete syntax; it is a partial illustration only.

One of the traditional ways for expressing syntax is via the following format:

Syntax_Element_Descriptor ::= Syntax_Element_Description

where the symbol '::=' denotes the assignment of a rule to a descriptor and the associated description consists of named segments punctuated with delimiters and grouping symbols:

[ ] signifies an optional segment that can occur once
{ } signifies an optional segment that can occur many times
| signifies a selection between two segments (i.e. a logical 'or')

The choice of descriptor names is entirely a matter of preference but should convey the purpose of the syntax element in a common-sense way.

Text that appears in bold font is treated as literal text, i.e. as it would actually appear in the finished code, e.g. **switch**. Text that appears in italic font indicates a variant component (i.e. something that must be supplied by the programmer), which would typically be a name of something, e.g. *Function_Name*. Also, italic text can be used as a prefix to a named item to indicate a variant of that named item, e.g. an Expression could be specialised as a *Logical_*Expression.

First of all, it is necessary to define a set of primitives, which provide the basic building blocks for the syntax development. In this case, it is sufficient to define the following:

- Variable
- Literal

These are interpreted as data elements without further elaboration. In MATLAB, data is created and used without prior declaration and so the language syntax does recognised the existence of data types; this is a feature that is managed internally to MATLAB and implemented in code via appropriate function calls.

B.3.1.1   Top-level Function Definition

Thus, a MATLAB function can be defined as the top-level syntax element, as follows:

```
Function_Declaration::=
        function [Function_Argument = |[ Function_Argument {,
           Function_Argument}]=]
                 Function_Name [( Function_Argument {, sFunction_Argument})]
        {Statement}
Function_Argument::=
        Variable
```

This shows clearly the 'function' keyword and the function name, together with optional single or multiple arguments. Note that the arguments are explicitly designated by a syntax element called Function_Argument, which is implemented as a Variable.

Statements (i.e. lines or blocks of code) can be categorised according to their purpose:

```
Statement::=
        Method_Statement | If_Statement | Switch_Statement |
            For_Statement | While_Statement | Try_Statement
```

B.3.1.2   *Method Statements*

The label 'Method_Statement' is used here to indicate general-purpose computation. This evaluates an expression (on the right) and, if required, assigns the result to one or more variables (on the left). The assignment is denoted by an *equals* sign.

```
Method_Statement::=
        [Variable = | [ Variable {, Variable} ] =] Expression
```

An Expression can exist in several forms:

```
Expression::=
              Data_Item | Algebraic_Expression | Bounded_Expression |
Concatenated_Expression
```

The simplest form is a data item, which can be a variable, a literal or a function call; the first two options represent *specified* data and the third represents *derived* data. The function call has the form that has been observed in previous examples, with a function name followed by arguments if required. The arguments here are designated by a syntax element called Actual_Argument, which can be implemented as a Variable or a Literal.

```
Data_Item::=
        Variable | Literal | Function_Call
Function_Call::=
        Function_Name [ ( Actual_Argument {, Actual_Argument} ) ]
Actual_Argument::=
        Variable | Literal
```

Note that the distinction between this and a Function_Argument; one pertains to the function instance while the other pertains to the function definition, respectively. When a function is invoked, a separate thread of calculation is created and separate memory is allocated; the current Actual_Argument values then have to be associated with corresponding Function_Argument variables through a process called *binding*. This ensures that the function has the correct information to commence.

Other forms of expression considered here are those that involve operators of various kinds (cf. Sections A.7 and A.8), those that involve parentheses (as a means of grouping calculations) and those that involve concatenation.

```
Algebraic_Expression::=
        [Operator] Expression {Operator Expression}
Bounded_Expression::=
      ( Expression )
Concatenated_Expression::=
              [ Expression { ; Expression} ] | [ Expression { , Expression} ]
```

Continued

```
Operator::=
        Arithmetic_Operator | Array_Operator | Logical_Operator |
           Relational_Operator
Arithmetic_Operator::=
        + | - | * | / | \ | ^ | ' | .'
Array_Operator::=
        .* | ./ | .\ | .^
Logical_Operator::=
        | | || | & | && | ~
Relational_Operator::=
        == | ~= | > | >= | < | <=
```

B.3.1.3   Control Statements

Various control mechanisms are provided in MATLAB, in common with most programming languages, which affect the order of calculation. In this interpretation of MATLAB syntax they are specified by specialised statements that organise other statements for the purposes of conditional execution, iterative execution or exception handling.

Conditional execution can be described using If_Statement and Switch_Statement syntax; an explanation is provided in Section B.3.2.

```
If_Statement::=
                if Logical_Expression
                    {Statement}
            { elseif Logical_Expression
                    {Statement}}
            [ else
                    {Statement}]
            end
```

```
Switch_Statement::=
                switch Variable
                case Option
                    {Statement}
            { case Option
                    {Statement}}
            [ otherwise
                    {Statement}}
            end
Option::=
        Variable | Literal
```

Iterative execution can be described using For_Statement and While_Statement syntax; an explanation is provided in Section B.3.3.

```
For_Statement::=
            for Number_Sequence
                {Statement}
        end
Number_Sequence::=
        Sequence_Variable | Sequence_Literal
```

```
While_Statement::=
            while Logical_Expression
                {Statement}
        end
```

Exception handling can be described using Try_Statement syntax; an explanation is provided in Section B.3.4.

```
Try_Statement::=
            try
                {Statement}
        catch
                {Statement}
        end
```

Although the explanation of these control statements is deferred, what should be clear is that these statements contain other statements, any of which could be a control statement. In this way, the computation can be decomposed into a nested hierarchy.

B.3.2   *Conditional Execution*

B.3.2.1   *'If' Statements*

Simple choices are made by determining *if* a particular condition exists or *else if* another condition exists, or *else* it is taken that some default condition exists; this is the *end* of decision process. So it is with program code; a simple combination of tests is applied, which has the general structure:

```
if condition1 is true
        perform task1
elseif condition2 is true
        perform task2
... continue testing other conditions
        ... and performing other tasks
else
        perform default_task
end
```

Without introducing any complexity, the basic principle is illustrated in the following example:

```
% Straight line y=x with saturation at |x|==1
if x>1
    y = 1 ;
elseif x<-1
    y = -1 ;
else
    y = x ;
end
```

Note that logical checks performed on character strings (i.e. *char* objects) are based on string comparisons and, for convenience, these should use the 'strcmp' function:

```
if strcmp(event,'arrival')
    greeting = 'hello'
else
    greeting = 'goodbye'
end
```

Without a string comparator, the implementation would be rather more tedious:

```
if length(event)==length('arrival') && event=='arrival'
    greeting = 'hello'
else
    greeting = 'goodbye'
end
```

which would probably become such a distraction that someone would write a string comparator, just like 'strcmp'!

B.3.2.2   'Switch' Statements

A somewhat neater method of organising decisions is via a '*switch*' statement. This operates on scalars and character strings and, depending on the value of the operand, control is transferred to the particular *case* that happens to be true, *otherwise* a default task is performed. Thus, in the program code; a simple combination of tests is applied, which has the general structure:

```
switch variable
        case value1
                perform task1
        case value2
                perform task2
... continue checking for other values
        ... and performing other tasks
```

```
otherwise
        perform default_task
end
```

This arrangement can have as many '*case*' segments as required to cater for defined conditions. A particular case can be associated with several values and, in MATLAB, this is established by a case statement followed by values, separated by commas and enclosed in braces (i.e. curly brackets).

The basic principle is illustrated in the following example:

```
switch carMoving
 case true
  switch trafficLights
   case {'red','amber'}
   action = 'stop' ;
  otherwise
   action = 'go' ;
  end
 otherwise
  switch trafficLights
   case 'green'
    action = 'go' ;
   otherwise
    action = 'stop' ;
  end
end
```

B.3.3   *Iterative Execution*

B.3.3.1   *'For' Statements*

Iterative (or repeated) calculations need to be performed in various places and programming languages provide looping mechanisms for this purpose. The standard in MATLAB and elsewhere is to specify a repetitive loop *for* some countable sequence of numbers and then bring the calculation to an *end*. In program code, this has the general structure:

```
for counter in a defined sequence
        perform task
        ... and keep repeating until the counter reaches the end
          of sequence
end
```

The basic principle is illustrated in the following example:

```
n = 1 ;
for iNumber = 1:10
  n = n * iNumber ;
end
```

This calculates the value of 10-factorial. Notice that the loop executes over a number sequence from 1 to ten and that this is specified in the form shown in Section A.6. In principle, any other sequence could be applied, such as:

```
sequence = 1:2:10 ;
n = 1 ;
for iNumber = sequence
 n = n * iNumber ;
end
```

Note that this is calculating something other than value of 10-factorial.

B.3.3.2 'While' Statements

An alternative, but entirely equivalent, form of iterative calculation is provided using a 'while' statement. *While* a given condition is true, the loop is executed repetitively until the condition is no longer true and the calculation is brought to an *end*. This requires a slightly different method for controlling the loop, namely initialising the enabling condition and checking that condition after each pass through the loop. This is different from a 'for' statement in that there is no predefined sequence of numbers driving the execution; so, the enabling condition must become a terminating condition at some point, otherwise the loop will continue indefinitely. In other words, a terminating condition must exist and this has to be verified.

In program code, this has the general structure:

```
set condition to true
while condition is true
        perform task
        check that condition is still true
        ... and if it is then repeat once more
end
```

The basic principle is illustrated in the following example:

```
n = 1 ;
iNumber = 0 ;
calculating = true ;
while calculating
 iNumber = iNumber + 1 ;
 n = n * iNumber ;
 calculating = iNumber 10 ;
end
```

This again calculates the value of 10-factorial. This appears to be messy when compared with the corresponding '*for*' loop and, in general, it is not the preferred way of performing iterations based on a simple counting scheme. However, in other circumstances, this can be far more appropriate as it allows decisions to be made based on the results of calculations (and often complicated calculation); thus, looping is completely generalised and context-dependent.

B.3.4   *Exception Handling*

Some languages provide the ability to jump between statements; a feature that was somewhat dubiously celebrated in FORTRAN and, from experience, one that could lead to a lot of entanglements. This is not recommended simply because it is too difficult to trace the order of execution of program code and, thus, the state of a calculation at any time is not predictable. MATLAB has a 'return' function, which causes a function to cease execution immediately; so it is a 'jump' and, again, this is not recommended.

However, many languages do support *exception handling*. This is a controlled jump that is triggered by an error condition (otherwise called an *exception*). In essence, a program is set up to *try* to execute a block of code and, if an exception is thrown, to *catch* that event and transfer control to a separate block of code (to implement recovery action) and then bring the particular calculation to an *end*. Importantly, this enables a calculation to terminate in a know state; under this regime, the program should be designed so that it does not just fail and stop.

In program code, this has the general structure:

```
try
    perform task
catch
    ... and perform another task in the event of an exception
end
```

An exception will be thrown if MATLAB encounters a condition that it cannot deal with (e.g. incorrect code or a numerical anomaly) or if the program executes an 'error' statement (installed by the programmer in order to respond to context-dependent anomalies).

A trivial example is as follows:

```
try
  x = a/b ;
catch
  x = [] ;
end
```

This approach would catch literally anything that MATLAB could not cope with, which would offer no discrimination between sources of error. As a matter of clarify, it would be preferable to use identifiers in the code and to take appropriate action based on where in the code the exception has occurred. With this in mind, another trivial example is:

```
a = double(a) ;
b = double(b) ;
try
    identifier = 'task1'
    x = a/b ;
    identifier = 'task2'
    y = b/a ;
catch
  switch identifier
```

Continued

```
case 'task1'
  error('b is zero?');
case 'task2'
  error('a is zero?');
end
```

The writing of exception handlers is tricky because they have to be able to mop up all manner of unintended events and to do sensible things when such events actually occur. This is very wide and involved topic and, while of interest, it is peripheral to the main thrust of this book and will not be considered any further.

B.3.5   Omissions

Having stated originally that this is not a complete syntax, it is reasonable to offer some indication of what is missing. First of all, note that the preceding sections are an author's interpretation of MATLAB syntax, simplified in order to emphasise the most important features. Next, note that this is not a textbook of computer programming and so it is not desirable to cover every detail of the MATLAB language.

Several minor omissions include:

1. *Comments*:
   These commence with the symbol '%' and truncate any line of code in which they appear.
2. *Colons*:
   Colons are used in the construction of number sequences (e.g. 1:5, as first introduced in Section A.6) and, strictly speaking, should be treated as operators.
3. *Keywords*:
   The complete range of MATLAB keywords can be obtained by typing 'iskeyword'; the syntax described up to now covers most of the keywords but not all.
4. *Continuations*:
   These can appear at the end of lines of code (before any comment) as a sequence of three period characters (i.e. full stops) '...', thereby allowing long lines to be split.

Also, the depth of explanation has not been particularly deep because the emphasis has been placed on familiarisation. Through usage, the design philosophy that underpins MATLAB should quickly become apparent.

In addition to syntax, there are other issues that arise in connection with program structure, namely user-defined functions, user-defined classes and practical implementation. These will be addresses briefly in the next three sections.

B.4   User-defined Functions

B.4.1   Interfacing

A common structure for a MATLAB function is as follows:

```
function output = thisFunction(input)
% ...followed by language statements
```

This is a function with a name (thisFunction), a single input and a single output. This can be extended to multiple inputs and/or multiple outputs and, in a more general form, a function would be introduced as:

```
function [output1,output2,output3] = thisFunction(input1,input2)
% ...followed by language statements
```

for a function with two inputs and three outputs. Functions might conclude with a 'return' statement but this is optional.

Functions can be defined with no inputs and/or no outputs, depending on specific circumstances. Typically, a function with no input would be used for specifying constants, as seen in the following example:

```
function R0 = universalGasConstant
R0 = 8314.33 ;
```

Functions are held in M-files and, as seen in Section A.9, an M-file can hold several functions; the primary function always comes first, with other functions serving as subfunctions. In accordance with normal scoping rules, subfunctions are only visible to other functions within the M-file, unless their existence is published to the outside world via functions handles. It should be stressed throughout that any function can call any other function, provided that it can be found on the MATLAB search path.

```
function f = applicationLibrary
f.sine = @sineFunction ;
f.cosine = @cosineFunction ;

function result = sineFunction(angle)
result = sin(angle*pi/180) ;

function result = cosineFunction(angle)
result = cosine(angle*pi/180) ;
```

The body of code that makes up the function content is essentially the same as a MATLAB script, except that data created during execution is local to that function (and as such cannot be seen or accessed from outside) and is cleared when execution is complete. The only exceptions are *global* data and *persistent* data.

Global data must be declared as *global* before use and can only be accessed from within functions (not the MATLAB command line) that declare the same data. This gives a source of shared data that is hidden from the function call and that can be accessed arbitrarily. Functions with access to global data require great care in their design and usage.

A suitable example is as follows:

```
function makeGlobal(parameter,value)
 global sharedData

sharedData.(parameter) = value ;
```

Persistent data must be declared as *persistent* before it is used and can only be accessed from the originating function during any invocation. Thus, data that are persistent can be changed whenever the function is called and the same data are available when the function is next called. This gives a source of private data that is hidden from the function call and that can be accessed arbitrarily.

A suitable example is as follows:

```
function latestCount = incrementCounter(thisIncrement)
 persistent counter

if isempty(counter)
 counter = thisIncrement ;
else
 counter = counter + thisIncrement ;
end
latestCount = counter ;
```

Functions with access to persistent data require great care is their design and usage, especially where it is possible to call one of these functions from different contexts, e.g. two different aircraft simulations called the same engine calculation. In all but the simplest of calculations, it is strongly recommended (if not essential) to treat this as a client-server application and, as such, call the function with some form of identification in order to maintain segregation between data for different clients.

A simple illustration of this approach is as follows:

```
function latestCount = incrementCounter(thisIncrement,clientNumber)
persistent counter

if isempty(counter(clientNumber))
  counter(clientNumber) = thisIncrement;
else
  counter(clientNumber) = counter(clientNumber) + thisIncrement;
end
latestCount = counter(clientNumber);
```

As a final comment in this section, it is noted that the data associated with graphics handles can be thought of as being both hidden and persistent. A similar degree of care is required in that setting also, as will be discussed in Chapter 6.

B.4.2   *Generic Functions*

Conditional execution offers a means of varying what a program does depending on the data that it knows about. Thus, different circumstances trigger different responses. More generally, functions can be configured such that they accept functions as arguments and even accept varying numbers of inputs and outputs. For instance, the Ada programming language provided this capability and, also, this is one of the intricacies much loved by C++ programmers. This is often denoted by the term *generic function* or *function template*. MATLAB provides the capability through the use of function handles and through generalised input and output variables.

Both the use of function handles and the use of multi-input/multi-output functions have been discussed already in Section B.4.1. Here is a simple example showing the combination of these two features:

```
function [out1,out2] = calculate(f,g,in1,out2)

out1 = f(in1) + g(in2) ;
out2 = f(in1) - g(in2) ;
```

An alternative implementation can make use of varargin and varargout, these being reserved names that denote cell arrays containing input and output arguments, respectively. The actual number of inputs and outputs in the function call is determined by the built-in functions nargin and nargout. These features enable arbitrary numbers of arguments to be specified and enable these numbers to be counted once the function has been invoked; in fact, using knowledge about data types from Section B.1, the class of each argument could easily be established, thereby allowing the function to verify the data that it been given before attempting to use it.

With this in mind, the previous example can be re-created as follows:

```
function varargout = calculate(varargin)

varargout{1} = [] ;
switch nargin
  case 4
   f = varargin{1} ;
   g = varargin{2} ;
   in1 = varargin{3} ;
   in2 = varargin{4} ;

   haveFunctions = isa(f,'function_handle') && isa(g,'function_handle') ;
   haveDoubles = isa(in1,'double') && isa(in2,'double') ;

   switch haveFunctions && haveDoubles
    case true
      varargout{1} = f(in1) + g(in2) ;
      if nargout>1
        varargout{2} = f(in1) - g(in2) ;
      end
      ok = true ;
    end
end
```

Application is demonstrated in the following few MATLAB commands:

```
>> calculate(@sin,@cos,0.1,0.2)
ans =
  1.0799
>> calculate(@sin,@sin,0.1,0.2)
ans =
  0.2985
>> [x,y] = calculate(@sin,@sin,0.1,0.2)
x =
  0.2985
y =
  -0.0988
```

Now, a generic function is a convenient way of packaging a common sequence of activities that can be implemented in different ways. As observed, it provides a template (which is simply a framework with place-holders for variant components). Typically functions like this would be used in combination with 'if' statements and 'switch' statements.

B.4.3   *Recursive Functions*

Iterative execution can be implemented using *recursive* functions as an alternative to 'for' and 'while' statements. This is a method based on a step-by-step functional decomposition, such that the result from step N can be applied directly as an input to step N+1. In practical terms, this means that a function calls itself; strictly speaking, the function definition contains a function call that references the same function definition. At run-time, when the actual function is invoked, it will encounter a call to the same function and, accordingly, the function is invoked again. This is not the same invocation; these are separate instances of the one function definition, with separate interfaces and separate workspace allocations. This continues until a condition is encountered that terminates this process ... which means that terminating condition must be designed into the function, otherwise the process would continue to infinity. When the process has terminated in the latest function instance, the whole process unwinds as every other function instance terminates and then the calculation is complete.

The calculation of factorials is one of the standard illustrations offered in textbooks and it seems appropriate to offer it here:

```
function result = factorial(number)
if number>0
  result = number * factorial(number-1) ;
else
  result = 1 ;
end
```

which calculates the value of N-factorial where N is the number given to the first instance of factorial. Now there is a potential problem with this function as it stands, namely it requires a positive integer (i.e. a natural number) in order to be able to terminate but there is nothing to guarantee that or even to check that. If the initial inputs are prechecked, this will not be a problem but, for arbitrary usage, it almost certainly will be a problem from time to time. Incidentally, exactly the same problem would arise with a 'while' statement.

Thus, a better attempt would be as follows:

```
function result = factorial(number)
number = round(number) ;
if number<0
   result = [] ;
elseif number = 0
   result = 1 ;
else
   result = number * factorial(number-1) ;
end
```

which forces the input argument to be rounded to the nearest integer and traps negative integers, thereby giving a sensible result or a null result.

B.4.4   *Private Functions*

Some degree of privacy is available in the definition of MATLAB functions, based on where M-files are located. This is an issue that was discussed in Section A.5 and it is important because of the way in which it uses its search path.

Any directory that is used to hold M-files can contain a directory called 'private' (e.g. C:\My Documents\MATLAB_functions\private or something like that). In this context, it is convenient to talk about the parent directory and the private subdirectory. A function whose M-file is located in the parent directory can make use of functions from the private subdirectory. Functions in any other directory cannot, unless they are part of a chain of execution that starts at a function is the parent directory. Nothing more really needs to be added to this explanation; it is a facility that might be useful.

B.5   User-defined Classes

Up to now, data has belonged to predefined types or classes and been manipulated directly by statements typed in at the MATLAB command line or contained within M-files. As a matter of preference and good management, applications should be designed based on functions. In this way, computation is well defined and packaged; more importantly, perhaps, data is managed locally for each function, thereby avoiding workspace clutter and/or accidental data corruption.

A completely different philosophy is available through the creation of application-specific classes. In other words, the developer can design an appropriate data structure and package it in such a way as to restrict access to it. The basic process is often referred to as encapsulation and this has the virtue of hiding information from general view. Having created a *class*, data access is granted via a set of functions that are defined explicitly for that class; generally these functions are called *methods*. Thus, encapsulated information is hidden except where suitable methods are provided to extract specific items.

In MATLAB the construction of objects of a given class is done via a constructor function, located in a directory of the same name prefixed with the symbol '@'. Thus, for a class called *vector*, new objects would be made by calling a function called vector located in a directory called '@vector' (which is referred to as a method directory). Note that method directories are not permitted on the MATLAB search path; so, this must be a subdirectory of a directory that is on the search path.

In essence, what the constructor function does is to encapsulate a data structure and associate with it a class name. This produces an *object* of a given *class*. What the developer of the constructor function has to do is to check that the object data that is supplied has the correct form and content to enable an object to be constructed. Once that is achieved the encapsulation is performed using a variation of the 'class' function that has appeared previously. Consider, for instance, the creation of a class called *vector* that, in its simplest manifestation, holds three coordinates. The constructor could be designed as follows:

```
function object = vector(objectData)

obj = [];
if ~isempty(objectData)
  switch class(objectData)
    case 'vector'
     object = objectData ;
    case 'double'
     if all(size(objectData)==[3 1]) && isreal(objectData)
       obj.components = objectData ;
     end
    case 'char'
     switch objectData
      case 'origin'
       obj.components = [ 0 0 0 ] ;
     end
    end
end

object = [];
if ~isempty(obj)
   object = class(obj,'vector') ;
end
```

If the input (called objectData) is already a *vector* then the output (called object) simply copies it. If it is *double* then, subject to it being a valid 3x1 vector of real numbers, it will be accepted as the set of vector components and added to the object data structure (called obj). If it is *char* then, provided that the input text is 'origin', it will be accepted as a specification for the origin of coordinates and these will be added to the object data structure. Finally, given that the object data structure was initialised as an empty array, it is tested to see if it now has content and, if so, it is encapsulated as a new object of class *vector*.

```
>> xyz = [1;3;6] ;
>> position = vector(xyz)
```

```
position =
  vector object : 1-by-1
>> position.components
??? Access to an object's fields is only permitted within its methods.
```

Having constructed an object, it can only be accessed and manipulated via method functions. Importantly, there is no chance of accidental overwriting any of the information content because no of it accessible without a specific method. A general-purpose method can be written to get the information content of an object:

```
function information = get (object, component)
information [] ;

switch nargin
  case 1
    information = fieldnames (object) ;
case 2
  switch class (component)
   case 'char'
    if isfield (object, component)
      information = object . (component) ;
    else
     information = [] ;
    end
  end
otherwise
 information = [] ;
end
```

```
>> get (position)
ans =
 'components'
>> get (position, 'components')
ans =
1
3
6
```

Here, interrogating the object gives a list of field names and, then, a particular field can be extracted. This same method could be applied to any encapsulated object; it happens to be a *vector* method because it resides in the appropriate method directory (@vector). When MATLAB invokes the 'get' function, if it finds a *vector* argument, it knows which method directory to use. (Issues of precedence will be discussed in Section B.6.3.)

In general, different classes can have methods with the same name (which is called method *overloading*) and MATLAB will select the correct one by context. In Sections A.7 and A.8, operators were related to equivalent functions; at the time, that probably seemed like a technicality but, here, it allows operators to be defined for any class. For instance, addition of vectors is conceptually simple but, without a suitable method, it would be rather clumsy. Consider the displacement of an physical object from one position to a new position:

```
>> newPosition = get (position, 'components') + get (displacement,
    'components') ;
```

It would be far more preferable to be able to perform addition in the conventional way, i.e. 'a+b'. For this to happen, a method is needed that overloads the symbol '+' for use with vector objects. This can be implemented as follows:

```
function resultant = plus (left, right) % Vector Addition

resultant = [] ;
switch class (right)
 case 'vector'
   sumVector = vector (left.components + right.components) ;
end
```

With this new capability, vector addition is much simpler:

```
>> newPosition = position + displacement ;
```

and much easier to understand!

In a similar way, vector cross products could be calculated using the operator '^':

```
function product = mpower (left, right)    % Vector cross product

product = [] ;
switch class (right)
 case 'vector'
   a = left.components ;
   b = right.components ;
   c (1) = a (2) *b (3) - a (3) *b (2) ;
   c (2) = a (3) *b (1) - a (1) *b (3) ;
   c (3) = a (1) *b (2) - a (2) *b (1) ;
   product = vector (c) ;
end
```

Thus, the basis vectors for a right-handed axis system could be interrelated as follows:

```
>> zAxis = xAxis ^ yAxis ;
```

B.6   Practical Implementation

Having gone through a wide selection of programming principles, it seems appropriate to offer some comments on practical implementation and to clarify some of the rules governing program execution. As with all matters to do with programming languages, there is never an ideal way of presenting information that satisfies everyone but, hopefully, what follows will be of some general assistance.

It should perhaps be obvious that there are many practical issues that concern programming style and that good style is an effective mitigation against serious design flaws. The rationale for this is that, if a piece of code is easily readable, then it is likely to be understandable and therefore it should be a straightforward task to establish the correctness of that code. Every developer will know only too well that time is a great leveller; what appeared obvious when code was being developed will appear progressively less obvious as time goes by. Even after a few months, questions might arise like 'why did I do that?' and 'what does that do?'. Beware!

B.6.1   *Naming Convention*

B.6.1.1   *Naming Variables*

Variables should always be given meaningful names; and, if that means using several words to achieve this, then concatenate several words together in order to achieve this. It is clearer if names are written in mixed case; it is common to see the first word in lower case, with each subsequent word starting with a capital letter (e.g. aircraftWingReferenceArea). Some developers prefer to distinguish data structures by starting the first word with capital letter. (e.g. Aircraft.Wing.referenceArea). In some situations it may be desirable to separate words with underscores '_' (e.g. aircraft_mass).

Following standard practice in data modelling (i.e. database design), variables should have singular names unless there is a good reason for making them plural. In this way, the expression 'object' or 'object(27)' signifies one object, while 'object(1:3)' signifies object(1) to object(3) inclusive. A quasi-plural name would be 'totalItems', which is a singular property (total) name referring to a plural target (Items).

Logical variables should have names that suggest a 'true' condition with a positive bias (e.g. isFinished). Bearing in mind that logical conditions can be 'false' (or not true), it is undesirable that tests names could give rise to double negatives (e.g. ~notFinished).

Continuing this theme, logical variables exist in order to identify activities and events, to establish conditions and to confirm attributes. This discussion could be refined and extended but this basic observation still holds. So, it is desirable that the variable names are appropriate to the logic; from experience, the following suggestions are offered:

- Associate *activities* with action verbs and subject nouns (e.g. trimmingAircraft)
- Associate *events* with predefined event names and highlight with an 'event_' prefix (e.g. event_AircraftTakeOff)
- Associate *conditions* with names prefixed by a variant of 'to be' (e.g. isComplete, canProceed, mustComply)
- Highlight *failed conditions* with names prefixed by 'not' (e.g. notComplete)
- Associate *attributes* with names prefixed by 'has' (e.g. hasFuel)

B.6.1.2   Naming Functions

Like variables, functions should always be given meaningful names. Given that there different types of functions, it is appropriate that there should some variation in the naming convention. Thus, the following suggestions are offered:

- Generally associate a *function* with an action verb and the intended result or activity, e.g. calculateAerodynamicDrag(...).
- Specifically associate a *simple function* (i.e. giving a single output) with the result of its calculation, e.g. temperature(...).
- Associate an *anonymous function* (i.e. a function handle for a one-line calculation) with the result of its calculation, e.g. radians(...).
- Associate a *constructor function* with a distinctive class name, e.g. perhaps define a 'fuelPump' rather than a 'pump' because it has more meaning and it is future-proof.
- Associate a *function library* with a structure of function handles, where the structure name relates to the library and the field names relate to the functional components, e.g. engine. compressor = @calculateCompressorPerformance.

B.6.1.3   Naming Constants

The C/C++ user community prefers to emphasise constants by using upper-case letters, which means that long names inevitably contain underscores (e.g. UNIVERSAL_GAS_CONSTANT). This is not to everyone's taste. An alternative approach would be to create structures of constants with meaningful names (e.g. Universal.gasConstant), thereby grouping constants together and applying a consistent naming convention in all aspects of program development.

Above all, it is important to recognise that MATLAB has no concept of a constant; a *constant* is just a variable whose value is not supposed to change. In critical applications, it may be desirable to construct a *universalConstant* class using the principles expounded in Section B.5:

```
function object = universalConstant

obj.gasConstant.value = 8314.33 ;
obj.gasConstant.units = {'J',1,'kg',-1,'K',-1} ;
obj.earthGgravity.value = 6.67e11 ;
obj.earthGgravity.units = {'m',3,'kg',-1,'s',-2} ;

object = class(obj,'universalConstant') ;
```

Simple methods could be provided to get values and units from a *universalConstant* object. In the absence of a method that would allow stored values to be changed, constants would remain constant!

B.6.1.4   Acronyms and Abbreviations

Be careful with acronyms! Incorporation of upper-case acronyms can render a name virtually unreadable. So, use mixed-case letters or apply the acronym as a suffix (perhaps separated from the remainder of the name by an underscore). Also, be sure that particular acronyms are widely understood in the domain where they are being applied.

Avoid abbreviations! They are more trouble that they are worth.

B.6.1.5   Counters

A recurring issue in program development is how best to count things inside iterative loops. Old practice was to apply (i,j,k, . . .) as iterative counters and (n,m, . . .) as cumulative totals, which was all rather anonymous. It is easy to run out of letters and it is not desirable to create more anonymity in the form (i1,i2, . . .). Note that, if 'i' and 'j' are used counters, they cannot be used as imaginary numbers (i.e. sqrt(−1)) because the default values will be overwritten.

Current practice is that counters and totals be given related names, such as *iLoop* and *nLoops*. The counter is singular and prefixed with 'i': the total is plural and prefixed with 'n'; if necessary an adjacent prefix can be used for counting (e.g. 'j' or 'k'). As an additional cue to the reader, it is often helpful to clarify the data object that is the subject of the particular loop; the data object might be buried deep inside a data structure and/or array and referencing it might be long-winded. An illustrative example is as follows:

```
inventory.installation.nut = 3 ;
inventory.installation.bolt = 2 ;
inventory.installation.spanner = 1 ;

itemList = fieldnames(inventory.installation) ;
nItems = length(itemList) ;

totalItems = 0 ;
for iItem = 1:nItems
   thisItem = itemList(iItem) ;
   thisValue = inventory.installation.(thisItem) ;
   totalItems = totalItems + ThisValue ;
end
```

B.6.2   Program Architecture

B.6.2.1   Top-Level Structure

Stating the obvious, the overall structure of a program should always be readable and every effort should be made to avoid cryptic code (which often implies overly compact code and/or poorly planned code). The top-level structure should be decomposed into computational steps, with each implemented by a dedicated function.

```
function object = performTask(varargin)
userInput = step1_getUserInput(varargin) ;
result = step2_performComputation(userInput) ;
```

This is 'modularisation'; it provides an overview of the program is intended to achieve and collects together activities in related task elements. In order to maintain cohesion between the program components, it is recommended that the program be held in an M-file, with the top-level structure defined as the primary function and the computational steps defined by subfunctions. As a matter of generality, minor computations should be provided by function libraries to the maximum extent possible. This will prevent re-implementation of minor functions for each major application and will establish a common standard for an end product.

Further levels of decomposition may be warranted for complex programs. If this is so, then perform another step-by-step decomposition but avoid too much decomposition in one place; always consider the natural breaks between task elements and decide whether one task should actually be implemented as several tasks. Also, avoid unbalanced decompositions as this probably belies an inconsistent approach to task management. These issues are especially important in modelling and simulation.

B.6.2.2   *Program Interfacing*

For small-scale functions it is sufficient to accept the sequence of input arguments as they as given. Such functions (especially if they belong to a predefined library) serve a well-defined and dedicated purpose and, in addition, they are likely to be used in a controlled way from inside a larger program.

For large-scale functions it is beneficial to treat the identification and verification of user input as a separate step, rather than just accepting input arguments at face value. Typically, there are three situations to be handled, namely a fixed interface definition, a variable interface definition and a named interface definition. It is recommended that each is handled appropriately.

A fixed interface definition is signified by an explicit set of input and output arguments in the function declaration. If this approach is adopted then it is appropriate to use the information as it is presented, i.e. change nothing! Specific arguments are then propagated to subfunctions as necessary. in this case, a typical declaration would be:

```
function [output1,output2] = doCalculation(input1,input2)
```

A variable interface definition is signified by the appearance of varargin and varargout in the function declaration. Rather than clutter up the top-level function with interpretive processing, it is strongly recommended that this be consigned to a dedicated subfunction. The main issues here will be ensuring that necessary and sufficient data are supplied to enable a given computation and to determine the context of that computation. Typically, this lacks any real sophistication and will be limited to counting numbers of arguments and checking a few data types. In this case, a typical declaration would be:

```
function varargout = doCalculation(varargin)
```

and an example of its usage might have the form:

```
result = doCalculation(x,y,z,2)
```

A named interface definition is a refinement of a variable interface definition and is illustrated by a function of the following format:

```
result = doCalculation('pressure',p,'temperature',t,'factor',2)
```

Thus, inputs appear in pairs, giving a name and a value. Recalling the top-level program structure illustrated in the previous section, this could be interpreted as follows:

```
function object = performTask(varargin)
userInput = step1_getUserInput(varargin) ;
result = step2_performComputation(userInput) ;

function userInput = step1_getUserInput(inputList)
userInput = [] ;

nInputs = length(inputList) ;
for iInput = 1:2:nInputs
  thisArgument = inputList{iInput} ;
  thisValue = inputList{iInput+1} ;

  if isvarname(thisArgument)
      userInput.(thisArgument) = thisValue ;
  end
end
```

and then step2_performComputation can proceed.

B.6.2.3   Statements

Statements follow syntax rules of the type shown in Section B.3.1. In particular, iterative execution, condition execution and exception handling are supported in a structured way and statements can be organised in nested hierarchies, theoretically, without limitation. Practically, it is not a good idea to bury things too deep and, arguably, a hierarchy of more than three or four levels is deep enough.

Hierarchical nesting should be accompanied by code indentation, as automatically inserted by the MATLAB editor. Deep hierarchies imply a lot of code indentation and it becomes difficult to find corresponding keywords within statement blocks (e.g. if-elseif-else-end, for-end and try-catch-end).

In general, statements should not contain more than is absolutely necessary. This implies a circular argument about not knowing what is needed but stopping when there is too much! The real intention here is to limit the length of text in any line of code. The MATLAB editor marks the page width at 75 characters; so it would be reasonable to limit the line length to perhaps 75 or 80 characters.

A related issue is that several statements can appear in one line of code. This is acceptable for very simple tasks, e.g. initialising variables, but is highly undesirable if any other context.

Specific comments are needed when considering logical expressions, e.g. in connection with 'if' statements and 'while' statements. Combinational logic can be intricate and difficult to follow and this is especially true when many logical (and/or relational) operators are consolidated in a single expression. In this situation, it is always recommended that the various parts of the expression be broken down using parentheses unless the logic is especially simple. This will serve to remove ambiguity in the way in which the logic interpreted and, more importantly, prevent errors arising from incorrect assumptions regarding operator precedence (cf. Section B.6.3). For particularly complicated logic, consider breaking the evaluation into separate statements and providing a combined decision at the end.

More general comments are needed when considering the logic associated with conditional execution, i.e. 'if' statements and 'switch' statements. Again, combinational logic can be intricate and difficult to follow . . . and this includes case selections. Humans are not naturally predisposed to logical analysis; even five or six logical conditions, when taken together, can cause confusion. Experience shows repeatedly that developers can be tempted into applying ad hoc logic, which really means that they only consider combinations that appear to be important without being exhaustive. If in doubt, make provision for every combination of logical conditions, no matter how long-winded that might be. Logic design can be challenging and adequate time and effort must be devoted to it.

For completeness, it should be pointed out that statements exist that enable jumps to be performed inside iterative and conditional statements, namely break and continue. Quite deliberately, these have been omitted from discussion because their use is controversial. However, they might be of use on some occasion but that is a decision for individuals.

As a common requirement, the role of comments should not be downplayed. Comments can occupy full lines of code or be appended after any text that defines a statement or part of a statement. It is possible to overburden code with too much comment so that the code is fragmented and its purpose is totally obscured. A sensible compromise is needed, such that variable names, function names, program structure and well-placed comments can provide a running commentary of what the code is doing. It is also important that program code is documented so that it does not stand alone in the wilderness.

B.6.3   *Precedence*

Before moving on from this somewhat lengthy chapter, the issue of *precedence* has to be addressed and clarified, simple because it lies in wait for the ill-informed. Rules of precedence apply in different ways to operators, classes and functions and, if the rules are not understood properly, mistakes will be made. Remember that computers do what they are instructed and do not have any independent intelligence that can identify errors of interpretation or errors of omission (at least, not yet!).

B.6.3.1   *Operator Precedence*

Expressions are composed of any combination of arithmetic, relational, and logical operators. The order of evaluation is determined by the precedence of the various operators that appear in a given expression. Where operators have equal precedence, they are evaluated from left to right. Where parts of an expression are enclosed by parentheses, this forces the order of evaluation; parentheses must be nested in order to make sense and, on this basis, evaluation proceeds from the innermost parentheses, working outwards.

MATLAB documentation lists the levels of precedence (listed from highest to lowest) as follows:

1. Parentheses ()
2. Transpose (.') Power (.^) Complex Conjugate Transpose (') Matrix Power (^)
3. Unary Plus (+) Unary Minus (−) Logical NOT (∼)
4. Multiplication (.*) Left Division (.\) Right Division (./) Matrix Multiplication (*) Matrix Left Division (\) Matrix Right Division (/)
5. Addition (+) subtraction (−)
6. Colon (:)
7. Less Than (<) Less Than or Equal (<=), Greater Than (>) Greater Than or Equal (>=) Equal (==) Not Equal (∼=)
8. Element-wise Logical AND (&)
9. Element-wise Logical OR (|)

10.   Short-circuit Logical AND (&&)
11.   Short-circuit Logical OR (||)

Although not mentioned, it is clear that brackets and concatenation operators have Level 1 precedence.

With eleven levels of precedence, the earlier comments about breaking logical expressions into fragments enclosed by parentheses should begin to resonate. As a matter of convenience, parentheses will save a lot of confusion.

B.6.3.2   Class Precedence

Class precedence is straightforward. User-defined classes have precedence over built-in classes. All user-defined classes have equal precedence unless the constructor function makes use of functions superiorto or inferiorto. This affects the choice of method function in case where input arguments are drawn from more than one user-defined class; the actual choice will follow *superior* objects.

B.6.3.3   Function Precedence

First of all, variables take precedence over functions. So, if a variable and a function were to have the same name, MATLAB would access the variable.

Functions must be *in scope* before they can be called; this means that a particular function has to be visible to a calling function or the current session (i.e. via the MATLAB path). For functions with the same name, located in different places, the following precedence applies:

- Subfunctions take precedence over all other functions that are on the path, even if the function is a recognised method function, i.e. in this case the method would be ignored.
- Private functions are called if there is no subfunction within the current scope and, in common with subfunctions, MATLAB will ignore method functions.
- Constructor functions take precedence over functions other than subfunctions and private functions, such that all other functions on the search path and the working directory will be ignored.
- Method functions take next precedence although the method that is actually called depends on the classes of objects in the argument list (and the applicable class precedence).
- Functions in the working directory then take precedence over functions elsewhere on the MATLAB path.
- Finally, functions elsewhere in the MATLAB path are called; the first function found (searching the path from top to bottom) will be the one that is called.

B.6.3.4   File Precedence

For completeness (as opposed to direct relevance to this book), it is noted that there are five file types recognised by MATLAB. For files with the same name appearing the same directory, the precedence is applied as follows:

MATLAB has five 'native' file types and uses file precedence to select between identically named functions in the same directory. For the file types covered in this chapter, built-in functions take precedence over M-files. The full order of precedence is as follows:

- built-in function;
- MEX-files (compiled/executable FORTRAN or C/C++ code);
- MDL file (Simulink® model);
- P-code file (pre-parsed M-file);
- M-file.

B.6.4   *Preferences*

General-purpose programming languages provide a framework for computational tasks and, within that framework, offer complete freedom to developers to do anything that they want. Chapters like this can only summarise the key language features and make helpful suggestions as to how programming style should be constrained. As this book is not dedicated to programming it is not appropriate to discuss these issues further, except to reiterate earlier comments about the readability of program code and the ease with which it should be understood by others. So, choose your style preferences carefully!

In order to handle data preferences, MATLAB does offer a mechanism for storing user data for global access during a session. Thus, data that are important in defining the context of what is being done can be held as a set of session preferences (using setpref) and accessed at any time, from anywhere (using getpref). These data are persistent and, having created them in one session, they will be available in future sessions. In principle, this provides the basis for tool preferences but, from experience, this needs to be designed and implemented with care.

B.7   Summary of Appendix B

MATLAB provides a range of data types (or classes) and allows user to construct their own object-based classes that are accessed via a set of method functions. Classes can be assigned different precedence levels relative to each other (using superiorto and inferiorto).

Functions represent the standard way of modularising programs and can be created in various forms, each contained within M-files. MATLAB provides flexibility in establishing the interface definition for a function; it also enables functions to be invoked via function handles, thereby providing flexibility in establishing generic functions. Function execution follows a set of precedence rules based on what is visible to the calling function or the MATLAB session and where particular functions are located. Note that functions can call themselves recursively.

Language statements conform to syntax rules and different types of statements can be built for manipulating data and controlling execution (i.e. iterative execution, conditional execution and exception handling). A wide range of arithmetic, logical and relational operators are provided for built-in classes and these can be overloaded for user-defined classes. These operators obey a set of precedence rules.

A collection of comments and recommendations were offered in connection with practical implementation, covering various issues of programming style and structure.

Appendix C

Organisations

C.1   Specialist Agencies of the United Nations

International Civil Aviation Organisation (ICAO)

> ICAO was established after World War II as a means of securing international cooperation in air transport, as well as fostering uniformity in regulations and standards (particularly in air navigation). This was mandated by the Convention on International Civil Aviation that became effective in 1947. More information can be found at **www.icao.org** or **www.icao.int**.
>
> Particular interest here relates to the widely used International Standard Atmosphere (ISA), as defined in ICAO Document 7488-CD (ICAO, 1993).

International Telecommunication Union (ITU)

> The International Telegraph Union was established in 1865 to address the need to interconnect national telegraph networks. Its role expanded into telephone and wireless communication and, during the 1920s, various consultative committees were established (such as the International Radio Consultative Committee (CCIR)). The current name was adopted in 1934. The current organisational structure is comprises three sectors, namely Telecommunications (ITU-T), Radiocommunication (ITU-R) and Development (ITU-D). More information can be found at **www.itu.int**.
>
> Particular interest here relates to the definition of Coordinated Universal Time (UTC) in CCIR Recommendation 460-4 (1986) (where CCIR subsequently became the ITU-R), which is available as ITU-R TF.460-4 'Standard-Frequency and Time-Signal Emissions'.

World Meteorological Organisation (WMO)

> WMO was established in 1950 (succeeding the International Meteorological Organisation founded in 1873) and became the specialised agency for meteorology, operational hydrology and related geophysical sciences. It is an intergovernmental organisation with 187 member states and

Continued

territories, with technical interests focused on the state and behaviour of the atmosphere, its interaction with the oceans, the climate it produces and the resulting distribution of water resources. More information can be found at **www.wmo.ch** and **www.worldweather.org**.

C.2   International Organisations

International Standardisation Organisation (ISO)

ISO is an organisation that was created after World War II in order 'to facilitate the international coordination and unification of industrial standards'. It officially commenced its work in 1947 and is now a network of national standards institutes, covering 156 countries, with a central secretariat located in Geneva, Switzerland. More information can be found at www.iso.ch.

TC20 *Technical Committee, Aircraft and space vehicles*
SC6 *Sub-Committee, Standard atmosphere*

ISO atmosphere models are coordinated by TC20/SC6 and specified by three principal publications:
ISO 2533:1975 Standard Atmosphere
ISO 5878:1982 Reference Atmospheres for Aerospace Use
ISO/TR 14618:1996 Middle atmosphere – Global model at altitudes between 30 km and 120 km, and wind model at altitudes above 30 km

ISO 2533:1975 specifies a standard atmosphere that can be used as a common basis for simulation, analysis and test. Importantly, it removes local variations in atmospheric properties, thereby permitting the unification of test results and the calibration of test instruments. *ISO 5878:1982* presents information on the temporal and spatial variations in atmospheric properties at levels between the surface and 80 km. This is achieved via a family of models, comprising the reference atmospheres tropical, subtropical, mid-latitude, subarctic and arctic. *ISO/TR 14618:1996* establishes a global model of monthly mean values of temperature, pressure, density and wind. These data cover the latitude range from 80° S to 80° N, extending from altitudes between 30 km and 120 km.

JTC1 *Joint Technical Committee, Information technology* (**www.jtc1.org**)
SC24 *Sub-Committee, Computer graphics, image processing and environmental data representation* (**www.bsi.org.uk/sc24**)
WG8 *Working group, Environmental representation* (**www.sedris.org/wg8home**)

JTC1 is a joint technical committee or ISO and IEC (the International Electrotechnical Commission). For present purposes, the major publication of JTC1/SC24/WG8 is: **ISO/IEC FDIS 18026:2006(E) Spatial Reference Model**. This defines the conceptual model and the methodologies that allow the description of geometric properties within spatial reference frames. It supports unambiguous specification of positions, directions, distances and times associated with spatial information. It defines algorithms for precise transformation of these properties between spatial reference frames. An electronic version of this document can be found at: **http://standards.sedris.org/18026_FDIS/C020811.html**.

TC211 *Technical Committee, on Geographic information/Geomatics*

The aim of TC211 is to establish standards for information concerning objects or phenomena that are associated with a location relative to the earth. These may specify methods, tools and services for the management, utilisation and transfer of such data, linking to appropriate standards for information technology and data. More information can be found at **www.isotc211.org**.

International Electrotechnical Commission (IEC)

IEC was founded in London in 1906, originally to consider standardisation of nomenclature and ratings of electrical apparatus and machinery. It is now a leading global organisation that prepares and publishes international standards for all electrical, electronic and related technologies. For present purposes the main interest lies in the joint technical activities undertaken with ISO on environmental representation. More information can be found at **www.iec.ch**.

International Earth Rotation and Reference Systems Service (IERS)

IERS[1] was established in 1987 by the International Astronomical Union (IAU) and the International Union of Geodesy and Geophysics. Its stated objectives are to provide data, publications and general information on Earth's rotation, the International Celestial Reference Frame (ICRF), the International Terrestrial Reference Frame (ITRF), plus geophysical data and conventions. More information can be found at **www.iers.org**.

International Council for Science (ICSU)

The International Council of Scientific Unions (ICSU) was founded in 1931 to promote international scientific activity and its application for the benefit of humanity. This is one of the oldest nongovernmental organisations in the world, with membership drawn from National Scientific Members (e.g. the Royal Society) and International Scientific Unions, who carry a weight of scientific expertise appropriate to major, international, interdisciplinary issues. In 1998, the name was changed to the **International Council for Science**, although the acronym (ICSU) was retained. More information can be found at **www.icsu.org**.

International Astronomical Union (IAU)

IAU was founded in 1919 to promote and safeguard the science of astronomy through international cooperation. It is one of the scientific unions within ICSU. Its individual members are professional

Continued

---

[1] Its original name was the International Earth Rotation Service.

astronomers from all over the world, who are active in professional research and education. Also there are 62 national members, which are local astronomical societies. It is the internationally recognised authority for assigning designations to celestial bodies and any surface features on them. Activities are run by the permanent IAU Secretariat, located at the Institut d'Astrophysique in Paris. More information can be found at **www.iau.org**.

International Union of Geodesy and Geophysics (IUGG)

The IUGG was founded in 1919 and describes itself as 'dedicated to the international promotion and coordination of scientific studies of Earth (. . .) and its environment in space'. It is one of the scientific unions within ICSU. Its membership comprises semi-autonomous *Associations* that pursue specific scientific themes and *Union Commissions* that foster links with the international geophysical community and promote interdisciplinary studies. More information can be found at **www.iugg.org**.

International Association of Geodesy (IAG)

For present purposes, the most relevant IUGG Association is the IAG, originally established in 1864. The main permanent activity is to monitor positions of points at the surface of the Earth through a global system of satellite tracking stations. The IAG support scientific activities such as improving and extending the IERS International Terrestrial Reference Frame (ITRF), monitoring variations in the earth rotation and mapping the gravitational field (and variations over time). More information can be found at **www.iag-aig.org**.

EUREF

EUREF is the Reference Frame Sub-Commission for Europe, integrated in the IAG Sub-Commission 1.3, Regional Reference Frames (under Commission 1, Reference Frames). It deals with the European Reference Frame, which is a geodetic infrastructure for multinational projects requiring precise geo-referencing, in cooperation with other parts of IAG activities and with Eurographics (the consortium of the national mapping agencies in Europe). More information can be found at **www.euref-iag.net**.

Bureau International des Poids et Mesures (BIPM)

The BIPM was established by the Metre Convention in order to provide the basis for a single, coherent measurement system across the world, traceable to the *Système International d'Unités* (SI). The Metre Convention was a treaty signed in 1875, which now has 51 member states. It gives authority to the General Conference on Weights and Measures (CGPM), the International Conference on Weights and Measures (CIPM) and the International Bureau on Weights and Measures (BIPM) to act on matters of world metrology. More information can be found at **tai.bipm.org**.

C.3   US Government Organisations

National Aeronautics and Space Administration (NASA)

> Building on the technical expertise of the National Advisory Committee on Aeronautics, founded in 1915, NASA was established in 1958 and is now so well known that it almost does not require an introduction. Interesting information on past, present and future activities can be found at **www.nasa.gov**.
>
> The Digital Earth Initiative is of particular interest, aimed at enabling access to geospatial information from multiple sources over the Internet. This requires interoperability between relevant information systems as well as a common set of geospatial concepts and associated standards. General information can be found at **www.digitalearth.gov**.
>
> The Jet Propulsion Laboratory (JPL) is a very good source for solar system dynamics and ephemeris data (i.e. position and velocity for orbiting bodies). It supports a wide range of research (including space science and astrodynamics) and it also builds and operates unmanned spacecraft. More information can be found at **www.jpl.nasa.gov**.

National Geospatial-Intelligence Agency (NGA)

> NGA is a major provider of imagery, maps, charts and environmental data, collectively termed 'geospatial intelligence'. The current name was adopted in 2003. Precursor agencies were the Defense Mapping Agency (DMA) (created in 1972 by merging most of the armed services' mapping and charting capabilities) and the National Imagery and Mapping Agency (NIMA) (created in 1996 by a similar rationalisation between the US Congress, the Central Intelligence Agency and the Department of Defense). Note that, because of the history, documents of interest may carry DMA, NIMA or NGA designations. More information can be found at **earth-info.nga.mil**.

National Oceanic and Atmospheric Administration (NOAA)

> NOAA is a research and data-gathering agency covering global oceans, atmosphere, space and sun. It conducts applied research in the areas of ecosystems, climate, weather and water, and commerce and transportation. A large amount of information can be accessed via **www.noaa.gov** but, for current purposes, three sources are of particular interest, namely:
>
> - National Geophysical Data Centre (NGDC) (**www.ngdc.noaa.gov**)
> - National Geodetic Survey (**www.ngs.noaa.gov**)
> - Aviation Digital Data Service (ADDS) (**adds.aviationweather.gov**).

US Geological Survey (USGS)

> As part of the US Department of the Interior, USGS is an impartial, multidisciplinary scientific organisation that focuses on biology, geography, geology, geospatial information and water. Useful information is available at **www.usgs.gov** and, particularly, via the Earth Resources Observation Systems (EROS) Data Center (or EDC) at **eros.usgs.gov**. Of particular interest here are the Digital Elevation Model (DEM) and the global topography (GTOPO30) databases.

US Naval Observatory (USNO)

> USNO was established in 1830 as the 'Depot of Charts and Instruments', its primary mission was to care for the chronometers, charts and navigational equipment used by the US Navy. Today, its role includes determining the positions and motions of the Earth, Sun, Moon, planets, stars and other celestial objects, providing astronomical data; determining precise time; measuring the Earth's rotation; and maintaining the Master Clock for the United States. General information can be found at **www.usno.navy.mil**; specific information regarding astronomical and nautical almanacs can be found at **aa.usno.navy.mil**.

Federal Geographic Data Committee (FGDC)

> FGDC is an interagency committee that is developing the National Spatial Data Infrastructure (NSDI), encompassing policies, standards, and procedures for organisations to cooperatively produce and share geographic data. This involves all levels of government, academia and the private sector. More information can be found at **www.fgdc.gov** and **www.geo-one-stop.gov**. Of particular note is the Geospatial Applications and Interoperability Working Group (**gai.fgdc.gov**).

Committee on Extension to the Standard Atmosphere (COESA)

> COESA was established in 1953 and produced various incarnations of the US Standard Atmosphere (most recently in 1976) in cooperation with NOAA, NASA and the US Air Force although, in total, there were thirty participant organisations. This work is significant because it represents atmospheric properties from sea level to 1000 km (identical to ISA[2] below an altitude of 32 km). More information can be found at: **modelweb.gsfc.nasa.gov/atmos/atmos_index.html**.

Defense Modeling and Simulation Office (DMSO)

> DMSO is the executive agency of the US Department of Defense responsible for providing environments for use in training service personnel, developing doctrine and tactics, formulating operational plans and assessing war-fighting situations. The modelling and simulation environments are intended to promote cooperation between participants, as well as offering a high degree of flexibility through the use of affordable, reusable components within an open systems architecture. More information can be found at **www.dmso.mil**.

Federal Aviation Authority (FAA)

> FAA is responsible for the safety of civil aviation in the United States. It pursues a wide range of activities including safety regulation, airspace and air traffic management, air navigation facilities and commercial space transportation. More information can be found at **www.faa.gov**.

---

[2] The ICAO standard atmosphere as described in Section 1.5.1.1.1.

C.4   UK Government Organisations

Civil Aviation Authority (CAA)

> CAA is a public corporation that acts as the UK's independent aviation regulator, with all civil aviation regulatory functions (economic regulation, airspace policy, safety regulation and consumer protection) integrated within a single specialist body. Of particular interest here is the Directorate of Airspace Policy (DAP), which is responsible for the planning and regulation of all UK airspace including the navigation and communications infrastructure to support safe and efficient operations. More information can be found at **www.caa.co.uk**.

Ordnance Survey

> Ordnance Survey is the UK national mapping agency and was founded in 1791 in response to the threat posed by France after the Napoleonic Revolution. In order to facilitate plans to repel a possible invasion, the Board of Ordnance (the then ministry of defence) undertook a comprehensive survey of the south coast of England. Eventually this led to the surveying and mapping of the whole country in detail and the development of considerable expertise in surveying and mapmaking. More information is available at **www.ordnancesurvey.co.uk**.

British Geological Survey

> BGS is responsible for advising the UK government on all aspects of geoscience as well as providing impartial geological advice to industry, academia and the public. More information can be found at **www.bgs.ac.uk**. Of particular interest here is the World Magnetic Model (WMM2005) produced jointly with the NOAA National Geophysical Data Center; this is the standard model in UK and US military navigation and attitude reference systems and is also used widely in civilian navigation systems. The model is revised every five years. More information on this is available at **www.geomag.bgs.ac.uk/navigation.html**.

HM Nautical Almanac Office (HMNAO)

> HMNAO is responsible for producing the annual volumes of *The Astronomical Almanac, The Nautical Almanac, Astronomical Phenomena, The Star Almanac* and *The UK Air Almanac* in the UK. Several of these publications are produced in collaboration with US Naval Observatory (USNO). HMNAO also produces astronomical data suitable for a wide range of users. More information can be found at **www.nao.rl.ac.uk**.

C.5   European Organisations

European Civil Aviation Conference

ECAC was founded in 1955 as an intergovernmental organisation, to promote the continued development of a safe, efficient and sustainable European air transport system. It seeks to harmonise civil aviation policies and practices amongst its member states. It provides a forum for discussion and maintains close working relationships with ICAO, EUROCONTROL, the Joint Aviation Authorities (JAA), the Council of Europe, the institutions of the European Union, plus the air transport industry. More information can be found at **www.ecac-ceac.org**.

Joint Aviation Authorities (JAA)

JAA is an associated body of ECAC, representing the civil aviation regulatory authorities of a number of European States who have agreed to co-operate in developing and implementing common regulatory standards and, through uniform application of those standards, to enable fair and equal competition between member states. Work is on-going to harmonise JAA regulations with those of the USA and also to achieve an efficient transition of activities to the European Aviation Safety Agency (EASA). More information can be found at **www.jaa.nl**.

European Aviation Safety Agency (EASA)

EASA is an agency of the European Union (EU) which has been given specific regulatory and executive tasks in the field of aviation safety. It was founded in 2003 in order to establish and maintain a high, uniform level of civil aviation safety and environmental protection in Europe. More information can be found at **europa.eu.int/agencies/easa/index-en.html** and **www.easa.eu.int**.

EUROCONTROL

EUROCONTROL is the European organisation for the safety of air navigation, with 36 member states. Its primary objective the development of a seamless, pan-European Air Traffic Management (ATM) system capable of handling the forecast growth in air traffic, while providing improvements in efficiency, economy, safety and environment. More information can be found at **www.eurocontrol.int**.

Following the adoption by ICAO of WGS84 as the global reference system, EUROCONTROL is coordinating its implementation within the ECAC area as part of EATCHIP (The European Air Traffic Control Harmonisation and Integration Programme). More information is available at **www.wgs84.com**.

The ECAC Navigation Strategy addresses the technical, operational and economic short-comings of the existing air navigation system. It aims to provide an integrated framework within which this system can evolve up to 2020 and beyond, taking account global interoperability. This strategy supports the developments towards a European Air Traffic Management System and is in line with the ICAO Global Air Navigation Plan. More information can be found at **www.ecacnav.com**.

EuroGeographics

EuroGeographics represents nearly all European National Mapping and Cadastral Agencies (NMCAs) (currently 46 organisations from 41 countries). The overall aim is to establish the European Spatial Data Infrastructure (ESDI), thereby achieving interoperability of European mapping and other geographical information. More information can be found at **www.eurogeographics.org**.

C.6   Open Projects and Consortia

The Virtual Terrain Project (VTP)

The VTP fosters the creation of tools for easily constructing any part of the real world in interactive, 3D digital form. This requires a convergence of Computer-Aided Design (CAD), Geographical Information Systems (GIS), visual simulation, surveying and remote sensing. VTP gathers information and tracks progress in areas such as procedural scene construction, feature extraction, and rendering algorithms. It creates and supports a set of software tools, which are shared freely. A lot more information can be found at **www.vterrain.org**.

Open Geospatial Consortium (OGC)

OGC is a nonprofit, international, voluntary organisation that is leading the development of standards for geospatial and location-based services. Via member-driven consensus programs, OGC works with government, industry and academia to create open and extensible software application programming interfaces for geographic information systems (GIS) and other mainstream technologies. Information can be found at **www.opengeospatial.org** or **www.opengis.org**.

Global Spatial Data Infrastructure (GSDI)

GSDI is an association of organisations, companies, agencies and individuals with an interest in promoting local, national and international spatial data initiatives so that countries can effectively address social, economic and environment issues. More information can be found at **www.gsdi.org**.

The Geographer's Craft Project

> The Geographer's Craft Project is one of the celebrated websites on the Internet, borne out of teaching courses at the University of Colorado. Although it states that the courses no longer run, the site has been maintained in order to provide resources for teaching and general interest. This material is highly recommended as an introduction to geographical concepts and information and can be found at **www.colorado.edu/geography/gcraft**.

SEDRIS

> The Synthetic Environment Data Representation and Interchange Standardisation (SEDRIS) project started as a US government-led program and, while there have been a variety of funding sources, DMSO has been the primary sponsor. Note that, in this context, 'environment' means air, land, sea and space, and 'interchange' means transfer between heterogeneous systems that are required to interoperate in networked and distributed applications. More information can be found at **www.sedris.org**.

Bibliography

Abramowitz M, Stegun IA (1965), *Handbook of Mathematical Functions*. New York: Dover.

Adams OS (1921/1949), *Latitude Developments Connected with Geodesy and Cartography with Tables, including a Table for Lambert Equal-Area Meridional Projection*, US Coast and Geodetic Survey Special Publication No. 67.

Aoki S, Guinot B, Kaplan GH, Kinoshita H, McCarthy DD, Seidelmann PK (1982), The new definition of universal time, *Astronomy and Astrophysics* 105: 359–61.

Arfken GB, Weber HJ (2001), *Mathematical Methods for Physicists*. New York: Academic Press.

Arias EF, Charlot P, Feissel M, Lestrade J-F (1995), The extragalactic reference system of the International Earth Rotation Service, ICRS, *Astronomy and Astrophysics* 303: 604–8.

Babister AW (1980), *Aircraft Dynamic Stability and Response*. Oxford: Pergamon Press.

Bennett BS (1995), *Simulation Fundamentals*. London: Prentice Hall.

Blakelock JH (1991), *Automatic Control of Aircraft and Missiles*, 2nd Edition. New York: John Wiley & Son, Inc.

Bomford G (1971), *Geodesy*. Oxford: Oxford University Press.

Borkowski KM (1989), Accurate algorithms to transform geodetic coordinates, *Bulletin Geodesique* 63: 50–6.

Bowring BR (1969), The further extension of the Gauss inverse problem, *Survey Review* XX(151): 40–3.

Bowring BR (1976), Transformation from spatial to geodetic coordinates, *Survey Review* 23, 181: 323–7.

Bowring BR (1985), The accuracy of geodetic latitude and altitude equations, Survey Review 28: 202–6.

British Standards Institution (1984), *Tables relating to altitudes, airspeed and Mach numbers for use in aeronautical instrument design and calibration*, British Standard 2G199.

Buccholz JJ, Bauschat J-M, Hahn K-U, Pausder HJ (1995), ATTAS & ATTHeS in-flight simulators: recent application experiences and future programs, AGARD CP-577. In Flight Vehicle Integration Panel Symposium, *Flight Simulation – Where are the Challenges?*, Braunschweig, Germany.

Capitaine N, Wallace PT, Chapront J (2003), Expressions for IAU 2000 precession quantities, *Astronomy and Astrophysics* 412: 567–86.

Capitaine N, Wallace PT, Chapront J (2005), Improvement of the IAU 2000 precession model, *Astronomy and Astrophysics* 432: 355–67.

Cellier FE (1991), *Continuous System Modelling*. Berlin: Springer Verlag.

Chobotov VA (ed.) (1991), *Orbital Mechanics*. American Institute of Aeronautics and Astronautics.

COESA (1976), *US Standard Atmosphere*. Committee on Extension to the Standard Atmosphere, US Government Printing Office.

Cook MV (1997), *Flight Dynamics Principles*, Oxford: Butterworth Heinemann.

Cunningham LE (1970), On the computation of the spherical harmonic terms needed during the numerical integration of the orbital motion of an artificial satellite, *Celestial Mechanics* 2: 207–16.

de Vries TJA (1994), *Conceptual design of controlled electro-mechanical systems: a modeling perspective*, PhD Thesis, Universiteit Enschede, Netherlands.

DMA (1989), TM8358.2: The Universal Grids: Universal Transverse Mercator (UTM) and Universal Polar Stereographic (UPS), US Defence Mapping Agency.

DMA (1990), *TM8358.1: Datums, Ellipsoids, Grids and Grid Reference Systems*, US Defence Mapping Agency.

Edwards CH, Penny DE (1986), *Calculus and Analytical Geometry*, 2nd Edition. Englewood Cliffs, NJ: Prentice-Hall.

ESDU 68046d (1992), *Atmospheric Data for Performance Calculations*. London: ESDU International plc.

ESDU 69026b (1992), *Airspeed Data for Performance Calculations*. London: ESDU International plc.

ESDU 72018 (1972), *International Standard Atmosphere*. London: ESDU International plc.

ESDU 72021b (2005), *Properties of a Standard Atmosphere*. London: ESDU International plc.

ESDU 72022b (1986), *Equations for Calculation of International Standard Atmosphere and Associated Off-Standard Atmospheres*. London: ESDU International plc.

ESDU 78008c (1992), *Physical Properties of Design Atmospheres*. London: ESDU International plc.

ESDU 78012b (1996), *Height Relationships for Non-Standard Atmospheres*. London: ESDU International plc.

Eshelby ME (2000), *Aircraft Performance: Theory and Practice*. London: Arnold.

Etkin B (1972), *Dynamics of Atmospheric Flight*. New York: John Wiley & Sons, Inc.

Etkin B, Reid LD (1996), *Dynamics of Flight: Stability and Control*, 3rd edition. New York: John Wiley & Sons, Inc.

Feissel M, Mignard F (1998), The adoption of ICRS on 1 January 1998: meaning and consequences, *Astronomy and Astrophysics* 331: L33–L36.

Filippone A (2006), *Flight performance of Fixed and Rotary Wing Aircraft*. Oxford: Elsevier.

Fok HS, Iz HB (2003), A comparative analysis of the performance of iterative and non-iterative solutions to the cartesian to geodetic coordinate transformation, *Journal of Geospatial Engineering* 5: 61–74.

Fricke W *et al.* (1988), *Fifth fundamental catalogue (FK5). Part 1. The Basic Fundamental Stars*. Veröffentlichungen Astronomisches Rechen-Insitut Heidelberg Nr 32, Verlag G Braun: Karlsruhe.

Gawthrop PJ, Smith L (1996), *Metamodelling: Bond Graphs and Dynamic Systems*. London: Prentice Hall.

Gawthrop PJ, Ballance DJ (1999), Symbolic computation for manipulation of hierarchical bond graphs. In N Munro (*1999*), *Symbolic Methods in Control Systems Analysis and Design*, IEE Control Engineering Series, Vol. 56.

Good RH, Nelson TJ (1971), *Classical Theory of Electric and Magnetic Fields*. New York: Academic Press.

Grossman N (1996), *The Sheer Joy of Celestial Mechanics*. Boston: Birkhäuser.

Hedgley DR Jr (1976), *An exact transformation from geocentric to geodetic coordinates for nonzero altitudes*, NASA-TR-458, March 1976.

Hildebrand FB (1967), *Advanced Calculus for Applications*, 2nd Edition. New York: Prentice Hall.

Hooijberg M (1997), *Practical Geodesy*, Berlin: Springer.

ICAO (1993), Manual of the ICAO Standard Atmosphere: extended to 80 kilometres (262500 feet), Document 7488-CD, 3rd Edition. ICAO.

ICD-GPS-200 (1994), Navstar GPS Space Segment/Navigation User Interfaces, Revision C, ARINC Research Corporation, El Segundo, CA.

Ismail IH, Bhinder FS (1991), Simulation of aircraft gas turbines, *J. Engineering for Gas Turbines and Power* 113: 95–9.

ISO 2533 (1975), *Standard Atmosphere*, International Standardisation Organisation.

ISO 5878 (1982), *Reference Atmospheres for Aerospace Use*, International Standardisation Organisation.

ISO/TR 14618 (1996), *Middle Atmosphere – Global Model at Altitudes between 30 km and 120 km, and Wind Model at Altitudes above 30 km*, International Standardisation Organisation.

Jackson JD (1975), *Classical Electrodynamics*, 2nd Edition. New York: John Wiley & Sons, Inc.

Jones GC (2002), New solutions for the geodetic coordinate transformation, *Bulletin Geodesique*, 76: 437–46.

Kaplan MH (1976), *Modern Spacecraft Dynamics and Control*. New York: John Wiley & Sons, Inc.

Karnopp DC, Margolis DL, Rosenberg RC (1990), *System Dynamics: A Unified Approach*. 2nd Edition. New York: John Wiley & Sons, Inc.

Kinoshita H (1977), Theory of the rotation of the rigid earth, *Celestial Mechanics* 15: 277.

Kinoshita H, Aoki S (1983), The definition of the ecliptic, *Celestial Mechanics* 31:329–38.

Konopliv AS, Asmar SW, Carranza E, Sjogren WL, Yuan DN (2001), Recent gravity models as a result of the Lunar Prospector Mission, *Icarus* 150: 1–18.

Kumar M (1993), World Geodetic System 1984: a reference system for global mapping, charting and geodetic applications, *Surveying and Land Information Systems* 53/1: 53–6.

Kuroishi, Yuki (1996), Precise gravimetric determination of the geoid in the vicinity of Japan, *Bulletin of the Geographical Survey Institute*, V.41, GSI, Tokyo, Japan, March 1995.[1]

Lemoine FG et al. (1996), EGM96, The NASA GSFC and NIMA Joint Geopotential Model, *Proceedings of the International Symposium on Gravity, Geoid, and Marine Geodesy*, Tokyo, Japan, 30 September–4 October 1996.[2]

Lemoine FG, Smith DE, Rowlands DD, Zuber MT, Meumann GA, Chinn DS (2001), An improved solution to the gravity field of Mars (GMM-2B) from Mars Global Surveyor, *J. Geophys. Res.* 106(E10), 23359–76.

Lieske JH, Lederle T, Fricke W, Morando B (1977), Expressions for the Precession Quantities based on the IAU (1976) System of Astronomical Constants, *Astronomy and Astrophysics* 73: 1.

List RJ (1971), *Smithsonian Meteorological Tables*, Smithsonian Institute Press.

McCarthy DD, Petit G (eds) (2004), International Earth Rotation and Reference Systems Service (IERS) Conventions (2003), IERS Technical Note no 32, Verlag des Bundesamts für Kartographie und Geodäsie, Frankfurt am Main.

McCormick BW (1995), *Aerodynamics, Aeronautics and Flight Mechanics*, 2nd Edition. New York: John Wiley & Sons, Inc.

MacFarlane AGJ (1970), *Dynamical System Models*. London: GG Harrap.

McLean S, McMillan S, Maus S, Lesur V, Thomson A, Dater D (2004), *The US/UK World Magnetic Model for 2005-2010*, NOAA Technical Report NESDIS/NGDC-1.

McMillan S, Quinn JM (2000a), *The Derivation of the World Magnetic Model 2000*, British Geological Survey Technical Report WM/00/17R.

McMillan S, Quinn JM (2000b), The 2000 Revision of the joint UK/US geomagnetic field models and the IGRF 2000 candidate model, *Earth Planets Space* 52: 1149–62.

McRuer D, Ashkenas I, Graham D (1973), *Aircraft Dynamics and Automatic Control*. Princeton, NJ: Princeton University Press.

Mohr PJ, Taylor BN (1998), *CODATA Recommended Values for the Fundamental Physical Constants: 1998*, National Institute of Standards and Technology, Gaithersburg, MD 20899-8401.

Montenbruck O, Gill E (2000), *Satellite Orbits: Models, Methods and Applications*. New York: Springer-Verlag.

Murray-Smith DJ (1995), *Advances in Simulation Model Validation: Theory, Software and Applications*, Proceedings EUROSIM '95, 75-84, Oxford: Elsevier.

Newcomb S (1898), Tables of the motion of the Earth on its axis and around the Sun, *Astronomical Papers of the American Ephemeris*, Vol. VI/1, 1–170, Washington.

NIMA (2004), *Department of Defense World Geodetic System 1984*, NIMA TR8350.2 Third Edition, Amendment 2, 23 June 2004, National Imagery and Mapping Agency, US Department of Defense.

Pamadi BN (2004), *Performance, Stability and Control of Airplanes*, 2nd edition, American Institute of Aeronautics and Astronautics.

Parish OO, Putnam TW (1977), *Equations for the Determination of Humidity from Dewpoint and Psychrometric Data*, NASA TN D-8401, January 1997.

Paynter KM (1961), *Analysis and Design of Engineering Systems*. Boston, MA: MIT Press.

Peel MC, Finlayson BL, McMahon TA (2007), Updated world map of the Köppen-Geiger climate classification, *Hydrol. Earth Syst. Sci.*, 11, 1633–1644.

Perkins C, Parry RB (1996), *Mapping the UK*, London: Bowker Saur.

Phillips WF (2004), *Mechanics of Flight*. New York: John Wiley & Sons, Inc.

Rainsford HF (1955), Long geodesics on the ellipsoid, *Bulletin Géodesique*, 37.

Reda I, Andreas A (2003), *Solar Position Algorithm for Solar Radiation Applications*, Technical Report NREL/TP-560-34302, National Renewable Energy Laboratory, US Department of Energy.

Regan FJ (1993), *Dynamics of Atmospheric Re-entry*, American Institute of Aeronautics and Astronautics.

Ridpath I (ed.) (2004), *Norton's Star Atlas and Reference Handbook*, 20th Edition, Pi Press.

Robinson AR (1958), *On the Use of Quaternions in Simulation of Rigid Body Motion*, Wright Air Development Center (WADC) Technical Report, No. 58-17.

[1] www.gsi-mc.go.jp/ENGLISH/RESEARCH/BULLETIN/vol-41/41-1.htm.

[2] ftp://cddisa.gsfc.nasa.gov/pub/egm96/general_info/tokyo_grageomar96_paper_web.ps.

Rogers GFC, Mayhew YR (1980), *Engineering Thermodynamics and Heat Transfer*, 3rd Edition. London: Longman.

Rolfe JM, Staples KJ (1986), *Flight Simulation*. Cambridge: Cambridge University Press.

Roncoli RB (2005), *Lunar Constants and Models Document*, JPL D-32296, Jet Propulsion Laboratory, California Institute of Technology.

Roy AE (1982), *Orbital Motion*, 2nd Edition. Bristol: Adam Hilger.

Ruijgrok GJJ (1990), *Elements of Airplane Performance*. Delft University Press.

Rumbaugh J, Blaha M, Premerlain W, Eddy F, Lorensen W (1991), *Object-oriented Modeling and Design*. Englewood-Cliffs, NJ: Prentice-Hall.

Seidelmann PK (1982), 1980 IAU Theory of Nutation: The Final Report of the IAU Working Group on Nutation, *Celestial Mechanics* 27: 79–106.

Seidelmann PK, Abalakin VK, Bursa M, *et al.* (2002), Report of the IAU/IAG Working Group on Cartographic Coordinates and Rotational Elements of the Planets and Satellites: 2000, *Celestial Mechanics and Dynamical Astronomy* 82: 83–110.

Sidi MJ (1997), *Spacecraft Dynamics and Control: A Practical Engineering Approach*. Cambridge: Cambridge University Press.

Smith DA (1998), There is no such thing as 'The' EGM96 geoid: Subtle points on the use of a global geopotential model, *IGeS Bulletin 8*, International Geoid Service, Milan, pp. 17–28.

Snyder JP (1987), *Map Projections – A Working Manual*, US Geological Survey Professional Paper 1395.

Standish EM (1995), Time scales in the JPL and CfA ephemerides, *Astronomy and Astrophysics* 336: 381–4.

Standish EM, Newhall XX, Williams JG, Folkner WM (1995), *JPL Planetary and Lunar Ephemerides*, DE403/LE403, JPL IOM 314.10-127, May 22, 1995.

Standish, E.M. (2001), *Approximate Mean Ecliptic Elements of the Lunar Orbit*, JPL IOM 312.F-01-004.

Stengel RF (2004), *Flight Dynamics*. Princeton, NJ: Princeton University Press.

Stevens BL, Lewis FL (2003), *Aircraft Control and Simulation*. New York: John Wiley & Sons, Inc.

Society for Computer Simulation, Technical Committee on Model Credibility (1979), Terminology for Model Credibility, *Simulation* 32: 103–4.

Sodano EM (1965), General non-iterative solution of the inverse and direct geodetic problems, *Bulletin Géodesique* 75.

Symon KR (1971), *Mechanics*. Reading, MA: Addison-Wesley.

Tamburro MB, Abbott AX, Townsend GE (1997), *Guidance, Flight Mechanics and Trajectory Optimisation, Volume1: Coordinate Systems and Time Measure*, NASA-CR-1000.

Thoma J (1975), *Introduction to Bond Graphs and their Applications*. Oxford: Pergamon.

Thoma J (1990), *Simulation by Bond Graphs*. Berlin: Springer-Verlag.

Tobey WM (1928), *Geodesy*, Geodetic Survey of Canada Publication No. 11.

Varsamidis T (1998), *Object-Oriented Modelling for Computer-Aided Control Engineering*, PhD thesis, University of Wales, Bangor.

Vincenty T (1975), Direct and inverse solutions of geodesics on the ellipsoid with application of nested equations, *Survey Review*, vol XXIII, no. 176.

Wellstead PE (1979), *Introduction to Physical System Modelling*. London: Academic Press.

Index